# FUNDAMENTALS OF PROBABILITY AND STATISTICS FOR ENGINEERS

# FUNDAMENTALS OF PROBABILITY AND STATISTICS FOR ENGINEERS

**T.T. Soong**
*State University of New York at Buffalo, Buffalo, New York, USA*

John Wiley & Sons, Ltd

### Other Wiley Editorial Offices

John Wiley & Sons Inc., 111 River Street, Hoboken, NJ 07030, USA

Jossey-Bass, 989 Market Street, San Francisco, CA 94103-1741, USA

Wiley-VCH Verlag GmbH, Boschstr. 12, D-69469 Weinheim, Germany

John Wiley & Sons Australia Ltd, 33 Park Road, Milton, Queensland 4064, Australia

John Wiley & Sons (Asia) Pte Ltd, 2 Clementi Loop #02-01, Jin Xing Distripark, Singapore
129809

John Wiley & Sons (Canada) Ltd, 22 Worcester Road, Etobicoke, Ontario M9W 1L1

Wiley also publishes its books in a variety of electronic formats. Some content that appears
in print may not be available in electronic books.

### British Library Cataloguing in Publication Data

A catalogue record for this book is available from the British Library

ISBN-10: 0–470–86813–9 (Cloth)
ISBN-13: 978–0470–86813–3 (Cloth)
ISBN-10: 0–470–86814–7 (Paper)
ISBN-13: 978–0470–86814–0 (Paper)

Typeset in 10/12pt Times LaTeX files supplied by the author, processed by
Integra Software Services Pvt. Ltd, Pondicherry, India

This book is printed on acid-free paper responsibly manufactured from sustainable forestry,
in which at least two trees are planted for each one used for paper production.

*To the memory of my parents*

# Contents

**APPENDIX A: TABLES                                                  365**

**APPENDIX B: COMPUTER SOFTWARE                                       375**

**APPENDIX C: ANSWERS TO SELECTED PROBLEMS                            379**

**SUBJECT INDEX                                                       389**

# Preface

This book was written for an introductory one-semester or two-quarter course in probability and statistics for students in engineering and applied sciences. No previous knowledge of probability or statistics is presumed but a good understanding of calculus is a prerequisite for the material.

The development of this book was guided by a number of considerations observed over many years of teaching courses in this subject area, including the following:

- As an introductory course, a sound and rigorous treatment of the basic principles is imperative for a proper understanding of the subject matter and for confidence in applying these principles to practical problem solving. A student, depending upon his or her major field of study, will no doubt pursue advanced work in this area in one or more of the many possible directions. How well is he or she prepared to do this strongly depends on his or her mastery of the fundamentals.
- It is important that the student develop an early appreciation for applications. Demonstrations of the utility of this material in nonsuperficial applications not only sustain student interest but also provide the student with stimulation to delve more deeply into the fundamentals.
- Most of the students in engineering and applied sciences can only devote one semester or two quarters to a course of this nature in their programs. Recognizing that the coverage is time limited, it is important that the material be self-contained, representing a reasonably complete and applicable body of knowledge.

The choice of the contents for this book is in line with the foregoing observations. The major objective is to give a careful presentation of the fundamentals in probability and statistics, the concept of probabilistic modeling, and the process of model selection, verification, and analysis. In this text, definitions and theorems are carefully stated and topics rigorously treated but care is taken not to become entangled in excessive mathematical details.

Practical examples are emphasized; they are purposely selected from many different fields and not slanted toward any particular applied area. The same objective is observed in making up the exercises at the back of each chapter.

Because of the self-imposed criterion of writing a comprehensive text and presenting it within a limited time frame, there is a tight continuity from one topic to the next. Some flexibility exists in Chapters 6 and 7 that include discussions on more specialized distributions used in practice. For example, extreme-value distributions may be bypassed, if it is deemed necessary, without serious loss of continuity. Also, Chapter 11 on linear models may be deferred to a follow-up course if time does not allow its full coverage.

It is a pleasure to acknowledge the substantial help I received from students in my courses over many years and from my colleagues and friends. Their constructive comments on preliminary versions of this book led to many improvements. My sincere thanks go to Mrs. Carmella Gosden, who efficiently typed several drafts of this book. As in all my undertakings, my wife, Dottie, cared about this project and gave me her loving support for which I am deeply grateful.

T.T. Soong
Buffalo, New York

# 1

# Introduction

At present, almost all undergraduate curricula in engineering and applied sciences contain at least one basic course in probability and statistical inference. The recognition of this need for introducing the ideas of probability theory in a wide variety of scientific fields today reflects in part some of the profound changes in science and engineering education over the past 25 years.

One of the most significant is the greater emphasis that has been placed upon complexity and precision. A scientist now recognizes the importance of studying scientific phenomena having complex interrelations among their components; these components are often not only mechanical or electrical parts but also 'soft-science' in nature, such as those stemming from behavioral and social sciences. The design of a comprehensive transportation system, for example, requires a good understanding of technological aspects of the problem as well as of the behavior patterns of the user, land-use regulations, environmental requirements, pricing policies, and so on.

Moreover, precision is stressed – precision in describing interrelationships among factors involved in a scientific phenomenon and precision in predicting its behavior. This, coupled with increasing complexity in the problems we face, leads to the recognition that a great deal of uncertainty and variability are inevitably present in problem formulation, and one of the mathematical tools that is effective in dealing with them is probability and statistics.

Probabilistic ideas are used in a wide variety of scientific investigations involving randomness. Randomness is an empirical phenomenon characterized by the property that the quantities in which we are interested do not have a predictable outcome under a given set of circumstances, but instead there is a statistical regularity associated with different possible outcomes. Loosely speaking, statistical regularity means that, in observing outcomes of an experiment a large number of times (say $n$), the ratio $m/n$, where $m$ is the number of observed occurrences of a specific outcome, tends to a unique limit as $n$ becomes large. For example, the outcome of flipping a coin is not predictable but there is statistical regularity in that the ratio $m/n$ approaches $\frac{1}{2}$ for either

*Fundamentals of Probability and Statistics for Engineers* T.T. Soong © 2004 John Wiley & Sons, Ltd
ISBNs: 0-470-86813-9 (HB)   0-470-86814-7 (PB)

heads or tails. Random phenomena in scientific areas abound: noise in radio signals, intensity of wind gusts, mechanical vibration due to atmospheric disturbances, Brownian motion of particles in a liquid, number of telephone calls made by a given population, length of queues at a ticket counter, choice of transportation modes by a group of individuals, and countless others. It is not inaccurate to say that randomness is present in any realistic conceptual model of a real-world phenomenon.

## 1.1 ORGANIZATION OF TEXT

This book is concerned with the development of basic principles in constructing probability models and the subsequent analysis of these models. As in other scientific modeling procedures, the basic cycle of this undertaking consists of a number of fundamental steps; these are schematically presented in Figure 1.1. A basic understanding of probability theory and random variables is central to the whole modeling process as they provide the required mathematical machinery with which the modeling process is carried out and consequences deduced. The step from B to C in Figure 1.1 is the induction step by which the structure of the model is formed from factual observations of the scientific phenomenon under study. Model verification and parameter estimation (E) on the basis of observed data (D) fall within the framework of statistical inference. A model

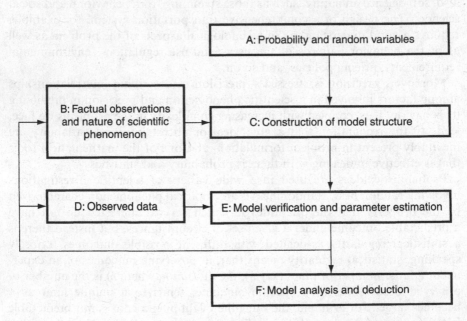

**Figure 1.1**   Basic cycle of probabilistic modeling and analysis

may be rejected at this stage as a result of inadequate inductive reasoning or insufficient or deficient data. A reexamination of factual observations or additional data may be required here. Finally, model analysis and deduction are made to yield desired answers upon model substantiation.

In line with this outline of the basic steps, the book is divided into two parts. Part A (Chapters 2–7) addresses probability fundamentals involved in steps $A \rightarrow C$, $B \rightarrow C$, and $E \rightarrow F$ (Figure 1.1). Chapters 2–5 provide these fundamentals, which constitute the foundation of all subsequent development. Some important probability distributions are introduced in Chapters 6 and 7. The nature and applications of these distributions are discussed. An understanding of the situations in which these distributions arise enables us to choose an appropriate distribution, or model, for a scientific phenomenon.

Part B (Chapters 8–11) is concerned principally with step $D \rightarrow E$ (Figure 1.1), the statistical inference portion of the text. Starting with data and data representation in Chapter 8, parameter estimation techniques are carefully developed in Chapter 9, followed by a detailed discussion in Chapter 10 of a number of selected statistical tests that are useful for the purpose of model verification. In Chapter 11, the tools developed in Chapters 9 and 10 for parameter estimation and model verification are applied to the study of linear regression models, a very useful class of models encountered in science and engineering.

The topics covered in Part B are somewhat selective, but much of the foundation in statistical inference is laid. This foundation should help the reader to pursue further studies in related and more advanced areas.

## 1.2  PROBABILITY TABLES AND COMPUTER SOFTWARE

The application of the materials in this book to practical problems will require calculations of various probabilities and statistical functions, which can be time consuming. To facilitate these calculations, some of the probability tables are provided in Appendix A. It should be pointed out, however, that a large number of computer software packages and spreadsheets are now available that provide this information as well as perform a host of other statistical calculations. As an example, some statistical functions available in Microsoft® Excel™ 2000 are listed in Appendix B.

## 1.3  PREREQUISITES

The material presented in this book is calculus-based. The mathematical prerequisite for a course using this book is a good understanding of differential and integral calculus, including partial differentiation and multidimensional integrals. Familiarity in linear algebra, vectors, and matrices is also required.

# Part A

# Probability and Random Variables

# 2

# Basic Probability Concepts

The mathematical theory of probability gives us the basic tools for constructing and analyzing mathematical models for random phenomena. In studying a random phenomenon, we are dealing with an experiment of which the outcome is not predictable in advance. Experiments of this type that immediately come to mind are those arising in games of chance. In fact, the earliest development of probability theory in the fifteenth and sixteenth centuries was motivated by problems of this type (for example, see Todhunter, 1949).

In science and engineering, random phenomena describe a wide variety of situations. By and large, they can be grouped into two broad classes. The first class deals with physical or natural phenomena involving uncertainties. Uncertainty enters into problem formulation through complexity, through our lack of understanding of all the causes and effects, and through lack of information. Consider, for example, weather prediction. Information obtained from satellite tracking and other meteorological information simply is not sufficient to permit a reliable prediction of what weather condition will prevail in days ahead. It is therefore easily understandable that weather reports on radio and television are made in probabilistic terms.

The second class of problems widely studied by means of probabilistic models concerns those exhibiting variability. Consider, for example, a problem in traffic flow where an engineer wishes to know the number of vehicles crossing a certain point on a road within a specified interval of time. This number varies unpredictably from one interval to another, and this variability reflects variable driver behavior and is inherent in the problem. This property forces us to adopt a probabilistic point of view, and probability theory provides a powerful tool for analyzing problems of this type.

It is safe to say that uncertainty and variability are present in our modeling of all real phenomena, and it is only natural to see that probabilistic modeling and analysis occupy a central place in the study of a wide variety of topics in science and engineering. There is no doubt that we will see an increasing reliance on the use of probabilistic formulations in most scientific disciplines in the future.

*Fundamentals of Probability and Statistics for Engineers* T.T. Soong © 2004 John Wiley & Sons, Ltd
ISBNs: 0-470-86813-9 (HB)   0-470-86814-7 (PB)

## 2.1   ELEMENTS OF SET THEORY

Our interest in the study of a random phenomenon is in the statements we can make concerning the events that can occur. Events and combinations of events thus play a central role in probability theory. The mathematics of events is closely tied to the theory of sets, and we give in this section some of its basic concepts and algebraic operations.

A *set* is a collection of objects possessing some common properties. These objects are called *elements* of the set and they can be of any kind with any specified properties. We may consider, for example, a set of numbers, a set of mathematical functions, a set of persons, or a set of a mixture of things. Capital letters $A, B, C, \Phi, \Omega, \ldots$ shall be used to denote sets, and lower-case letters $a, b, c, \phi, \omega, \ldots$ to denote their elements. A set is thus described by its elements. Notationally, we can write, for example,

$$A = \{1, 2, 3, 4, 5, 6\},$$

which means that set $A$ has as its elements integers 1 through 6. If set $B$ contains two elements, success and failure, it can be described by

$$B = \{s, f\},$$

where $s$ and $f$ are chosen to represent success and failure, respectively. For a set consisting of all nonnegative real numbers, a convenient description is

$$C = \{x : x \geq 0\}.$$

We shall use the convention

$$a \in A \qquad\qquad (2.1)$$

to mean 'element $a$ belongs to set $A$'.

A set containing no elements is called an *empty* or *null* set and is denoted by $\emptyset$. We distinguish between sets containing a finite number of elements and those having an infinite number. They are called, respectively, *finite sets* and *infinite sets*. An infinite set is called *enumerable* or *countable* if all of its elements can be arranged in such a way that there is a one-to-one correspondence between them and all positive integers; thus, a set containing all positive integers 1, 2, ... is a simple example of an enumerable set. A *nonenumerable* or *uncountable* set is one where the above-mentioned one-to-one correspondence cannot be established. A simple example of a nonenumerable set is the set $C$ described above.

If every element of a set $A$ is also an element of a set $B$, the set $A$ is called a *subset* of $B$ and this is represented symbolically by

$$A \subset B \quad \text{or} \quad B \supset A. \qquad\qquad (2.2)$$

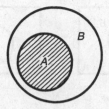

**Figure 2.1**   Venn diagram for $A \subset B$

**Example 2.1.** Let $A = \{2, 4\}$ and $B = \{1, 2, 3, 4\}$. Then $A \subset B$, since every element of $A$ is also an element of $B$. This relationship can also be presented graphically by using a Venn diagram, as shown in Figure 2.1. The set $B$ occupies the interior of the larger circle and $A$ the shaded area in the figure.

It is clear that an empty set is a subset of any set. When both $A \subset B$ and $B \subset A$, set $A$ is then *equal* to $B$, and we write

$$A = B. \tag{2.3}$$

We now give meaning to a particular set we shall call *space*. In our development, we consider only sets that are subsets of a fixed (nonempty) set. This 'largest' set containing all elements of all the sets under consideration is called *space* and is denoted by the symbol $S$.

Consider a subset $A$ in $S$. The set of all elements in $S$ that are not elements of $A$ is called the *complement* of $A$, and we denote it by $\overline{A}$. A Venn diagram showing $A$ and $\overline{A}$ is given in Figure 2.2 in which space $S$ is shown as a rectangle and $\overline{A}$ is the shaded area. We note here that the following relations clearly hold:

$$\overline{S} = \emptyset, \quad \overline{\emptyset} = S, \quad \overline{\overline{A}} = A. \tag{2.4}$$

### 2.1.1   SET OPERATIONS

Let us now consider some algebraic operations of sets $A, B, C, \ldots$ that are subsets of space $S$.

The *union* or *sum* of $A$ and $B$, denoted by $A \cup B$, is the set of all elements belonging to $A$ or $B$ or both.

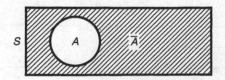

**Figure 2.2**   $A$ and $\overline{A}$

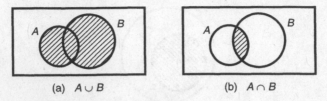

(a)  $A \cup B$                    (b)  $A \cap B$

**Figure 2.3**   (a) Union and (b) intersection of sets $A$ and $B$

The *intersection* or *product* of $A$ and $B$, written as $A \cap B$, or simply $AB$, is the set of all elements that are common to $A$ and $B$.

In terms of Venn diagrams, results of the above operations are shown in Figures 2.3(a) and 2.3(b) as sets having shaded areas.

If $AB = \emptyset$, sets $A$ and $B$ contain no common elements, and we call $A$ and $B$ *disjoint*. The symbol '+' shall be reserved to denote the union of two disjoint sets when it is advantageous to do so.

**Example 2.2.** Let $A$ be the set of all men and $B$ consist of all men and women over 18 years of age. Then the set $A \cup B$ consists of all men as well as all women over 18 years of age. The elements of $A \cap B$ are all men over 18 years of age.

**Example 2.3.** Let $S$ be the space consisting of a real-line segment from 0 to 10 and let $A$ and $B$ be sets of the real-line segments from 1–7 and 3–9 respectively. Line segments belonging to $A \cup B$, $A \cap B$, $\overline{A}$, and $\overline{B}$ are indicated in Figure 2.4. Let us note here that, by definition, a set and its complement are always disjoint.

The definitions of union and intersection can be directly generalized to those involving any arbitrary number (finite or countably infinite) of sets. Thus, the set

$$A_1 \cup A_2 \ldots \cup A_n = \bigcup_{j=1}^{n} A_j \qquad (2.5)$$

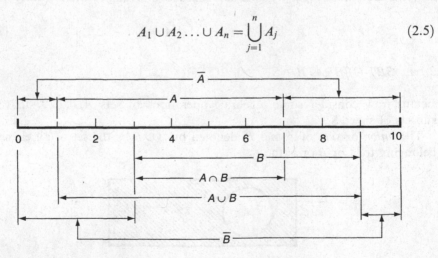

**Figure 2.4**   Sets defined in Example 2.3

stands for the set of all elements belonging to one or more of the sets $A_j$, $j = 1, 2, \ldots, n$. The intersection

$$A_1 A_2 \ldots A_n = \bigcap_{j=1}^{n} A_j \tag{2.6}$$

is the set of all elements common to all $A_j, j = 1, 2, \ldots, n$. The sets $A_j, j = 1, 2, \ldots, n$, are disjoint if

$$A_i A_j = \emptyset, \quad \text{for every } i, j \ (i \neq j). \tag{2.7}$$

Using Venn diagrams or analytical procedures, it is easy to verify that union and intersection operations are associative, commutative, and distributive; that is,

$$\left.\begin{aligned}
&(A \cup B) \cup C = A \cup (B \cup C) = A \cup B \cup C, \\
&A \cup B = B \cup A, \\
&(AB)C = A(BC) = ABC, \\
&AB = BA, \\
&A(B \cup C) = (AB) \cup (AC).
\end{aligned}\right\} \tag{2.8}$$

Clearly, we also have

$$\left.\begin{aligned}
&A \cup A = AA = A, \\
&A \cup \emptyset = A, \\
&A\emptyset = \emptyset, \\
&A \cup S = S, \\
&AS = A, \\
&A \cup \overline{A} = S, \\
&A\overline{A} = \emptyset.
\end{aligned}\right\} \tag{2.9}$$

Moreover, the following useful relations hold, all of which can be easily verified using Venn diagrams:

$$\left.\begin{aligned}
&A \cup (BC) = (A \cup B)(A \cup C), \\
&A \cup B = A \cup (\overline{A}B) = A + (\overline{A}B), \\
&\overline{(A \cup B)} = \overline{A}\,\overline{B}, \\
&\overline{(AB)} = \overline{A} \cup \overline{B}, \\
&\overline{\left(\bigcup_{j=1}^{n} A_j\right)} = \bigcap_{j=1}^{n} \overline{A}_j, \\
&\overline{\left(\bigcap_{j=1}^{n} A_j\right)} = \bigcup_{j=1}^{n} \overline{A}_j.
\end{aligned}\right\} \tag{2.10}$$

The second relation in Equations (2.10) gives the union of two sets in terms of the union of two disjoint sets. As we will see, this representation is useful in probability calculations. The last two relations in Equations (2.10) are referred to as *DeMorgan's laws*.


## 2.2  SAMPLE SPACE AND PROBABILITY MEASURE

In probability theory, we are concerned with an experiment with an outcome depending on chance, which is called a *random experiment*. It is assumed that all possible distinct outcomes of a random experiment are known and that they are elements of a fundamental set known as the *sample space*. Each possible outcome is called a *sample point*, and an *event* is generally referred to as a subset of the sample space having one or more sample points as its elements.

It is important to point out that, for a given random experiment, the associated sample space is not unique and its construction depends upon the point of view adopted as well as the questions to be answered. For example, $100\,\Omega$ resistors are being manufactured by an industrial firm. Their values, owing to inherent inaccuracies in the manufacturing and measurement processes, may range from 99 to $101\,\Omega$. A measurement taken of a resistor is a random experiment for which the possible outcomes can be defined in a variety of ways depending upon the purpose for performing such an experiment. On the one hand, if, for a given user, a resistor with resistance range of 99.9–100.1 $\Omega$ is considered acceptable, and unacceptable otherwise, it is adequate to define the sample space as one consisting of two elements: 'acceptable' and 'unacceptable'. On the other hand, from the viewpoint of another user, possible outcomes may be the ranges 99–99.5 $\Omega$, 99.5–100 $\Omega$, 100–100.5 $\Omega$, and 100.5–101 $\Omega$. The sample space in this case has four sample points. Finally, if each possible reading is a possible outcome, the sample space is now a real line from 99 to 101 on the ohm scale; there is an uncountably infinite number of sample points, and the sample space is a nonenumerable set.

To illustrate that a sample space is not fixed by the action of performing the experiment but by the point of view adopted by the observer, consider an energy negotiation between the United States and another country. From the point of view of the US government, success and failure may be looked on as the only possible outcomes. To the consumer, however, a set of more direct possible outcomes may consist of price increases and decreases for gasoline purchases.

The description of sample space, sample points, and events shows that they fit nicely into the framework of set theory, a framework within which the analysis of outcomes of a random experiment can be performed. All relations between outcomes or events in probability theory can be described by sets and set operations. Consider a space $S$ of elements $a, b, c, \ldots$, and with subsets

**Table 2.1**  Corresponding statements in set theory and probability

| Set theory | Probability theory |
|---|---|
| Space, $S$ | Sample space, sure event |
| Empty set, $\emptyset$ | Impossible event |
| Elements $a, b, \ldots$ | Sample points $a, b, \ldots$ (or simple events) |
| Sets $A, B, \ldots$ | Events $A, B, \ldots$ |
| $A$ | Event $A$ occurs |
| $\overline{A}$ | Event $A$ does not occur |
| $A \cup B$ | At least one of $A$ and $B$ occurs |
| $AB$ | Both $A$ and $B$ occur |
| $A \subset B$ | $A$ is a subevent of $B$ (i.e. the occurrence of $A$ necessarily implies the occurrence of $B$) |
| $AB = \emptyset$ | $A$ and $B$ are mutually exclusive (i.e. they cannot occur simultaneously) |

$A, B, C, \ldots$. Some of these corresponding sets and probability meanings are given in Table 2.1. As Table 2.1 shows, the empty set $\emptyset$ is considered an impossible event since no possible outcome is an element of the empty set. Also, by 'occurrence of an event' we mean that the observed outcome is an element of that set. For example, event $A \cup B$ is said to occur if and only if the observed outcome is an element of $A$ or $B$ or both.

**Example 2.4.** Consider an experiment of counting the number of left-turning cars at an intersection in a group of 100 cars. The possible outcomes (possible numbers of left-turning cars) are $0, 1, 2, \ldots, 100$. Then, the sample space $S$ is $S = \{0, 1, 2, \ldots, 100\}$. Each element of $S$ is a sample point or a possible outcome. The subset $A = \{0, 1, 2, \ldots, 50\}$ is the event that there are 50 or fewer cars turning left. The subset $B = \{40, 41, \ldots, 60\}$ is the event that between 40 and 60 (inclusive) cars take left turns. The set $A \cup B$ is the event of 60 or fewer cars turning left. The set $A \cap B$ is the event that the number of left-turning cars is between 40 and 50 (inclusive). Let $C = \{80, 81, \ldots, 100\}$. Events $A$ and $C$ are *mutually exclusive* since they cannot occur simultaneously. Hence, disjoint sets are mutually exclusive events in probability theory.

## 2.2.1  AXIOMS OF PROBABILITY

We now introduce the notion of a *probability function*. Given a random experiment, a finite number $P(A)$ is assigned to every event $A$ in the sample space $S$ of all possible events. The number $P(A)$ is a function of set $A$ and is assumed to be defined for all sets in $S$. It is thus a set function, and $P(A)$ is called the *probability measure* of $A$ or simply the *probability* of $A$. It is assumed to have the following properties (axioms of probability):

- Axiom 1: $P(A) \geq 0$ (nonnegative).
- Axiom 2: $P(S) = 1$ (normed).
- Axiom 3: for a countable collection of mutually exclusive events $A_1, A_2, \ldots$ in $S$,

$$P(A_1 \cup A_2 \cup \ldots) = P\left(\sum_j A_j\right) = \sum_j P(A_j) \quad \text{(additive).} \tag{2.11}$$

These three axioms define a countably additive and nonnegative set function $P(A)$, $A \subset S$. As we shall see, they constitute a sufficient set of postulates from which all useful properties of the probability function can be derived. Let us give below some of these important properties.

First, $P(\emptyset) = 0$. Since $S$ and $\emptyset$ are disjoint, we see from Axiom 3 that

$$P(S) = P(S + \emptyset) = P(S) + P(\emptyset).$$

It then follows from Axiom 2 that

$$1 = 1 + P(\emptyset)$$

or

$$P(\emptyset) = 0.$$

Second, if $A \subset C$, then $P(A) \leq P(C)$. Since $A \subset C$, one can write

$$A + B = C,$$

where $B$ is a subset of $C$ and disjoint with $A$. Axiom 3 then gives

$$P(C) = P(A + B) = P(A) + P(B).$$

Since $P(B) \geq 0$ as required by Axiom 1, we have the desired result.

Third, given two arbitrary events $A$ and $B$, we have

$$\boxed{P(A \cup B) = P(A) + P(B) - P(AB).} \tag{2.12}$$

In order to show this, let us write $A \cup B$ in terms of the union of two mutually exclusive events. From the second relation in Equations (2.10), we write

$$A \cup B = A + \overline{A}B.$$

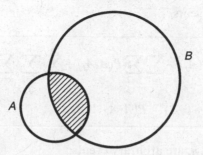

**Figure 2.5**   Venn diagram for derivation of Equation (2.12)

Hence, using Axiom 3,

$$P(A \cup B) = P(A + \overline{A}B) = P(A) + P(\overline{A}B). \tag{2.13}$$

Furthermore, we note

$$\overline{A}B + AB = B.$$

Hence, again using Axiom 3,

$$P(\overline{A}B) + P(AB) = P(B),$$

or

$$P(\overline{A}B) = P(B) - P(AB).$$

Substitution of this equation into Equation (2.13) yields Equation (2.12).

Equation (2.12) can also be verified by inspecting the Venn diagram in Figure 2.5. The sum $P(A) + P(B)$ counts twice the events belonging to the shaded area $AB$. Hence, in computing $P(A \cup B)$, the probability associated with one $AB$ must be subtracted from $P(A) + P(B)$, giving Equation (2.12) (see Figure 2.5).

The important result given by Equation (2.12) can be immediately generalized to the union of three or more events. Using the same procedure, we can show that, for arbitrary events $A$, $B$, and $C$,

$$P(A \cup B \cup C) = P(A) + P(B) + P(C) - P(AB) - P(AC)$$
$$- P(BC) + P(ABC). \tag{2.14}$$

and, in the case of $n$ events,

$$P\left(\bigcup_{j=1}^{n} A_j\right) = \sum_{j=1}^{n} P(A_j) - \sum_{i=1}^{n}\sum_{\substack{j=2 \\ i<j}}^{n} P(A_iA_j) + \sum_{i=1}^{n}\sum_{\substack{j=2 \\ i<j<k}}^{n}\sum_{k=3}^{n} P(A_iA_jA_k)$$
$$- \cdots + (-1)^{n-1} P(A_1A_2\ldots A_n),$$

(2.15)

where $A_j, j = 1, 2, \ldots, n$, are arbitrary events.

**Example 2.5.** Let us go back to Example 2.4 and assume that probabilities $P(A)$, $P(B)$, and $P(C)$ are known. We wish to compute $P(A \cup B)$ and $P(A \cup C)$.

Probability $P(A \cup C)$, the probability of having either 50 or fewer cars turn-ing left or between 80 to 100 cars turning left, is simply $P(A) + P(C)$. This follows from Axiom 3, since $A$ and $C$ are mutually exclusive. However, $P(A \cup B)$, the probability of having 60 or fewer cars turning left, is found from

$$P(A \cup B) = P(A) + P(B) - P(AB)$$

The information given above is thus not sufficient to determine this probability and we need the additional information, $P(AB)$, which is the probability of having between 40 and 50 cars turning left.

With the statement of three axioms of probability, we have completed the mathematical description of a random experiment. It consists of three funda-mental constituents: a sample space $S$, a collection of events $A, B, \ldots$, and the probability function $P$. These three quantities constitute a *probability space* associated with a random experiment.

### 2.2.2  ASSIGNMENT OF PROBABILITY

The axioms of probability define the properties of a probability measure, which are consistent with our intuitive notions. However, they do not guide us in assigning probabilities to various events. For problems in applied sciences, a natural way to assign the probability of an event is through the observation of *relative frequency*. Assuming that a random experiment is performed a large number of times, say $n$, then for any event $A$ let $n_A$ be the number of occurrences of $A$ in the $n$ trials and define the ratio $n_A/n$ as the relative frequency of $A$. Under stable or statistical regularity conditions, it is expected that this ratio will tend to a unique limit as $n$ becomes large. This limiting value of the relative frequency clearly possesses the properties required of the probability measure and is a natural candidate for the probability of $A$. This interpretation is used, for example, in saying that the probability of 'heads' in flipping a coin is $1/2$. The relative frequency approach to probability assignment is objective and consistent with the axioms stated in Section 2.2.1 and is one commonly adopted in science and engineering.

Another common but more subjective approach to probability assignment is that of *relative likelihood*. When it is not feasible or is impossible to perform an experiment a large number of times, the probability of an event may be assigned as a result of subjective judgement. The statement 'there is a 40% probability of rain tomorrow' is an example in this interpretation, where the number 0.4 is assigned on the basis of available information and professional judgement.

In most problems considered in this book, probabilities of some simple but basic events are generally assigned by using either of the two approaches. Other probabilities of interest are then derived through the theory of probability. Example 2.5 gives a simple illustration of this procedure where the probabilities of interest, $P(A \cup B)$ and $P(A \cup C)$, are derived upon assigning probabilities to simple events $A$, $B$, and $C$.

## 2.3  STATISTICAL INDEPENDENCE

Let us pose the following question: given individual probabilities $P(A)$ and $P(B)$ of two events $A$ and $B$, what is $P(AB)$, the probability that both $A$ and $B$ will occur? Upon little reflection, it is not difficult to see that the knowledge of $P(A)$ and $P(B)$ is not sufficient to determine $P(AB)$ in general. This is so because $P(AB)$ deals with joint behavior of the two events whereas $P(A)$ and $P(B)$ are probabilities associated with individual events and do not yield information on their joint behavior. Let us then consider a special case in which the occurrence or nonoccurrence of one does not affect the occurrence or nonoccurrence of the other. In this situation events $A$ and $B$ are called *statistically independent* or simply *independent* and it is formalized by Definition 2.1.

**Definition 2.1.** Two events $A$ and $B$ are said to be *independent* if and only if

$$P(AB) = P(A)P(B). \tag{2.16}$$

To show that this definition is consistent with our intuitive notion of independence, consider the following example.

**Example 2.6.** In a large number of trials of a random experiment, let $n_A$ and $n_B$ be, respectively, the numbers of occurrences of two outcomes $A$ and $B$, and let $n_{AB}$ be the number of times both $A$ and $B$ occur. Using the relative frequency interpretation, the ratios $n_A/n$ and $n_B/n$ tend to $P(A)$ and $P(B)$, respectively, as $n$ becomes large. Similarly, $n_{AB}/n$ tends to $P(AB)$. Let us now confine our attention to only those outcomes in which $A$ is realized. If $A$ and $B$ are independent, we expect that the ratio $n_{AB}/n_A$ also tends to $P(B)$ as $n_A$ becomes large. The independence assumption then leads to the observation that

$$\frac{n_{AB}}{n_A} \cong P(B) \cong \frac{n_B}{n}.$$

This then gives

$$\frac{n_{AB}}{n} \cong \left(\frac{n_A}{n}\right)\left(\frac{n_B}{n}\right),$$

or, in the limit as $n$ becomes large,

$$P(AB) = P(A)P(B),$$

which is the definition of independence introduced above.

**Example 2.7.** In launching a satellite, the probability of an unsuccessful launch is $q$. What is the probability that two successive launches are unsuccessful? Assuming that satellite launchings are independent events, the answer to the above question is simply $q^2$. One can argue that these two events are not really completely independent, since they are manufactured by using similar processes and launched by the same launcher. It is thus likely that the failures of both are attributable to the same source. However, we accept this answer as reasonable because, on the one hand, the independence assumption is acceptable since there are a great deal of unknowns involved, any of which can be made accountable for the failure of a launch. On the other hand, the simplicity of computing the joint probability makes the independence assumption attractive. In physical problems, therefore, the independence assumption is often made whenever it is considered to be reasonable.

Care should be exercised in extending the concept of independence to more than two events. In the case of three events, $A_1$, $A_2$, and $A_3$, for example, they are mutually independent if and only if

$$P(A_j A_k) = P(A_j)P(A_k), \quad j \neq k, \quad j,k = 1,2,3, \tag{2.17}$$

and

$$P(A_1 A_2 A_3) = P(A_1)P(A_2)P(A_3). \tag{2.18}$$

Equation (2.18) is required because pairwise independence does not generally lead to mutual independence. Consider, for example, three events $A_1$, $A_2$, and $A_3$ defined by

$$A_1 = B_1 \cup B_2, \quad A_2 = B_1 \cup B_3, \quad A_3 = B_2 \cup B_3,$$

where $B_1$, $B_2$, and $B_3$ are mutually exclusive, each occurring with probability $\frac{1}{4}$. It is easy to calculate the following:

$$P(A_1) = P(B_1 \cup B_2) = P(B_1) + P(B_2) = \frac{1}{2},$$

$$P(A_2) = P(A_3) = \frac{1}{2},$$

$$P(A_1 A_2) = P[(B_1 \cup B_2) \cap (B_1 \cup B_3)] = P(B_1) = \frac{1}{4},$$

$$P(A_1 A_3) = P(A_2 A_3) = \frac{1}{4},$$

$$P(A_1 A_2 A_3) = P[(B_1 \cup B_2) \cap (B_1 \cup B_3) \cap (B_2 \cup B_3)] = P(\emptyset) = 0.$$

We see that Equation (2.17) is satisfied for every $j$ and $k$ in this case, but Equation (2.18) is not. In other words, events $A_1$, $A_2$, and $A_3$ are pairwise independent but they are not mutually independent.

In general, therefore, we have Definition 2.2 for mutual independence of $n$ events.

**Definition 2.2.** Events $A_1, A_2, \ldots, A_n$ are mutually independent if and only if, with $k_1, k_2, \ldots, k_m$ being any set of integers such that $1 \leq k_1 < k_2 \ldots < k_m \leq n$ and $m = 2, 3, \ldots, n$,

$$\boxed{P(A_{k_1} A_{k_2} \ldots A_{k_m}) = P(A_{k_1}) P(A_{k_2}) \ldots P(A_{k_m}).} \tag{2.19}$$

The total number of equations defined by Equation (2.19) is $2^n - n - 1$.

**Example 2.8.** Problem: a system consisting of five components is in working order only when each component is functioning ('good'). Let $S_i, i = 1, \ldots, 5$, be the event that the $i$th component is good and assume $P(S_i) = p_i$. What is the probability $q$ that the system fails?

Answer: assuming that the five components perform in an independent manner, it is easier to determine $q$ through finding the probability of system success $p$. We have from the statement of the problem

$$p = P(S_1 S_2 S_3 S_4 S_5).$$

Equation (2.19) thus gives, due to mutual independence of $S_1, S_2, \ldots, S_5$,

$$p = P(S_1) P(S_2) \ldots P(S_5) = p_1 p_2 p_3 p_4 p_5. \tag{2.20}$$

Hence,

$$q = 1 - p = 1 - p_1 p_2 p_3 p_4 p_5. \tag{2.21}$$

An expression for $q$ may also be obtained by noting that the system fails if any one or more of the five components fail, or

$$q = P(\overline{S}_1 \cup \overline{S}_2 \cup \overline{S}_3 \cup \overline{S}_4 \cup \overline{S}_5), \tag{2.22}$$

where $\overline{S}_i$ is the complement of $S_i$ and represents a bad $i$th component. Clearly, $P(\overline{S}_i) = 1 - p_i$. Since events $\overline{S}_i, i = 1, \ldots, 5$, are not mutually exclusive, the calculation of $q$ with use of Equation (2.22) requires the use of Equation (2.15). Another approach is to write the unions in Equation (2.22) in terms of unions of mutually exclusive events so that Axiom 3 (Section 2.2.1) can be directly utilized. The result is, upon applying the second relation in Equations (2.10),

$$\overline{S}_1 \cup \overline{S}_2 \cup \overline{S}_3 \cup \overline{S}_4 \cup \overline{S}_5 = \overline{S}_1 + S_1\overline{S}_2 + S_1S_2\overline{S}_3 + S_1S_2S_3\overline{S}_4 + S_1S_2S_3S_4\overline{S}_5,$$

where the '$\cup$' signs are replaced by '$+$' signs on the right-hand side to stress the fact that they are mutually exclusive events. Axiom 3 then leads to

$$q = P(\overline{S}_1) + P(S_1\overline{S}_2) + P(S_1S_2\overline{S}_3) + P(S_1S_2S_3\overline{S}_4) + P(S_1S_2S_3S_4\overline{S}_5),$$

and, using statistical independence,

$$\begin{aligned} q &= (1 - p_1) + p_1(1 - p_2) + p_1p_2(1 - p_3) + p_1p_2p_3(1 - p_4) \\ &\quad + p_1p_2p_3p_4(1 - p_5) \end{aligned} \tag{2.23}$$

Some simple algebra will show that this result reduces to Equation (2.21).

Let us mention here that probability $p$ is called the *reliability* of the system in systems engineering.


## 2.4  CONDITIONAL PROBABILITY

The concept of conditional probability is a very useful one. Given two events $A$ and $B$ associated with a random experiment, probability $P(A|B)$ is defined as the *conditional probability* of $A$, given that $B$ has occurred. Intuitively, this probability can be interpreted by means of relative frequencies described in Example 2.6, except that events $A$ and $B$ are no longer assumed to be independent. The number of outcomes where both $A$ and $B$ occur is $n_{AB}$. Hence, given that event $B$ has occurred, the relative frequency of $A$ is then $n_{AB}/n_B$. Thus we have, in the limit as $n_B$ becomes large,

$$P(A|B) \cong \frac{n_{AB}}{n_B} = \frac{n_{AB}}{n} \Big/ \frac{n_B}{n} \cong \frac{P(AB)}{P(B)}$$

This relationship leads to Definition 2.3.

**Definition 2.3.** The conditional probability of $A$ given that $B$ has occurred is given by

$$P(A|B) = \frac{P(AB)}{P(B)}, \quad P(B) \neq 0. \tag{2.24}$$

Definition 2.3 is meaningless if $P(B) = 0$.

It is noted that, in the discussion of conditional probabilities, we are dealing with a contracted sample space in which $B$ is known to have occurred. In other words, $B$ replaces $S$ as the sample space, and the conditional probability $P(A|B)$ is found as the probability of $A$ with respect to this new sample space.

In the event that $A$ and $B$ are independent, it implies that the occurrence of $B$ has no effect on the occurrence or nonoccurrence of $A$. We thus expect $P(A|B) = P(A)$, and Equation (2.24) gives

$$P(A) = \frac{P(AB)}{P(B)},$$

or

$$P(AB) = P(A)P(B),$$

which is precisely the definition of independence.

It is also important to point out that conditional probabilities are probabilities (i.e. they satisfy the three axioms of probability). Using Equation (2.24), we see that the first axiom is automatically satisfied. For the second axiom we need to show that

$$P(S|B) = 1.$$

This is certainly true, since

$$P(S|B) = \frac{P(SB)}{P(B)} = \frac{P(B)}{P(B)} = 1.$$

As for the third axiom, if $A_1, A_2, \ldots$ are mutually exclusive, then $A_1B, A_2B, \ldots$ are also mutually exclusive. Hence,

$$\begin{aligned}
P[(A_1 \cup A_2 \cup \ldots)|B] &= \frac{P[(A_1 \cup A_2 \cup \ldots)B]}{P(B)} \\
&= \frac{P(A_1B \cup A_2B \cup \ldots)}{P(B)} \\
&= \frac{P(A_1B)}{P(B)} + \frac{P(A_2B)}{P(B)} + \cdots \\
&= P(A_1|B) + P(A_2|B) + \cdots,
\end{aligned}$$

and the third axiom holds.

The definition of conditional probability given by Equation (2.24) can be used not only to compute conditional probabilities but also to compute joint probabilities, as the following examples show.

**Example 2.9.** Problem: let us reconsider Example 2.8 and ask the following question: what is the conditional probability that the first two components are good given that (a) the first component is good and (b) at least one of the two is good?

Answer: the event $S_1 S_2$ means both are good components, and $S_1 \cup S_2$ is the event that at least one of the two is good. Thus, for question (a) and in view of Equation (2.24),

$$P(S_1 S_2 | S_1) = \frac{P(S_1 S_2 S_1)}{P(S_1)} = \frac{P(S_1 S_2)}{P(S_1)} = \frac{p_1 p_2}{p_1} = p_2.$$

This result is expected since $S_1$ and $S_2$ are independent. Intuitively, we see that this question is equivalent to one of computing $P(S_2)$.

For question (b), we have

$$P(S_1 S_2 | S_1 \cup S_2) = \frac{P[S_1 S_2 (S_1 \cup S_2)]}{P(S_1 \cup S_2)}.$$

Now, $S_1 S_2 (S_1 \cup S_2) = S_1 S_2$. Hence,

$$P(S_1 S_2 | S_1 \cup S_2) = \frac{P(S_1 S_2)}{P(S_1 \cup S_2)} = \frac{P(S_1 S_2)}{P(S_1) + P(S_2) - P(S_1 S_2)}$$

$$= \frac{p_1 p_2}{p_1 + p_2 - p_1 p_2}.$$

**Example 2.10.** Problem: in a game of cards, determine the probability of drawing, without replacement, two aces in succession.

Answer: let $A_1$ be the event that the first card drawn is an ace, and similarly for $A_2$. We wish to compute $P(A_1 A_2)$. From Equation (2.24) we write

$$P(A_1 A_2) = P(A_2 | A_1) P(A_1). \tag{2.25}$$

Now, $P(A_1) = 4/52$ and $P(A_2 | A_1) = 3/51$ (there are 51 cards left and three of them are aces). Therefore,

$$P(A_1 A_2) = \frac{3}{51} \left( \frac{4}{52} \right) = \frac{1}{221}.$$

Equation (2.25) is seen to be useful for finding joint probabilities. Its extension to more than two events has the form

$$P(A_1 A_2 \ldots A_n) = P(A_1)P(A_2|A_1)P(A_3|A_1 A_2) \ldots P(A_n|A_1 A_2 \ldots A_{n-1}). \quad (2.26)$$

where $P(A_i) > 0$ for all $i$. This can be verified by successive applications of Equation (2.24).

In another direction, let us state a useful theorem relating the probability of an event to conditional probabilities.

**Theorem 2.1:** *theorem of total probability.* Suppose that events $B_1, B_2, \ldots,$ and $B_n$ are mutually exclusive and exhaustive (i.e. $S = B_1 + B_2 + \cdots + B_n$). Then, for an arbitrary event $A$,

$$
\begin{aligned}
P(A) &= P(A|B_1)P(B_1) + P(A|B_2)P(B_2) + \cdots + P(A|B_n)P(B_n) \\
&= \sum_{j=1}^{n} P(A|B_j)P(B_j).
\end{aligned}
\qquad (2.27)
$$

**Proof of Theorem 2.1:** referring to the Venn diagram in Figure 2.6, we can clearly write $A$ as the union of mutually exclusive events $AB_1, AB_2, \ldots, AB_n$ (i.e. $A = AB_1 + AB_2 + \cdots + AB_n$). Hence,

$$P(A) = P(AB_1) + P(AB_2) + \cdots + P(AB_n),$$

which gives Equation (2.27) on application of the definition of conditional probability.

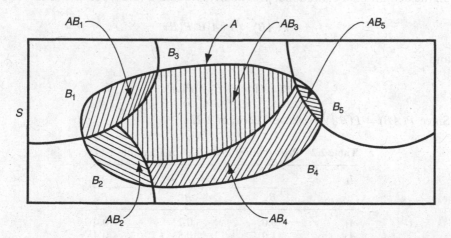

**Figure 2.6**  Venn diagram associated with total probability

The utility of this result rests with the fact that the probabilities in the sum in Equation (2.27) are often more readily obtainable than the probability of $A$ itself.

**Example 2.11.** Our interest is in determining the probability that a critical level of peak flow rate is reached during storms in a storm-sewer system on the basis of separate meteorological and hydrological measurements.

Let $B_i, i = 1, 2, 3$, be the different levels (low, medium, high) of precipitation caused by a storm and let $A_j, j = 1, 2$, denote, respectively, critical and non-critical levels of peak flow rate. Then probabilities $P(B_i)$ can be estimated from meteorological records and $P(A_j|B_i)$ can be estimated from runoff analysis. Since $B_1, B_2$, and $B_3$ constitute a set of mutually exclusive and exhaustive events, the desired probability, $P(A_1)$, can be found from

$$P(A_1) = P(A_1|B_1)P(B_1) + P(A_1|B_2)P(B_2) + P(A_1|B_3)P(B_3).$$

Assume the following information is available:

$$P(B_1) = 0.5, \quad P(B_2) = 0.3, \quad P(B_3) = 0.2,$$

and that $P(A_j|B_i)$ are as shown in Table 2.2. The value of $P(A_1)$ is given by

$$P(A_1) = 0(0.5) + 0.2(0.3) + 0.6(0.2) = 0.18.$$

Let us observe that in Table 2.2, the sum of the probabilities in each column is 1.0 by virtue of the conservation of probability. There is, however, no such requirement for the sum of each row.

A useful result generally referred to as Bayes' theorem can be derived based on the definition of conditional probability. Equation (2.24) permits us to write

$$P(AB) = P(A|B)P(B)$$

and

$$P(BA) = P(B|A)P(A).$$

Since $P(AB) = P(BA)$, we have Theorem 2.2.

**Table 2.2**   Probabilities $P(A_j|B_i)$, for Example 2.11

| $A_j$ | $B_i$ | | |
| --- | --- | --- | --- |
| | $B_1$ | $B_2$ | $B_3$ |
| $A_1$ | 0.0 | 0.2 | 0.6 |
| $A_2$ | 1.0 | 0.8 | 0.4 |

**Theorem 2.2:** *Bayes' theorem.* Let $A$ and $B$ be two arbitrary events with $P(A) \neq 0$ and $P(B) \neq 0$. Then:

$$P(B|A) = \frac{P(A|B)P(B)}{P(A)}. \qquad (2.28)$$

Combining this theorem with the total probability theorem we have a useful consequence:

$$P(B_i|A) = P(A|B_i)P(B_i) \left/ \sum_{j=1}^{n} [P(A|B_j)P(B_j)] \right. . \qquad (2.29)$$

for any $i$ where events $B_j$ represent a set of mutually exclusive and exhaustive events.

The simple result given by Equation (2.28) is called Bayes' theorem after the English philosopher Thomas Bayes and is useful in the sense that it permits us to evaluate *a posteriori* probability $P(B|A)$ in terms of *a priori* information $P(B)$ and $P(A|B)$, as the following examples illustrate.

**Example 2.12.** Problem: a simple binary communication channel carries messages by using only two signals, say 0 and 1. We assume that, for a given binary channel, 40% of the time a 1 is transmitted; the probability that a transmitted 0 is correctly received is 0.90, and the probability that a transmitted 1 is correctly received is 0.95. Determine (a) the probability of a 1 being received, and (b) given a 1 is received, the probability that 1 was transmitted.

Answer: let

$$A = \text{event that 1 is transmitted,}$$
$$\overline{A} = \text{event that 0 is transmitted,}$$
$$B = \text{event that 1 is received,}$$
$$\overline{B} = \text{event that 0 is received.}$$

The information given in the problem statement gives us

$$P(A) = 0.4, \qquad P(\overline{A}) = 0.6;$$
$$P(B|A) = 0.95, \qquad P(\overline{B}|A) = 0.05;$$
$$P(\overline{B}|\overline{A}) = 0.90, \qquad P(B|\overline{A}) = 0.10.$$

and these are represented diagrammatically in Figure 2.7.

For part (a) we wish to find $P(B)$. Since $A$ and $\overline{A}$ are mutually exclusive and exhaustive, it follows from the theorem of total probability [Equation (2.27)]

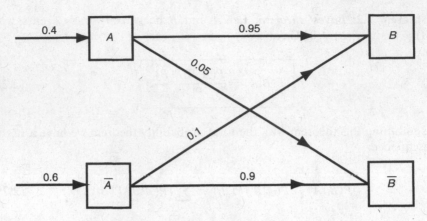

**Figure 2.7**   Probabilities associated with a binary channel, for Example 2.12

that

$$P(B) = P(B|A)P(A) + P(B|\overline{A})P(\overline{A}) = 0.95(0.4) + 0.1(0.6) = 0.44.$$

The probability of interest in part (b) is $P(A|B)$, and this can be found using Bayes' theorem [Equation (2.28)]. It is given by:

$$P(A|B) = \frac{P(B|A)P(A)}{P(B)} = \frac{0.95(0.4)}{0.44} = 0.863.$$

It is worth mentioning that $P(B)$ in this calculation is found by means of the total probability theorem. Hence, Equation (2.29) is the one actually used here in finding $P(A|B)$. In fact, probability $P(A)$ in Equation (2.28) is often more conveniently found by using the total probability theorem.

**Example 2.13.** Problem: from Example 2.11, determine $P(B_2|A_2)$, the probability that a noncritical level of peak flow rate will be caused by a medium-level storm. Answer: from Equations (2.28) and (2.29) we have

$$
\begin{aligned}
P(B_2|A_2) &= \frac{P(A_2|B_2)P(B_2)}{P(A_2)} \\
&= \frac{P(A_2|B_2)P(B_2)}{P(A_2|B_1)P(B_1) + P(A_2|B_2)P(B_2) + P(A_2|B_3)P(B_3)} \\
&= \frac{0.8(0.3)}{1.0(0.5) + 0.8(0.3) + 0.4(0.2)} = 0.293.
\end{aligned}
$$

In closing, let us introduce the use of tree diagrams for dealing with more complicated experiments with 'limited memory'. Consider again Example 2.12

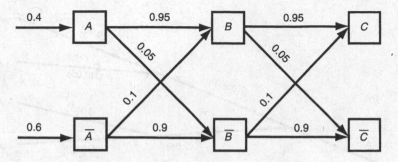

**Figure 2.8** A two-stage binary channel

by adding a second stage to the communication channel, with Figure 2.8 showing all the associated probabilities. We wish to determine $P(C)$, the probability of receiving a 1 at the second stage.

Tree diagrams are useful for determining the behavior of this system when the system has a 'one-stage' memory; that is, when the outcome at the second stage is dependent only on what has happened at the first stage and not on outcomes at stages prior to the first. Mathematically, it follows from this property that

$$P(C|BA) = P(C|B), \quad P(\overline{C}|\overline{B}A) = P(\overline{C}|\overline{B}), \quad \text{etc.} \tag{2.30}$$

The properties described above are commonly referred to as *Markovian* properties. Markov processes represent an important class of probabilistic process that are studied at a more advanced level.

Suppose that Equations (2.30) hold for the system described in Figure 2.8. The tree diagram gives the flow of conditional probabilities originating from the source. Starting from the transmitter, the tree diagram for this problem has the appearance shown in Figure 2.9. The top branch, for example, leads to the probability of the occurrence of event $ABC$, which is, according to Equations (2.26) and (2.30),

$$P(ABC) = P(A)P(B|A)P(C|BA)$$
$$= P(A)P(B|A)P(C|B)$$
$$= 0.4(0.95)(0.95) = 0.361.$$

The probability of $C$ is then found by summing the probabilities of all events that end with $C$. Thus,

$$P(C) = P(ABC) + P(A\overline{B}C) + P(\overline{A}BC) + P(\overline{A}\,\overline{B}C)$$
$$= 0.95(0.95)(0.4) + 0.1(0.05)(0.4) + 0.95(0.1)(0.6) + 0.1(0.9)(0.6)$$
$$= 0.472.$$

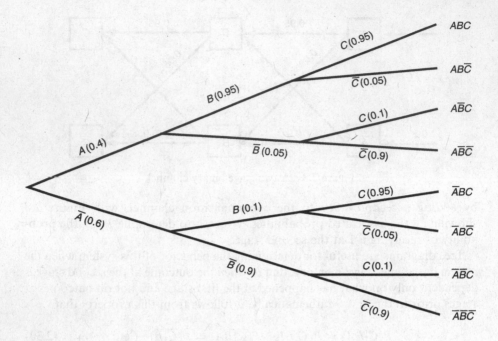

**Figure 2.9** A tree diagram

# REFERENCE

Todhunter, I., 1949, *A History of the Mathematical Theory of Probability from the Time of Pascal to Laplace*, Chelsea, New York.

# FURTHER READING

More accounts of early development of probability theory related to gambling can be found in

David, F.N., 1962, *Games, Gods, and Gambling*, Hafner, New York.

# PROBLEMS

2.1 Let *A*, *B*, and *C* be arbitrary sets. Determine which of the following relations are correct and which are incorrect:
   (a)  $ABC = AB(C \cup B)$.
   (b)  $\overline{AB} = \overline{A} \cup \overline{B}$.
   (c)  $\overline{A \cup B} = \overline{A}\,\overline{B}$.
   (d)  $\overline{(A \cup B)}C = \overline{A}\,\overline{B}C$.

(e) $A\overline{B} \subset A \cup B$.

(f) $(AB)(\overline{A}C) = \emptyset$.

2.2 The second relation in Equations (2.10) expresses the union of two sets as the union of two disjoint sets (i.e. $A \cup B = A + \overline{A}B$). Express $A \cup B \cup C$ in terms of the union of disjoint sets where $A$, $B$, and $C$ are arbitrary sets.

2.3 Verify DeMorgan's laws, given by the last two equations of Equations (2.10).

2.4 Let $S = \{1, 2, \ldots, 10\}$, $A = \{1, 3, 5\}$, $B = \{1, 4, 6\}$, and $C = \{2, 5, 7\}$. Determine elements of the following sets:

(a) $S \cup C$.

(b) $A \cup B$.

(c) $\overline{A}C$.

(d) $\overline{A} \cup (BC)$.

(e) $\overline{ABC}$.

(f) $\overline{A}\,\overline{B}$.

(g) $(AB) \cup (BC) \cup (CA)$.

2.5 Repeat Problem 2.4 if $S = \{x : 0 \le x \le 10\}$, $A = \{x : 1 \le x \le 5\}$, $B = \{x : 1 \le x \le 6\}$, and $C = \{x : 2 \le x \le 7\}$.

2.6 Draw Venn diagrams of events $A$ and $B$ representing the following situations:

(a) $A$ and $B$ are arbitrary.

(b) If $A$ occurs, $B$ must occur.

(c) If $A$ occurs, $B$ cannot occur.

(d) $A$ and $B$ are independent.

2.7 Let $A$, $B$, and $C$ be arbitrary events. Find expressions for the events that of $A$, $B$, $C$:

(a) None occurs.

(b) Only $A$ occurs.

(c) Only one occurs.

(d) At least one occurs.

(e) $A$ occurs and either $B$ or $C$ occurs but not both.

(f) $B$ and $C$ occur, but $A$ does not occur.

(g) Two or more occur.

(h) At most two occur.

(i) All three occur.

2.8 Events $A$, $B$, and $C$ are independent, with $P(A) = a$, $P(B) = b$, and $P(C) = c$. Determine the following probabilities in terms of $a$, $b$, and $c$:

(a) $P(AB)$.

(b) $P(A \cup B)$.

(c) $P(A \cup B|B)$.

(d) $P(A \cup B|C)$.

2.9 An engineering system has two components. Let us define the following events:

A : first component is good; $\overline{A}$: first component is defective.

B : second component is good; $\overline{B}$: second component is defective.

Describe the following events in terms of $A$, $\overline{A}$, $B$, and $\overline{B}$:

(a) At least one of the components is good.

(b) One is good and one is defective.

2.10 For the two components described in Problem 2.9, tests have produced the following result:

$$P(A) = 0.8, \quad P(B|A) = 0.85, \quad P(B|\overline{A}) = 0.75.$$

Determine the probability that:
(a) The second component is good.
(b) At least one of the components is good.
(c) The first component is good given that the second is good.
(d) The first component is good given that at most one component is good.

  For the two events $A$ and $B$:
(e) Are they independent? Verify your answer.
(f) Are they mutually exclusive? Verify your answer.

2.11 A satellite can fail for many possible reasons, two of which are computer failure and engine failure. For a given mission, it is known that:

The probability of engine failure is 0.008.

The probability of computer failure is 0.001.

Given engine failure, the probability of satellite failure is 0.98.

Given computer failure, the probability of satellite failure is 0.45.

Given any other component failure, the probability of satellite failure is zero.

(a) Determine the probability that a satellite fails.
(b) Determine the probability that a satellite fails and is due to engine failure.
(c) Assume that engines in different satellites perform independently. Given a satellite has failed as a result of engine failure, what is the probability that the same will happen to another satellite?

2.12 Verify Equation (2.14).

2.13 Show that, for arbitrary events $A_1, A_2, \ldots, A_n$,

$$P(A_1 \cup A_2 \cup \ldots \cup A_n) \leq P(A_1) + P(A_2) + \cdots + P(A_n)$$

This is known as *Boole's inequality*.

2.14 A box contains 20 parts, of which 5 are defective. Two parts are drawn at random from the box. What is the probability that:
(a) Both are good?
(b) Both are defective?
(c) One is good and one is defective?

2.15 An automobile braking device consists of three subsystems, all of which must work for the device to work. These systems are an electronic system, a hydraulic system, and a mechanical activator. In braking, the reliabilities (probabilities of success) of these units are 0.96, 0.95, and 0.95, respectively. Estimate the system reliability assuming that these subsystems function independently.
   *Comment*: systems of this type can be graphically represented as shown in Figure 2.10, in which subsystems $A$ (electronic system), $B$ (hydraulic system), and

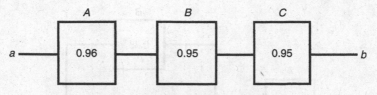

**Figure 2.10**    Figure for Problem 2.15

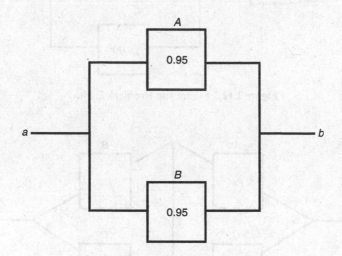

**Figure 2.11**    Figure for Problem 2.17

*C* (mechanical activator) are arranged in series. Consider the path $a \rightarrow b$ as the 'path to success'. A breakdown of any or all of *A*, *B*, or *C* will block the path from *a* to *b*.

2.16  A spacecraft has 1000 components in series. If the required reliability of the spacecraft is 0.9 and if all components function independently and have the same reliability, what is the required reliability of each component?

2.17  Automobiles are equipped with redundant braking circuits; their brakes fail only when *all* circuits fail. Consider one with two redundant braking circuits, each having a reliability of 0.95. Determine the system reliability assuming that these circuits act independently.

   *Comment*: systems of this type are graphically represented as in Figure 2.11, in which the circuits (*A* and *B*) have a parallel arrangement. The path to success is broken only when breakdowns of both *A* and *B* occur.

2.18  On the basis of definitions given in Problems 2.15 and 2.17 for series and parallel arrangements of system components, determine reliabilities of the systems described by the block diagrams as follows.
   (a)  The diagram in Figure 2.12.
   (b)  The diagram in Figure 2.13.

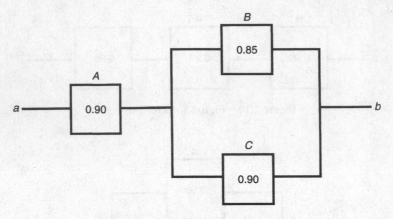

**Figure 2.12**   Figure for Problem 2.18(a)

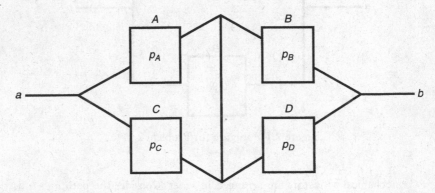

**Figure 2.13**   Figure for Problem 2.18(b)

2.19 A rifle is fired at a target. Assuming that the probability of scoring a hit is 0.9 for each shot and that the shots are independent, compute the probability that, in order to score a hit:
(a) It takes more than two shots.
(b) The number of shots required is between four and six (inclusive).

2.20 Events $A$ and $B$ are mutually exclusive. Can they also be independent? Explain.

2.21 Let $P(A) = 0.4$, and $P(A \cup B) = 0.7$. What is $P(B)$ if:
(a) $A$ and $B$ are independent?
(b) $A$ and $B$ are mutually exclusive?

2.22 Let $P(A \cup B) = 0.75$, and $P(AB) = 0.25$. Is it possible to determine $P(A)$ and $P(B)$? Answer the same question if, in addition:
(a) $A$ and $B$ are independent.
(b) $A$ and $B$ are mutually exclusive.

2.23 Events $A$ and $B$ are mutually exclusive. Determine which of the following relations are true and which are false:
   (a) $P(A|B) = P(A)$.
   (b) $P(A \cup B|C) = P(A|C) + P(B|C)$.
   (c) $P(A) = 0, P(B) = 0$, or both.
   (d) $\dfrac{P(A|B)}{P(B)} = \dfrac{P(B|A)}{P(A)}$.
   (e) $P(AB) = P(A)P(B)$.

   Repeat the above if events $A$ and $B$ are independent.

2.24 On a stretch of highway, the probability of an accident due to human error in any given minute is $10^{-5}$, and the probability of an accident due to mechanical breakdown in any given minute is $10^{-7}$. Assuming that these two causes are independent:
   (a) Find the probability of the occurrence of an accident on this stretch of highway during any minute.
   (b) In this case, can the above answer be approximated by $P$(accident due to human error) $+ P$(accident due to mechanical failure)? Explain.
   (c) If the events in succeeding minutes are mutually independent, what is the probability that there will be no accident at this location in a year?

2.25 Rapid transit trains arrive at a given station every five minutes and depart after stopping at the station for one minute to drop off and pick up passengers. Assuming trains arrive every hour on the hour, what is the probability that a passenger will be able to board a train immediately if he or she arrives at the station at a random instant between 7:54 a.m. and 8:06 a.m.?

2.26 A telephone call occurs at random in the interval $(0, t)$. Let $T$ be its time of occurrence. Determine, where $0 \leq t_0 \leq t_1 \leq t$:
   (a) $P(t_0 \leq T \leq t_1)$.
   (b) $P(t_0 \leq T \leq t_1 | T \geq t_0)$.

2.27 For a storm-sewer system, estimates of annual maximum flow rates (AMFR) and their likelihood of occurrence [assuming that a maximum of 12 cfs (cubic feet per second) is possible] are given as follows:

   Event $A = (5$ to $10\,\text{cfs})$,    $P(A) = 0.6$.
   Event $B = (8$ to $12\,\text{cfs})$,    $P(B) = 0.6$.
   Event $C = A \cup B$,          $P(C) = 0.7$.

   Determine:
   (a) $P(8 \leq \text{AMFR} \leq 10)$, the probability that the AMFR is between 8 and 10 cfs.
   (b) $P(5 \leq \text{AMFR} \leq 12)$.
   (c) $P(10 \leq \text{AMFR} \leq 12)$.
   (d) $P(8 \leq \text{AMFR} \leq 10 | 5 \leq \text{AMFR} \leq 10)$.
   (e) $P(5 \leq \text{AMFR} \leq 10 | \text{AMFR} \geq 5)$.

2.28 At a major and minor street intersection, one finds that, out of every 100 gaps on the major street, 65 are acceptable, that is, large enough for a car arriving on the minor street to cross. When a vehicle arrives on the minor street:
   (a) What is the probability that the first gap is not an acceptable one?
   (b) What is the probability that the first two gaps are both unacceptable?
   (c) The first car has crossed the intersection. What is the probability that the second will be able to cross at the very next gap?

2.29 A machine part may be selected from any of three manufacturers with probabilities $p_1 = 0.25$, $p_2 = 0.50$, and $p_3 = 0.25$. The probabilities that it will function properly during a specified period of time are 0.2, 0.3, and 0.4, respectively, for the three manufacturers. Determine the probability that a randomly chosen machine part will function properly for the specified time period.

2.30 Consider the possible failure of a transportation system to meet demand during rush hour.
  (a) Determine the probability that the system will fail if the probabilities shown in Table 2.3 are known.

**Table 2.3**  Probabilities of demand levels and of system failures for the given demand level, for Problem 2.30

| Demand level | P(level) | P(system failure\|level) |
|---|---|---|
| Low | 0.6 | 0 |
| Medium | 0.3 | 0.1 |
| High | 0.1 | 0.5 |

  (b) If system failure was observed, find the probability that a 'medium' demand level was its cause.

2.31 A cancer diagnostic test is 95% accurate both on those who have cancer and on those who do not. If 0.005 of the population actually does have cancer, compute the probability that a particular individual has cancer, given that the test indicates he or she has cancer.

2.32 A quality control record panel of transistors gives the results shown in Table 2.4 when classified by manufacturer and quality.
  Let one transistor be selected at random. What is the probability of it being:
  (a) From manufacturer A and with acceptable quality?
  (b) Acceptable given that it is from manufacturer C?
  (c) From manufacturer B given that it is marginal?

**Table 2.4**  Quality control results, for Problem 2.32

| Manufacturer | Quality | | | |
|---|---|---|---|---|
| | Acceptable | Marginal | Unacceptable | Total |
| A | 128 | 10 | 2 | 140 |
| B | 97 | 5 | 3 | 105 |
| C | 110 | 5 | 5 | 120 |

2.33 Verify Equation (2.26) for three events.

2.34 In an elementary study of synchronized traffic lights, consider a simple four-light system. Suppose that each light is red for 30 seconds of a 50-second cycle, and suppose

$$P(S_{j+1}|S_j) = 0.15$$

and

$$P(S_{j+1}|\overline{S}_j) = 0.40$$

for $j = 1, 2, 3$, where $S_j$ is the event that a driver is stopped by the $j$th light. We assume a 'one-light' memory for the system. By means of the tree diagram, determine the probability that a driver:

(a)  Will be delayed by all four lights.

(b)  Will not be delayed by any of the four lights.

(c)  Will be delayed by at most one light.

# 3

# Random Variables and Probability Distributions

We have mentioned that our interest in the study of a random phenomenon is in the statements we can make concerning the events that can occur, and these statements are made based on probabilities assigned to simple outcomes. Basic concepts have been developed in Chapter 2, but a systematic and unified procedure is needed to facilitate making these statements, which can be quite complex. One of the immediate steps that can be taken in this unifying attempt is to require that each of the possible outcomes of a random experiment be represented by a real number. In this way, when the experiment is performed, each outcome is identified by its assigned real number rather than by its physical description. For example, when the possible outcomes of a random experiment consist of success and failure, we arbitrarily assign the number one to the event 'success' and the number zero to the event 'failure'. The associated sample space has now $\{1, 0\}$ as its sample points instead of success and failure, and the statement 'the outcome is 1' means 'the outcome is success'.

This procedure not only permits us to replace a sample space of arbitrary elements by a new sample space having only real numbers as its elements but also enables us to use arithmetic means for probability calculations. Furthermore, most problems in science and engineering deal with quantitative measures. Consequently, sample spaces associated with many random experiments of interest are already themselves sets of real numbers. The real-number assignment procedure is thus a natural unifying agent. On this basis, we may introduce a variable $X$, which is used to represent real numbers, the values of which are determined by the outcomes of a random experiment. This leads to the notion of a random variable, which is defined more precisely below.

## 3.1 RANDOM VARIABLES

Consider a random experiment to which the outcomes are elements of sample space $S$ in the underlying probability space. In order to construct a model for

---

*Fundamentals of Probability and Statistics for Engineers* T.T. Soong © 2004 John Wiley & Sons, Ltd
ISBNs: 0-470-86813-9 (HB)    0-470-86814-7 (PB)

a random variable, we assume that it is possible to assign a real number $X(s)$ for each outcome $s$ following a certain set of rules. We see that the 'number' $X(s)$ is really a real-valued *point function* defined over the domain of the basic probability space (see Definition 3.1).

**Definition 3.1.** The point function $X(s)$ is called a *random variable* if (a) it is a finite real-valued function defined on the sample space $S$ of a random experiment for which the probability function is defined, and (b) for every real number $x$, the set $\{s : X(s) \leq x\}$ is an event. The relation $X = X(s)$ takes every element $s$ in $S$ of the probability space onto a point $X$ on the real line $R^1 = (-\infty, \infty)$.

Notationally, the dependence of random variable $X(s)$ on $s$ will be omitted for convenience.

The second condition stated in Definition 3.1 is the so-called 'measurability condition'. It ensures that it is meaningful to consider the probability of event $X \leq x$ for every $x$, or, more generally, the probability of any finite or countably infinite combination of such events.

To see more clearly the role a random variable plays in the study of a random phenomenon, consider again the simple example where the possible outcomes of a random experiment are success and failure. Let us again assign number one to the event success and zero to failure. If $X$ is the random variable associated with this experiment, then $X$ takes on two possible values: 1 and 0. Moreover, the following statements are equivalent:

- The outcome is success.
- The outcome is 1.
- $X = 1$.

The random variable $X$ is called a *discrete* random variable if it is defined over a sample space having a finite or a countably infinite number of sample points. In this case, random variable $X$ takes on discrete values, and it is possible to enumerate all the values it may assume. In the case of a sample space having an uncountably infinite number of sample points, the associated random variable is called a *continuous* random variable, with its values distributed over one or more continuous intervals on the real line. We make this distinction because they require different probability assignment considerations. Both types of random variables are important in science and engineering and we shall see ample evidence of this in the subsequent chapters.

In the following, all random variables will be written in capital letters, $X, Y, Z, \ldots$. The value that a random variable $X$ can assume will be denoted by corresponding lower-case letters such as $x, y, z$, or $x_1, x_2, \ldots$.

We will have many occasions to consider a sequence of random variables $X_j, j = 1, 2, \ldots, n$. In these cases we assume that they are defined on the same probability space. The random variables $X_1, X_2, \ldots, X_n$ will then map every element $s$ of $S$ in the probability space onto a point in the $n$-dimensional

Euclidian space $R^n$. We note here that an analysis involving $n$ random variables is equivalent to considering a *random vector* having the $n$ random variables as its components. The notion of a random vector will be used frequently in what follows, and we will denote them by bold capital letters $X$, $Y$, $Z$, . . . .

## 3.2 PROBABILITY DISTRIBUTIONS

The behavior of a random variable is characterized by its probability distribution, that is, by the way probabilities are distributed over the values it assumes. A probability distribution function and a probability mass function are two ways to characterize this distribution for a discrete random variable. They are equivalent in the sense that the knowledge of either one completely specifies the random variable. The corresponding functions for a continuous random variable are the probability distribution function, defined in the same way as in the case of a discrete random variable, and the probability density function. The definitions of these functions now follow.

### 3.2.1  PROBABILITY DISTRIBUTION FUNCTION

Given a random experiment with its associated random variable $X$ and given a real number $x$, let us consider the probability of the event $\{s : X(s) \leq x\}$, or, simply, $P(X \leq x)$. This probability is clearly dependent on the assigned value $x$. The function

$$F_X(x) = P(X \leq x), \qquad (3.1)$$

is defined as the *probability distribution function* (PDF), or simply the *distribution function*, of $X$. In Equation (3.1), subscript $X$ identifies the random variable. This subscript is sometimes omitted when there is no risk of confusion. Let us repeat that $F_X(x)$ is simply $P(A)$, the probability of an event $A$ occurring, the event being $X \leq x$. This observation ties what we do here with the development of Chapter 2.

The PDF is thus the probability that $X$ will assume a value lying in a subset of $S$, the subset being point $x$ and all points lying to the 'left' of $x$. As $x$ increases, the subset covers more of the real line, and the value of PDF increases until it reaches 1. The PDF of a random variable thus accumulates probability as $x$ increases, and the name *cumulative distribution function* (CDF) is also used for this function.

In view of the definition and the discussion above, we give below some of the important properties possessed by a PDF.

- It exists for discrete and continuous random variables and has values between 0 and 1.
- It is a nonnegative, continuous-to-the-left, and nondecreasing function of the real variable $x$. Moreover, we have

$$F_X(-\infty) = 0, \quad \text{and} \quad F_X(+\infty) = 1. \tag{3.2}$$

- If $a$ and $b$ are two real numbers such that $a < b$, then

$$P(a < X \le b) = F_X(b) - F_X(a). \tag{3.3}$$

This relation is a direct result of the identity

$$P(X \le b) = P(X \le a) + P(a < X \le b).$$

We see from Equation (3.3) that the probability of $X$ having a value in an arbitrary interval can be represented by the difference between two values of the PDF. Generalizing, probabilities associated with any sets of intervals are derivable from the PDF.

**Example 3.1.** Let a discrete random variable $X$ assume values $-1, 1, 2$, and $3$, with probabilities $\frac{1}{4}, \frac{1}{8}, \frac{1}{8}$, and $\frac{1}{2}$, respectively. We then have

$$F_X(x) = \begin{cases} 0, & \text{for } x < -1; \\ \frac{1}{4}, & \text{for } -1 \le x < 1; \\ \frac{3}{8}, & \text{for } 1 \le x < 2; \\ \frac{1}{2}, & \text{for } 2 \le x < 3; \\ 1, & \text{for } x \ge 3. \end{cases}$$

This function is plotted in Figure 3.1. It is typical of PDFs associated with discrete random variables, increasing from 0 to 1 in a 'staircase' fashion.

A continuous random variable assumes a nonenumerable number of values over the real line. Hence, the probability of a continuous random variable assuming any particular value is zero and therefore no discrete jumps are possible for its PDF. A typical PDF for continuous random variables is shown in Figure 3.2. It has no jumps or discontinuities as in the case of the discrete random variable. The probability of $X$ having a value in a given interval is found by using Equation (3.3), and it makes sense to speak only of this kind of probability for continuous random variables. For example, in Figure 3.2.

$$P(-1 < X \le 1) = F_X(1) - F_X(-1) = 0.8 - 0.4 = 0.4.$$

Clearly, $P(X = a) = 0$ for any $a$.

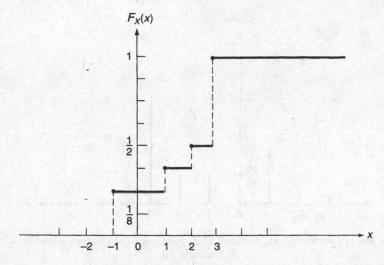

**Figure 3.1**   Probability distribution function of $X$, $F_X(x)$, for Example 3.1

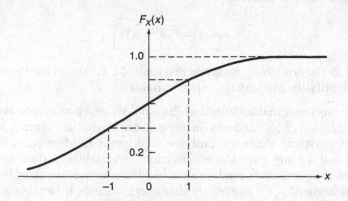

**Figure 3.2**   Probability distribution function of a continuous random variable $X$, $F_X(x)$

## 3.2.2   PROBABILITY MASS FUNCTION FOR DISCRETE RANDOM VARIABLES

Let $X$ be a discrete random variable that assumes at most a countably infinite number of values $x_1, x_2, \ldots$ with nonzero probabilities. If we denote $P(X = x_i) = p(x_i), i = 1, 2, \ldots$, then, clearly,

$$\left. \begin{array}{l} 0 < p(x_i) \leq 1, \text{ for all i;} \\ \sum_i p(x_i) = 1. \end{array} \right\} \qquad (3.4)$$

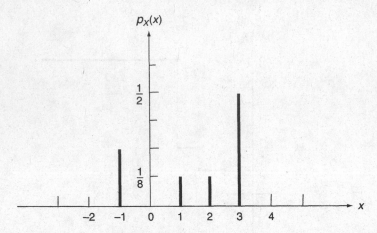

**Figure 3.3**   Probability mass function of $X$, $p_X(x)$, for the random variable defined in Example 3.1

**Definition 3.2.** The function

$$p_X(x) = P(X = x).$$

(3.5)

is defined as the *probability mass function* (pmf) of $X$. Again, the subscript $X$ is used to identify the associated random variable.

For the random variable defined in Example 3.1, the pmf is zero everywhere except at $x_i, i = 1, 2, \ldots,$ and has the appearance shown in Figure 3.3.

This is a typical shape of pmf for a discrete random variable. Since $P(X = x) = 0$ for any $x$ for continuous random variables, it does not exist in the case of the continuous random variable. We also observe that, like $F_X(x)$, the specification of $p_X(x)$ completely characterizes random variable $X$; furthermore, these two functions are simply related by:

$$p_X(x_i) = F_X(x_i) - F_X(x_{i-1}),$$

(3.6)

$$F_X(x) = \sum_{i=1}^{i:x_i \le x} p_X(x_i),$$

(3.7)

(assuming $x_1 < x_2 < \ldots$).

The upper limit for the sum in Equation (3.7) means that the sum is taken over all $i$ satisfying $x_i \le x$. Hence, we see that the PDF and pmf of a discrete random variable contain the same information; each one is recoverable from the other.

One can also give PDF and pmf a useful physical interpretation. In terms of the distribution of one unit of mass over the real line $-\infty < x < \infty$, the PDF of a random variable at $x$, $F_X(x)$, can be interpreted as the total mass associated with point $x$ and all points lying to the left of $x$. The pmf, in contrast, shows the distribution of this unit of mass over the real line; it is distributed at discrete points $x_i$ with the amount of mass equal to $p_X(x_i)$ at $x_i, i = 1, 2, \ldots$.

**Example 3.2.** A discrete distribution arising in a large number of physical models is the *binomial distribution*. Much more will be said of this important distribution in Chapter 6 but, at present, let us use it as an illustration for graphing the PDF and pmf of a discrete random variable.

A discrete random variable $X$ has a binomial distribution when

$$p_X(k) = \binom{n}{k} p^k (1-p)^{n-k}, \quad k = 0, 1, 2, \ldots, n, \tag{3.8}$$

where $n$ and $p$ are two parameters of the distribution, $n$ being a positive integer, and $0 < p < 1$. The binomial coefficient

$$\binom{n}{k}$$

is defined by

$$\binom{n}{k} = \frac{n!}{k!(n-k)!}. \tag{3.9}$$

The pmf and PDF of $X$ for $n = 10$ and $p = 0.2$ are plotted in Figure 3.4.

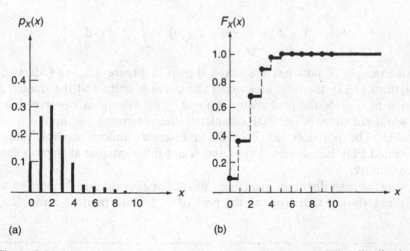

**Figure 3.4** (a) Probability mass function, $p_X(x)$, and (b) probability distribution function, $F_X(x)$, for the discrete random variable $X$ described in Example 3.2

### 3.2.3   PROBABILITY DENSITY FUNCTION FOR CONTINUOUS RANDOM VARIABLES

For a continuous random variable $X$, its PDF, $F_X(x)$, is a continuous function of $x$, and the derivative

$$f_X(x) = \frac{\mathrm{d}F_X(x)}{\mathrm{d}x},$$
(3.10)

exists for all $x$. The function $f_X(x)$ is called the *probability density function* (pdf), or simply the *density function*, of $X$.[1]

Since $F_X(x)$ is monotone nondecreasing, we clearly have

$$f_X(x) \geq 0 \quad \text{for all } x.$$
(3.11)

Additional properties of $f_X(x)$ can be derived easily from Equation (3.10); these include

$$F_X(x) = \int_{-\infty}^{x} f_X(u)\,\mathrm{d}u,$$
(3.12)

and

$$\left.\begin{aligned}
&\int_{-\infty}^{\infty} f_X(x)\,\mathrm{d}x = 1, \\
&P(a < X \leq b) = F_X(b) - F_X(a) = \int_a^b f_X(x)\,\mathrm{d}x.
\end{aligned}\right\}$$
(3.13)

An example of pdfs has the shape shown in Figure 3.5. As indicated by Equations (3.13), the total area under the curve is unity and the shaded area from $a$ to $b$ gives the probability $P(a < X \leq b)$. We again observe that the knowledge of either pdf or PDF completely characterizes a continuous random variable. The pdf does not exist for a discrete random variable since its associated PDF has discrete jumps and is not differentiable at these points of discontinuity.

Using the mass distribution analogy, the pdf of a continuous random variable plays exactly the same role as the pmf of a discrete random variable. The

---

[1] Note the use of upper-case and lower-case letters, PDF and pdf, to represent the distribution and density functions, respectively.

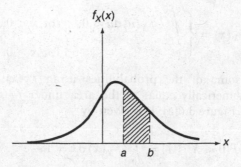

**Figure 3.5**   A probability density function, $f_X(x)$

function $f_X(x)$ can be interpreted as the mass density (mass per unit length). There are no masses attached to discrete points as in the discrete random variable case. The use of the term *density function* is therefore appropriate here for $f_X(x)$.

**Example 3.3.** A random variable $X$ for which the density function has the form ($a > 0$):

$$f_X(x) = \begin{cases} ae^{-ax}, & \text{for } x \geq 0; \\ 0, & \text{elsewhere;} \end{cases} \qquad (3.14)$$

is said to be *exponentially distributed*. We can easily check that all the conditions given by Equations (3.11)–(3.13) are satisfied. The pdf is presented graphically in Figure 3.6(a), and the associated PDF is shown in Figure 3.6(b). The functional form of the PDF as obtained from Equation (3.12) is

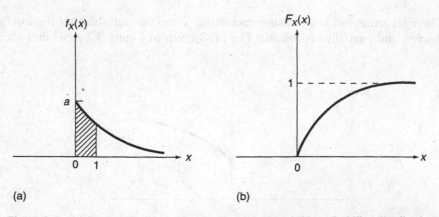

(a)                                                              (b)

**Figure 3.6**   (a) Probability density function, $f_X(x)$, and (b) probability distribution function, $F_X(x)$, for random variable $X$ in Example 3.3

$$F_X(x) = \begin{cases} \int_{-\infty}^{x} f_X(u)\, du = 0, & \text{for } x < 0; \\ 1 - e^{-ax}, & \text{for } x \geq 0. \end{cases} \qquad (3.15)$$

Let us compute some of the probabilities using $f_X(x)$. The probability $P(0 < X \leq 1)$ is numerically equal to the area under $f_X(x)$ from $x = 0$ to $x = 1$, as shown in Figure 3.6(a). It is given by

$$P(0 < X \leq 1) = \int_{0}^{1} f_X(x)\, dx = 1 - e^{-a}.$$

The probability $P(X > 3)$ is obtained by computing the area under $f_X(x)$ to the right of $x = 3$. Hence,

$$P(X > 3) = \int_{3}^{\infty} f_X(x)\, dx = e^{-3a}.$$

The same probabilities can be obtained from $F_X(x)$ by taking appropriate differences, giving:

$$P(0 < X \leq 1) = F_X(1) - F_X(0) = (1 - e^{-a}) - 0 = 1 - e^{-a},$$
$$P(X > 3) = F_X(\infty) - F_X(3) = 1 - (1 - e^{-3a}) = e^{-3a}.$$

Let us note that there is no numerical difference between $P(0 < X \leq 1)$ and $P(0 \leq X \leq 1)$ for continuous random variables, since $P(X = 0) = 0$.

### 3.2.4   MIXED-TYPE DISTRIBUTION

There are situations in which one encounters a random variable that is partially discrete and partially continuous. The PDF given in Figure 3.7 represents such

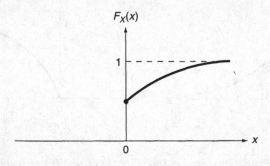

**Figure 3.7**   A mixed-type probability distribution function, $F_X(x)$

a case in which random variable $X$ is continuously distributed over the real line except at $X = 0$, where $P(X = 0)$ is a positive quantity. This situation may arise when, for example, random variable $X$ represents the waiting time of a customer at a ticket counter. Let $X$ be the time interval from time of arrival at the ticket counter to the time being served. It is reasonable to expect that $X$ will assume values over the interval $X \geq 0$. At $X = 0$, however, there is a finite probability of not having to wait at all, giving rise to the situation depicted in Figure 3.7.

Strictly speaking, neither a pmf nor a pdf exists for a random variable of the mixed type. We can, however, still use them separately for different portions of the distribution, for computational purposes. Let $f_X(x)$ be the pdf for the *continuous* portion of the distribution. It can be used for calculating probabilities in the positive range of $x$ values for this example. We observe that the total area under the pdf curve is no longer 1 but is equal to $1 - P(X = 0)$.

**Example 3.4.** Problem: since it is more economical to limit long-distance telephone calls to three minutes or less, the PDF of $X$ – the duration in minutes of long-distance calls – may be of the form

$$F_X(x) = \begin{cases} 0, & \text{for } x < 0; \\ 1 - e^{-x/3}, & \text{for } 0 \leq x < 3; \\ 1 - \frac{e^{-x/3}}{2}, & \text{for } x \geq 3. \end{cases}$$

Determine the probability that $X$ is (a) more than two minutes and (b) between two and six minutes.

Answer: the PDF of $X$ is plotted in Figure 3.8, showing that $X$ has a mixed-type distribution. The desired probabilities can be found from the PDF as before. Hence, for part (a),

$$P(X > 2) = 1 - P(X \leq 2) = 1 - F_X(2)$$
$$= 1 - (1 - e^{-2/3}) = e^{-2/3}.$$

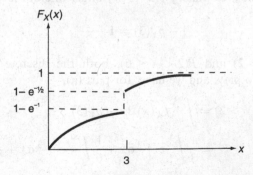

**Figure 3.8** Probability distribution function, $F_X(x)$, of $X$, as described in Example 3.4

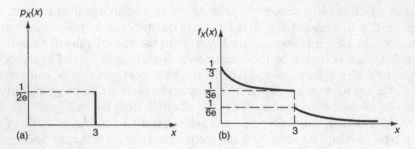

**Figure 3.9** (a) Partial probability mass function, $p_X(x)$, and (b) partial probability density function, $f_X(x)$, of $X$, as described in Example 3.4

For part (b),

$$P(2 < X \le 6) = F_X(6) - F_X(2)$$

$$= \left(1 - \frac{e^{-2}}{2}\right) - (1 - e^{-2/3}) = e^{-2/3} - \frac{e^{-2}}{2}.$$

Figure 3.9 shows $p_X(x)$ for the discrete portion and $f_X(x)$ for the continuous portion of $X$. They are given by:

$$p_X(x) = \begin{cases} \dfrac{1}{2e}, & \text{at } x = 3; \\ 0, & \text{elsewhere}; \end{cases}$$

and

$$f_X(x) = \frac{dF_X(x)}{dx} = \begin{cases} 0, & \text{for } x < 0; \\ \dfrac{1}{3} e^{-x/3}, & \text{for } 0 \le x < 3; \\ \dfrac{1}{6} e^{-x/3}, & \text{for } x \ge 3. \end{cases}$$

Note again that the area under $f_X(x)$ is no longer one but is

$$1 - p_X(3) = 1 - \frac{1}{2e}.$$

To obtain $P(X > 2)$ and $P(2 < X \le 6)$, both the discrete and continuous portions come into play, and we have, for part (a),

$$P(X > 2) = \int_2^\infty f_X(x)\,dx + p_X(3)$$

$$= \frac{1}{3} \int_2^3 e^{-x/3}\,dx + \frac{1}{6} \int_3^\infty e^{-x/3}\,dx + \frac{1}{2e}$$

$$= e^{-2/3}$$

and, for part (b),

$$P(2 < X \le 6) = \int_2^6 f_X(x)\,dx + p_X(3)$$

$$= \frac{1}{3}\int_2^3 e^{-x/3}\,dx + \frac{1}{6}\int_3^6 e^{-x/3}\,dx + \frac{1}{2e}$$

$$= e^{-2/3} - \frac{e^{-2}}{2}$$

These results are, of course, the same as those obtained earlier using the PDF.

## 3.3 TWO OR MORE RANDOM VARIABLES

In many cases it is more natural to describe the outcome of a random experiment by two or more numerical numbers simultaneously. For example, the characterization of both weight and height in a given population, the study of temperature and pressure variations in a physical experiment, and the distribution of monthly temperature readings in a given region over a given year. In these situations, two or more random variables are considered jointly and the description of their joint behavior is our concern.

Let us first consider the case of two random variables $X$ and $Y$. We proceed analogously to the single random variable case in defining their joint probability distributions. We note that random variables $X$ and $Y$ can also be considered as components of a two-dimensional random vector, say $Z$. Joint probability distributions associated with two random variables are sometimes called *bivariate distributions*.

As we shall see, extensions to cases involving more than two random variables, or *multivariate distributions*, are straightforward.

### 3.3.1 JOINT PROBABILITY DISTRIBUTION FUNCTION

The *joint probability distribution* function (JPDF) of random variables $X$ and $Y$, denoted by $F_{XY}(x, y)$, is defined by

$$\boxed{F_{XY}(x, y) = P(X \le x \cap Y \le y),} \qquad (3.16)$$

for all $x$ and $y$. It is the probability of the intersection of two events; random variables $X$ and $Y$ thus induce a probability distribution over a two-dimensional Euclidean plane.

Using again the mass distribution analogy, let one unit of mass be distributed over the $(x, y)$ plane in such a way that the mass in any given region $R$ is equal to the probability that $X$ and $Y$ take values in $R$. Then JPDF $F_{XY}(x, y)$ represents the total mass in the quadrant to the left and below the point $(x, y)$, inclusive of the boundaries. In the case where both $X$ and $Y$ are discrete, all the mass is concentrated at a finite or countably infinite number of points in the $(x, y)$ plane as point masses. When both are continuous, the mass is distributed continuously over the $(x, y)$ plane.

It is clear from the definition that $F_{XY}(x, y)$ is nonnegative, nondecreasing in $x$ and $y$, and continuous to the left with respect to $x$ and $y$. The following properties are also a direct consequence of the definition:

$$\left.\begin{aligned}
F_{XY}(-\infty, -\infty) &= F_{XY}(-\infty, y) = F_{XY}(x, -\infty) = 0, \\
F_{XY}(+\infty, +\infty) &= 1, \\
F_{XY}(x, +\infty) &= F_X(x), \\
F_{XY}(+\infty, y) &= F_Y(y).
\end{aligned}\right\} \qquad (3.17)$$

For example, the third relation above follows from the fact that the joint event $X \leq x \cap Y \leq +\infty$ is the same as the event $X \leq x$, since $Y \leq +\infty$ is a sure event. Hence,

$$F_{XY}(x, +\infty) = P(X \leq x \cap Y \leq +\infty) = P(X \leq x) = F_X(x).$$

Similarly, we can show that, for any $x_1, x_2, y_1$, and $y_2$ such that $x_1 < x_2$ and $y_1 < y_2$, the probability $P(x_1 < X \leq x_2 \cap y_1 < Y \leq y_2)$ is given in terms of $F_{XY}(x, y)$ by

$$\begin{aligned}
P(x_1 < X \leq x_2 \cap y_1 < Y \leq y_2) &= F_{XY}(x_2, y_2) - F_{XY}(x_1, y_2) \\
&\quad - F_{XY}(x_2, y_1) + F_{XY}(x_1, y_1),
\end{aligned} \qquad (3.18)$$

which shows that all probability calculations involving random variables $X$ and $Y$ can be made with the knowledge of their JPDF.

Finally, we note that the last two equations in Equations (3.17) show that distribution functions of individual random variables are directly derivable from their joint distribution function. The converse, of course, is not true. In the context of several random variables, these individual distribution functions are called *marginal distribution functions*. For example, $F_X(x)$ is the marginal distribution function of $X$.

The general shape of $F_{XY}(x, y)$ can be visualized from the properties given in Equations (3.17). In the case where $X$ and $Y$ are discrete, it has the appearance of a corner of an irregular staircase, something like that shown in Figure 3.10. It rises from zero to the height of one in the direction moving from the third quadrant to the

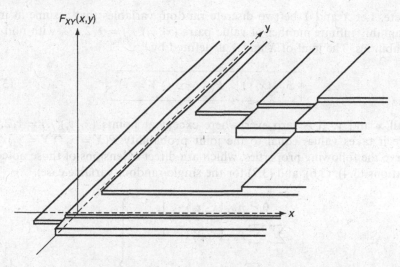

**Figure 3.10**   A joint probability distribution function of $X$ and $Y$, $F_{XY}(x,y)$, when $X$ and $Y$ are discrete

first quadrant. When both $X$ and $Y$ are continuous, $F_{XY}(x, y)$ becomes a smooth surface with the same features. It is a staircase type in one direction and smooth in the other if one of the random variables is discrete and the other continuous.

The joint probability distribution function of more than two random variables is defined in a similar fashion. Consider $n$ random variables $X_1, X_2, \ldots, X_n$. Their JPDF is defined by

$$F_{X_1 X_2 \ldots X_n}(x_1, x_2, \ldots, x_n) = P(X_1 \leq x_1 \cap X_2 \leq x_2 \cap \ldots \cap X_n \leq x_n). \qquad (3.19)$$

These random variables induce a probability distribution in an $n$-dimensional Euclidean space. One can deduce immediately its properties in parallel to those noted in Equations (3.17) and (3.18) for the two-random-variable case.

As we have mentioned previously, a finite number of random variables $X_j, j = 1, 2, \ldots n$, may be regarded as the components of an $n$-dimensional random vector $X$. The JPDF of $X$ is identical to that given above but it can be written in a more compact form, namely, $F_X(x)$, where $x$ is the vector, with components $x_1, x_2, \ldots, x_n$.

### 3.3.2   JOINT PROBABILITY MASS FUNCTION

The *joint probability mass function* (jpmf) is another, and more direct, characterization of the joint behavior of two or more random variables when they are discrete. Let $X$ and $Y$ be two discrete random variables that assume at most a countably infinite number of value pairs $(x_i, y_j), i, j = 1, 2, \ldots$, with nonzero probabilities. The jpmf of $X$ and $Y$ is defined by

$$\boxed{p_{XY}(x, y) = P(X = x \cap Y = y),}$$ (3.20)

for all $x$ and $y$. It is zero everywhere except at points $(x_i, y_j), i, j = 1, 2, \ldots$, where it takes values equal to the joint probability $P(X = x_i \cap Y = y_j)$. We observe the following properties, which are direct extensions of those noted in Equations (3.4), (3.6), and (3.7) for the single-random-variable case:

$$\left.\begin{aligned}
0 &< p_{XY}(x_i, y_j) \leq 1, \\
\sum_i \sum_j & p_{XY}(x_i, y_j) = 1, \\
\sum_i & p_{XY}(x_i, y) = p_Y(y), \\
\sum_j & p_{XY}(x, y_j) = p_X(x),
\end{aligned}\right\}$$ (3.21)

where $p_X(x)$ and $p_Y(y)$ are now called *marginal probability mass functions*. We also have

$$\boxed{F_{XY}(x, y) = \sum_{i=1}^{i:x_i \leq x} \sum_{j=1}^{j:y_j \leq y} p_{XY}(x_i, y_j).}$$ (3.22)

**Example 3.5.** Problem: consider a simplified version of a two-dimensional 'random walk' problem. We imagine a particle that moves in a plane in unit steps starting from the origin. Each step is one unit in the positive direction, with probability $p$ along the $x$ axis and probability $q$ ($p + q = 1$) along the $y$ axis. We further assume that each step is taken independently of the others. What is the probability distribution of the position of this particle after five steps?

Answer: since the position is conveniently represented by two coordinates, we wish to establish $p_{XY}(x, y)$ where random variable $X$ represents the $x$ coordinate of the position after five steps and where $Y$ represents the $y$ coordinate. It is clear that jpmf $p_{XY}(x, y)$ is zero everywhere except at those points satisfying $x + y = 5$ and $x, y \geq 0$. Invoking the independence of events of taking successive steps, it follows from Section 3.3 that $p_{XY}(5, 0)$, the probability of the particle being at $(5, 0)$ after five steps, is the product of probabilities of taking five successive steps in the positive $x$ direction. Hence

$$p_{XY}(5,0) = p^5.$$

For $p_{XY}(4,1)$, there are five distinct ways of reaching that position (4 steps in the $x$ direction and 1 in $y$; 3 in the $x$ direction, 1 in $y$, and 1 in the $x$ direction; and so on), each with a probability of $p^4q$. We thus have

$$p_{XY}(4,1) = 5p^4q.$$

Similarly, other nonvanishing values of $p_{XY}(x,y)$ are easily calculated to be

$$p_{XY}(x,y) = \begin{cases} 10p^3q^2, & \text{for } (x,y) = (3,2); \\ 10p^2q^3, & \text{for } (x,y) = (2,3); \\ 5pq^4, & \text{for } (x,y) = (1,4); \\ q^5, & \text{for } (x,y) = (0,5). \end{cases}$$

The jpmf $p_{XY}(x,y)$ is graphically presented in Figure 3.11 for $p = 0.4$ and $q = 0.6$. It is easy to check that the sum of $p_{XY}(x,y)$ over all $x$ and $y$ is 1, as required by the second of Equations (3.21).

Let us note that the marginal probability mass functions of $X$ and $Y$ are, following the last two expressions in Equations (3.21),

$$p_X(x) = \sum_j p_{XY}(x,y_j) = \begin{cases} q^5, & \text{for } x = 0; \\ 5pq^4, & \text{for } x = 1; \\ 10p^2q^3, & \text{for } x = 2; \\ 10p^3q^2, & \text{for } x = 3; \\ 5p^4q, & \text{for } x = 4; \\ p^5, & \text{for } x = 5; \end{cases}$$

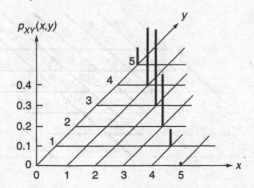

**Figure 3.11** The joint probability mass function, $p_{XY}(x,y)$, for Example 3.5, with $p = 0.4$ and $q = 0.6$

and

$$p_Y(y) = \sum_i p_{XY}(x_i, y) = \begin{cases} p^5, & \text{for } y = 0; \\ 5p^4q, & \text{for } y = 1; \\ 10p^3q^2, & \text{for } y = 2; \\ 10p^2q^3, & \text{for } y = 3; \\ 5pq^4, & \text{for } y = 4; \\ q^5, & \text{for } y = 5. \end{cases}$$

These are marginal pmfs of $X$ and $Y$.

The joint probability distribution function $F_{XY}(x, y)$ can also be constructed, by using Equation (3.22). Rather than showing it in three-dimensional form, Figure 3.12 gives this function by indicating its value in each of the dividing regions. One should also note that the arrays of indicated numbers beyond $y = 5$ are values associated with the marginal distribution function $F_X(x)$. Similarly, $F_Y(y)$ takes those values situated beyond $x = 5$. These observations are also indicated on the graph.

The knowledge of the joint probability mass function permits us to make all probability calculations of interest. The probability of any event being realized involving $X$ and $Y$ is found by determining the pairs of values of $X$ and $Y$ that give rise to this event and then simply summing over the values of $p_{XY}(x, y)$ for all such pairs. In Example 3.5, suppose we wish to determine the probability of $X > Y$; it is given by

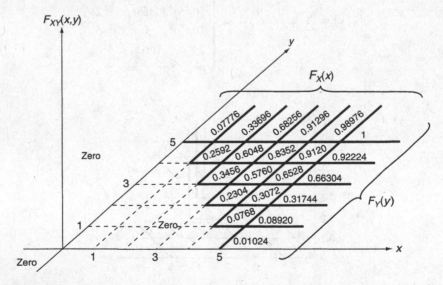

**Figure 3.12** The joint probability distribution function, $F_{XY}(x,y)$, for Example 3.5, with $p = 0.4$ and $q = 0.6$

**Table 3.1**  Joint probability mass function for low, medium, and high precipitation levels ($x = 1$, 2, and 3, respectively) and critical and noncritical peak flow rates ($y = 1$ and 2, respectively), for Example 3.6

| $y$ | $x$ | | |
|---|---|---|---|
|  | 1 | 2 | 3 |
| 1 | 0.0 | 0.06 | 0.12 |
| 2 | 0.5 | 0.24 | 0.08 |

$$P(X > Y) = P(X = 5 \cap Y = 0) + P(X = 4 \cap Y = 1) + P(X = 3 \cap Y = 2)$$
$$= 0.01024 + 0.0768 + 0.2304 = 0.31744.$$

**Example 3.6.** Let us discuss again Example 2.11 in the context of random variables. Let $X$ be the random variable representing precipitation levels, with values 1, 2, and 3 indicating low, medium, and high, respectively. The random variable $Y$ will be used for the peak flow rate, with the value 1 when it is critical and 2 when noncritical. The information given in Example 2.11 defines jpmf $p_{XY}(x, y)$, the values of which are tabulated in Table 3.1.

In order to determine the probability of reaching the critical level of peak flow rate, for example, we simply sum over all $p_{XY}(x, y)$ satisfying $y = 1$, regardless of $x$ values. Hence, we have

$$P(Y = 1) = p_{XY}(1, 1) + p_{XY}(2, 1) + p_{XY}(3, 1) = 0.0 + 0.06 + 0.12 = 0.18.$$

The definition of jpmf for more than two random variables is a direct extension of that for the two-random-variable case. Consider $n$ random variables $X_1, X_2, \ldots, X_n$. Their jpmf is defined by

$$\boxed{p_{X_1 X_2 \ldots X_n}(x_1, x_2, \ldots, x_n) = P(X_1 = x_1 \cap X_2 = x_2 \cap \ldots \cap X_n = x_n),} \quad (3.23)$$

which is the probability of the intersection of $n$ events. Its properties and utilities follow directly from our discussion in the two-random-variable case. Again, a more compact form for the jpmf is $p_X(x)$ where $X$ is an $n$-dimensional random vector with components $X_1, X_2, \ldots, X_n$.

### 3.3.3  JOINT PROBABILITY DENSITY FUNCTION

As in the case of single random variables, probability density functions become appropriate when the random variables are continuous. The *joint probability density function* (jpdf) of two random variables, $X$ and $Y$, is defined by the partial derivative

$$f_{XY}(x, y) = \frac{\partial^2 F_{XY}(x, y)}{\partial x \partial y}. \tag{3.24}$$

Since $F_{XY}(x, y)$ is monotone nondecreasing in both $x$ and $y$, $f_{XY}(x, y)$ is nonnegative for all $x$ and $y$. We also see from Equation (3.24) that

$$F_{XY}(x, y) \equiv P(X \leq x \cap Y \leq y) = \int_{-\infty}^{y} \int_{-\infty}^{x} f_{XY}(u, v) \, du \, dv. \tag{3.25}$$

Moreover, with $x_1 < x_2$, and $y_1 < y_2$,

$$P(x_1 < X \leq x_2 \cap y_1 < Y \leq y_2) = \int_{y_1}^{y_2} \int_{x_1}^{x_2} f_{XY}(x, y) \, dx \, dy. \tag{3.26}$$

The jpdf $f_{XY}(x, y)$ defines a surface over the $(x, y)$ plane. As indicated by Equation (3.26), the probability that random variables $X$ and $Y$ fall within a certain region $R$ is equal to the volume under the surface of $f_{XY}(x, y)$ and bounded by that region. This is illustrated in Figure 3.13.

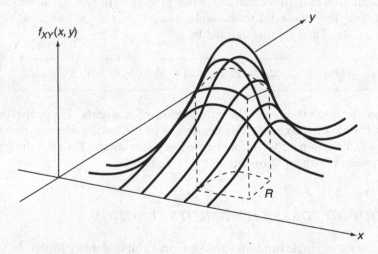

**Figure 3.13**  A joint probability density function, $f_{XY}(x, y)$

We also note the following important properties:

$$\int_{-\infty}^{\infty} \int_{-\infty}^{\infty} f_{XY}(x,y) \mathrm{d}x \mathrm{d}y = 1, \tag{3.27}$$

$$\int_{-\infty}^{\infty} f_{XY}(x,y) \mathrm{d}y = f_X(x), \tag{3.28}$$

$$\int_{-\infty}^{\infty} f_{XY}(x,y) \mathrm{d}x = f_Y(y). \tag{3.29}$$

Equation (3.27) follows from Equation (3.25) by letting $x, y \to +\infty, +\infty$, and this shows that the total volume under the $f_{XY}(x,y)$ surface is unity. To give a derivation of Equation (3.28), we know that

$$F_X(x) = F_{XY}(x, +\infty) = \int_{-\infty}^{\infty} \int_{-\infty}^{x} f_{XY}(u,y) \mathrm{d}u \mathrm{d}y.$$

Differentiating the above with respect to $x$ gives the desired result immediately. The density functions $f_X(x)$ and $f_Y(y)$ in Equations (3.28) and (3.29) are now called the *marginal density functions* of $X$ and $Y$, respectively.

**Example 3.7.** Problem: a boy and a girl plan to meet at a certain place between 9 a.m. and 10 a.m., each not waiting more than 10 minutes for the other. If all times of arrival within the hour are equally likely for each person, and if their times of arrival are independent, find the probability that they will meet.

Answer: for a single continuous random variable $X$ that takes all values over an interval $a$ to $b$ with equal likelihood, the distribution is called a *uniform* distribution and its density function $f_X(x)$ has the form

$$f_X(x) = \begin{cases} \dfrac{1}{b-a}, & \text{for } a \le x \le b; \\ 0, & \text{elsewhere.} \end{cases} \tag{3.30}$$

The height of $f_X(x)$ over the interval $(a, b)$ must be $1/(b-a)$ in order that the area is 1 below the curve (see Figure 3.14). For a two-dimensional case as described in this example, the joint density function of two independent uniformly distributed random variables is a flat surface within prescribed bounds. The volume under the surface is unity.

Let the boy arrive at $X$ minutes past 9 a.m. and the girl arrive at $Y$ minutes past 9 a.m. The jpdf $f_{XY}(x,y)$ thus takes the form shown in Figure 3.15 and is given by

$$f_{XY}(x,y) = \begin{cases} \dfrac{1}{3600}, & \text{for } 0 \le x \le 60, \text{and } 0 \le y \le 60; \\ 0, & \text{elsewhere.} \end{cases}$$

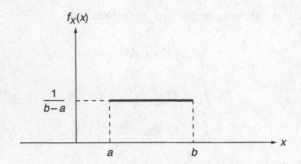

**Figure 3.14**  A uniform density function, $f_X(x)$

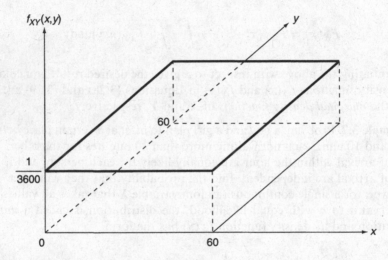

**Figure 3.15**  The joint probability density function $f_{XY}(x,y)$, for Example 3.7

The probability we are seeking is thus the volume under this surface over an appropriate region $R$. For this problem, the region $R$ is given by

$$R : |X - Y| \leq 10$$

and is shown in Figure 3.16 in the $(x, y)$ plane.

The volume of interest can be found by inspection in this simple case. Dividing $R$ into three regions as shown, we have

$$P(\text{they will meet}) = P(|X - Y| \leq 10)$$
$$= [2(5)(10) + 10\sqrt{2}(50\sqrt{2})]/3600 = \frac{11}{36}$$

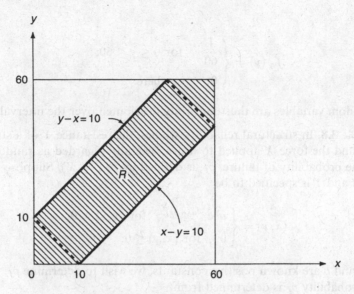

**Figure 3.16**  Region $R$ in Example 3.7

Note that, for a more complicated jpdf, one needs to carry out the volume integral $\iint_R f_{XY}(x,y)\mathrm{d}x\mathrm{d}y$ for volume calculations.

As an exercise, let us determine the joint probability distribution function and the marginal density functions of random variables $X$ and $Y$ defined in Example 3.7.

The JPDF of $X$ and $Y$ is obtained from Equation (3.25). It is clear that

$$F_{XY}(x,y) = \begin{cases} 0, & \text{for } (x,y) < (0,0); \\ 1, & \text{for } (x,y) > (60,60). \end{cases}$$

Within the region $(0,0) \le (x,y) \le (60,60)$, we have

$$F_{XY}(x,y) = \int_0^y \int_0^x \left(\frac{1}{3600}\right) \mathrm{d}x\mathrm{d}y = \frac{xy}{3600}.$$

For marginal density functions, Equations (3.28) and (3.29) give us

$$f_X(x) = \begin{cases} \displaystyle\int_0^{60} \left(\frac{1}{3600}\right) \mathrm{d}y = \frac{1}{60}, & \text{for } 0 \le x \le 60; \\ 0, & \text{elsewhere.} \end{cases}$$

Similarly,

$$f_Y(y) = \begin{cases} \dfrac{1}{60}, & \text{for } 0 \le y \le 60; \\ 0, & \text{elsewhere.} \end{cases}$$

Both random variables are thus uniformly distributed over the interval $(0, 60)$.

**Example 3.8.** In structural reliability studies, the resistance $Y$ of a structural element and the force $X$ applied to it are generally regarded as random variables. The probability of failure, $p_f$, is defined by $P(Y < X)$. Suppose that the jpdf of $X$ and $Y$ is specified to be

$$f_{XY}(x, y) = \begin{cases} abe^{-(ax+by)}, & \text{for } (x, y) > 0; \\ 0, & \text{for } (x, y) \le 0; \end{cases}$$

where $a$ and $b$ are known positive constants, we wish to determine $p_f$.

The probability $p_f$ is determined from

$$p_f = \iint_R f_{XY}(x, y)\mathrm{d}x\mathrm{d}y,$$

where $R$ is the region satisfying $Y \le X$. Since $X$ and $Y$ take only positive values, the region $R$ is that shown in Figure 3.17. Hence,

$$p_f = \int_0^\infty \int_y^\infty abe^{-(ax+by)}\mathrm{d}x\mathrm{d}y = \frac{b}{a+b}.$$

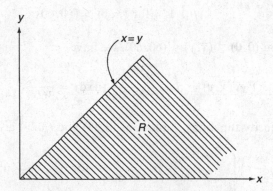

**Figure 3.17**   Region $R$ in Example 3.8

In closing this section, let us note that generalization to the case of many random variables is again straightforward. The joint distribution function of $n$ random variables $X_1, X_2, \ldots, X_n$, or $X$, is given, by Equation (3.19), as

$$F_X(x) = P(X_1 \leq x_1 \cap X_2 \leq x_2 \ldots \cap X_n \leq x_n). \qquad (3.31)$$

The corresponding joint density function, denoted by $f_X(x)$, is then

$$f_X(x) = \frac{\partial^n F_X(x)}{\partial x_1 \partial x_2 \ldots \partial x_n}, \qquad (3.32)$$

if the indicated partial derivatives exist. Various properties possessed by these functions can be readily inferred from those indicated for the two-random-variable case.

### 3.4 CONDITIONAL DISTRIBUTION AND INDEPENDENCE

The important concepts of conditional probability and independence introduced in Sections 2.2 and 2.4 play equally important roles in the context of random variables. The *conditional distribution function* of a random variable $X$, given that another random variable $Y$ has taken a value $y$, is defined by

$$F_{XY}(x|y) = P(X \leq x | Y = y). \qquad (3.33)$$

Similarly, when random variable $X$ is discrete, the definition of *conditional mass function* of $X$ given $Y = y$ is

$$p_{XY}(x|y) = P(X = x | Y = y). \qquad (3.34)$$

Using the definition of conditional probability given by Equation (2.24), we have

$$p_{XY}(x|y) = P(X = x | Y = y) = \frac{P(X = x \cap Y = y)}{P(Y = y)},$$

or

$$p_{XY}(x|y) = \frac{p_{XY}(x, y)}{p_Y(y)}, \text{ if } p_Y(y) \neq 0, \qquad (3.35)$$

which is expected. It gives the relationship between the joint jpmf and the conditional mass function. As we will see in Example 3.9, it is sometimes more convenient to derive joint mass functions by using Equation (3.35), as conditional mass functions are more readily available.

If random variables $X$ and $Y$ are independent, then the definition of independence, Equation (2.16), implies

$$p_{XY}(x|y) = p_X(x), \tag{3.36}$$

and Equation (3.35) becomes

$$p_{XY}(x,y) = p_X(x)p_Y(y). \tag{3.37}$$

Thus, when, and only when, random variables $X$ and $Y$ are independent, their jpmf is the product of the marginal mass functions.

Let $X$ be a continuous random variable. A consistent definition of the conditional density function of $X$ given $Y = y, f_{XY}(x|y)$, is the derivative of its corresponding conditional distribution function. Hence,

$$\boxed{f_{XY}(x|y) = \frac{dF_{XY}(x|y)}{dx},} \tag{3.38}$$

where $F_{XY}(x|y)$ is defined in Equation (3.33). To see what this definition leads to, let us consider

$$P(x_1 < X \le x_2 | y_1 < Y \le y_2) = \frac{P(x_1 < X \le x_2 \cap y_1 < Y \le y_2)}{P(y_1 < Y \le y_2)}. \tag{3.39}$$

In terms of jpdf $f_{XY}(x,y)$, it is given by

$$P(x_1 < X \le x_2 | y_1 < Y \le y_2) = \int_{y_1}^{y_2} \int_{x_1}^{x_2} f_{XY}(x,y)\,dx\,dy \Big/ \int_{y_1}^{y_2} \int_{-\infty}^{\infty} f_{XY}(x,y)\,dx\,dy$$

$$= \int_{y_1}^{y_2} \int_{x_1}^{x_2} f_{XY}(x,y)\,dx\,dy \Big/ \int_{y_1}^{y_2} f_Y(y)\,dy. \tag{3.40}$$

By setting $x_1 = -\infty, x_2 = x, y_1 = y$, and $y_2 = y + \Delta y$, and by taking the limit $\Delta y \to 0$, Equation (3.40) reduces to

$$F_{XY}(x|y) = \frac{\int_{-\infty}^{x} f_{XY}(u,y)\,du}{f_Y(y)}, \tag{3.41}$$

provided that $f_Y(y) \ne 0$.

Now we see that Equation (3.38) leads to

$$f_{XY}(x|y) = \frac{\mathrm{d}F_{XY}(x|y)}{\mathrm{d}x} = \frac{f_{XY}(x,y)}{f_Y(y)}, \quad f_Y(y) \neq 0, \tag{3.42}$$

which is in a form identical to that of Equation (3.35) for the mass functions – a satisfying result. We should add here that this relationship between the conditional density function and the joint density function is obtained at the expense of Equation (3.33) for $F_{XY}(x|y)$. We say 'at the expense of' because the definition given to $F_{XY}(x|y)$ does not lead to a convenient relationship between $F_{XY}(x|y)$ and $F_{XY}(x, y)$, that is,

$$F_{XY}(x|y) \neq \frac{F_{XY}(x,y)}{F_Y(y)}. \tag{3.43}$$

This inconvenience, however, is not a severe penalty as we deal with density functions and mass functions more often.

When random variables $X$ and $Y$ are independent, $F_{XY}(x|y) = F_X(x)$ and, as seen from Equation (3.42),

$$f_{XY}(x|y) = f_X(x), \tag{3.44}$$

and

$$f_{XY}(x,y) = f_X(x)f_Y(y), \tag{3.45}$$

which shows again that the joint density function is equal to the product of the associated marginal density functions when $X$ and $Y$ are independent.

Finally, let us note that, when random variables $X$ and $Y$ are discrete,

$$F_{XY}(x|y) = \sum_{i=1}^{i:x_i \leq x} p_{XY}(x_i|y), \tag{3.46}$$

and, in the case of a continuous random variable,

$$F_{XY}(x|y) = \int_{-\infty}^{x} f_{XY}(u|y)\mathrm{d}u. \tag{3.47}$$

Comparison of these equations with Equations (3.7) and (3.12) reveals they are identical to those relating these functions for $X$ alone.

Extensions of the above results to the case of more than two random variables are again straightforward. Starting from

$$P(ABC) = P(A|BC)P(B|C)P(C)$$

[see Equation (2.26)], for three events $A$, $B$, and $C$, we have, in the case of three random variables $X$, $Y$, and $Z$,

$$\left.\begin{aligned} p_{XYZ}(x,y,z) &= p_{XYZ}(x|y,z)p_{YZ}(y|z)p_Z(z) \\ f_{XYZ}(x,y,z) &= f_{XYZ}(x|y,z)f_{YZ}(y|z)f_Z(z) \end{aligned}\right\} \tag{3.48}$$

Hence, for the general case of $n$ random variables, $X_1, X_2, \ldots, X_n$, or $X$, we can write

$$\left.\begin{aligned} p_X(x) &= p_{X_1 X_2 \ldots X_n}(x_1|x_2,\ldots,x_n)p_{X_2 \ldots X_n}(x_2|x_3,\ldots,x_n)\ldots p_{X_{n-1}X_n}(x_{n-1}|x_n)p_{X_n}(x_n); \\ f_X(x) &= f_{X_1 X_2 \ldots X_n}(x_1|x_2,\ldots,x_n)f_{X_2 \ldots X_n}(x_2|x_3,\ldots,x_n)\ldots f_{X_{n-1}X_n}(x_{n-1}|x_n)f_{X_n}(x_n). \end{aligned}\right\}$$

$$\tag{3.49}$$

In the event that these random variables are mutually independent, Equations (3.49) become

$$\left.\begin{aligned} p_X(x) &= p_{X_1}(x_1)p_{X_2}(x_2)\ldots p_{X_n}(x_n); \\ f_X(x) &= f_{X_1}(x_1)f_{X_2}(x_2)\ldots f_{X_n}(x_n). \end{aligned}\right\} \tag{3.50}$$

**Example 3.9.** To show that joint mass functions are sometimes more easily found by finding first the conditional mass functions, let us consider a traffic problem as described below.

Problem: a group of $n$ cars enters an intersection from the south. Through prior observations, it is estimated that each car has the probability $p$ of turning east, probability $q$ of turning west, and probability $r$ of going straight on ($p + q + r = 1$). Assume that drivers behave independently and let $X$ be the number of cars turning east and $Y$ the number turning west. Determine the jpmf $p_{XY}(x,y)$.

Answer: since

$$p_{XY}(x,y) = p_{XY}(x|y)p_Y(y),$$

we proceed by determining $p_{XY}(x|y)$ and $p_Y(y)$. The marginal mass function $p_Y(y)$ is found in a way very similar to that in the random walk situation described in Example 3.5. Each car has two alternatives: turning west, and not turning west. By enumeration, we can show that it has a binomial distribution (to be more fully justified in Chapter 6)

$$p_Y(y) = \binom{n}{y}q^y(1-q)^{n-y}, \quad y = 1, 2, \ldots. \tag{3.51}$$

Consider now the conditional mass function $p_{XY}(x|y)$. With $Y = y$ having happened, the situation is again similar to that for determining $p_Y(y)$ except that the number of cars available for taking possible eastward turns is now $n - y$; also, here, the probabilities $p$ and $r$ need to be renormalized so that they sum to 1. Hence, $p_{XY}(x|y)$ takes the form

$$p_{XY}(x|y) = \binom{n-y}{x} \left(\frac{p}{r+p}\right)^x \left(1 - \frac{p}{r+p}\right)^{n-y-x}, \quad x = 0, 1, \ldots, n-y, \; y = 0, 1, \ldots, n.$$

(3.52)

Finally, we have $p_{XY}(x, y)$ as the product of the two expressions given by Equations (3.51) and (3.52). The ranges of values for $x$ and $y$ are $x = 0, 1, \ldots, n-y$, and $y = 0, 1, \ldots, n$.

Note that $p_{XY}(x, y)$ has a rather complicated expression that could not have been derived easily in a direct way. This also points out the need to exercise care in determining the limits of validity for $x$ and $y$.

**Example 3.10.** Problem: resistors are designed to have a resistance $R$ of $50 \pm 2\,\Omega$. Owing to imprecision in the manufacturing process, the actual density function of $R$ has the form shown by the solid curve in Figure 3.18. Determine the density function of $R$ after screening – that is, after all the resistors having resistances beyond the 48–52 $\Omega$ range are rejected.

Answer: we are interested in the conditional density function, $f_R(r|A)$, where $A$ is the event $\{48 \leq R \leq 52\}$. This is not the usual conditional density function but it can be found from the basic definition of conditional probability.

We start by considering

$$F_R(r|A) = P(R \leq r|48 \leq R \leq 52) = \frac{P(R \leq r \cap 48 \leq R \leq 52)}{P(48 \leq R \leq 52)}.$$

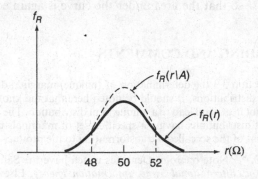

**Figure 3.18** The actual, $f_R(r)$, and conditional, $f_R(r|A)$, for Example 3.10

However,

$$R \le r \cap 48 \le R \le 52 = \begin{cases} \emptyset, & \text{for } r < 48; \\ 48 \le R \le r, & \text{for } 48 \le r \le 52; \\ 48 \le R \le 52, & \text{for } r > 52. \end{cases}$$

Hence,

$$F_R(r|A) = \begin{cases} 0, & \text{for } r < 48; \\ \dfrac{P(48 \le R \le r)}{P(48 \le R \le 52)} = \dfrac{\int_{48}^{r} f_R(r)\mathrm{d}r}{c}, & \text{for } 48 \le r \le 52; \\ 1, & \text{for } r > 52; \end{cases}$$

where

$$c = \int_{48}^{52} f_R(r)\mathrm{d}r.$$

is a constant.

The desired $f_R(r|A)$ is then obtained from the above by differentiation. We obtain

$$f_R(r|A) = \frac{\mathrm{d}F_R(r|A)}{\mathrm{d}r} = \begin{cases} \dfrac{f_R(r)}{c}, & \text{for } 48 \le r \le 52 \\ 0, & \text{elsewhere} \end{cases}$$

It can be seen from Figure 3.18 (dashed line) that the effect of screening is essentially a truncation of the tails of the distribution beyond the allowable limits. This is accompanied by an adjustment within the limits by a multiplicative factor $1/c$ so that the area under the curve is again equal to 1.

## FURTHER READING AND COMMENTS

We discussed in Section 3.3 the determination of (unique) marginal distributions from a knowledge of joint distributions. It should be noted here that the knowledge of marginal distributions does not in general lead to a unique joint distribution. The following reference shows that all joint distributions having a specified set of marginals can be obtained by repeated applications of the so-called $\theta$ transformation to the product of the marginals:

Becker, P.W., 1970, "A Note on Joint Densities which have the Same Set of Marginal Densities", in *Proc. International Symp. Information Theory*, Elsevier Scientific Publishers, The Netherlands.

## PROBLEMS

3.1 For each of the functions given below, determine constant $a$ so that it possesses all the properties of a probability distribution function (PDF). Determine, in each case, its associated probability density function (pdf) or probability mass function (pmf) if it exists and sketch all functions.
(a) Case 1:

$$F(x) = \begin{cases} 0, & \text{for } x < 5; \\ a, & \text{for } x \geq 5. \end{cases}$$

(b) Case 2:

$$F(x) = \begin{cases} 0, & \text{for } x < 5; \\ \dfrac{1}{3}, & \text{for } 5 \leq x < 7; \\ a, & \text{for } x \geq 7. \end{cases}$$

(c) Case 3:

$$F(x) = \begin{cases} 0, & \text{for } x < 1; \\ \displaystyle\sum_{j=1}^{k} 1/a^j, & \text{for } k \leq x < k+1, \text{ and } k = 1, 2, 3, \ldots. \end{cases}$$

(d) Case 4:

$$F(x) = \begin{cases} 0, & \text{for } x \leq 0; \\ 1 - e^{-ax}, & \text{for } x > 0. \end{cases}$$

(e) Case 5:

$$F(x) = \begin{cases} 0, & \text{for } x < 0; \\ x^a, & \text{for } 0 \leq x \leq 1; \\ 1, & \text{for } x > 1. \end{cases}$$

(f) Case 6:

$$F(x) = \begin{cases} 0, & \text{for } x < 0; \\ a\sin^{-1}\sqrt{x}, & \text{for } 0 \leq x \leq 1; \\ 1, & \text{for } x > 0. \end{cases}$$

(g) Case 7:

$$F(x) = \begin{cases} 0, & \text{for } x < 0; \\ a(1 - e^{-x/2}) + \dfrac{1}{2}, & \text{for } x \geq 0. \end{cases}$$

3.2 For each part of Problem 3.1, determine:
(a) $P(X \leq 6)$;
(b) $P(\frac{1}{2} < X \leq 7)$.

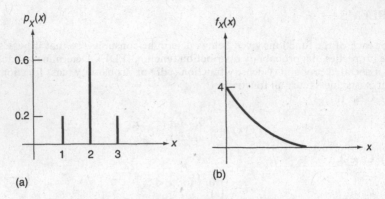

**Figure 3.19**  The probability mass function, $p_X(x)$, and probability density function, $f_X(x)$, for Problem 3.3

3.3  For $p_X(x)$ and $f_X(x)$ in Figure 3.19(a) and 3.19(b) respectively, sketch roughly in scale the corresponding PDF $F_X(x)$ and show on all graphs the procedure for finding $P(2 < X < 4)$.

3.4  For each part, find the corresponding PDF for random variable $X$.
  (a)  Case 1:

$$f_X(x) = \begin{cases} 0.1, & \text{for } 90 \leq x < 100; \\ 0, & \text{elsewhere.} \end{cases}$$

  (b)  Case 2:

$$f_X(x) = \begin{cases} 2(1-x), & \text{for } 0 \leq x < 1; \\ 0, & \text{elsewhere.} \end{cases}$$

  (c)  Case 3:

$$f_X(x) = \frac{1}{\pi(1+x^2)}, \quad \text{for } -\infty < x < \infty.$$

3.5  The pdf of $X$ is shown in Figure 3.20.
  (a)  Determine the value of $a$.
  (b)  Graph $F_X(x)$ approximately.
  (c)  Determine $P(X \geq 2)$.
  (d)  Determine $P(X \geq 2 | X \geq 1)$.

3.6  The life $X$, in hours, of a certain kind of electronic component has a pdf given by

$$f_X(x) = \begin{cases} 0, & \text{for } x < 100; \\ \dfrac{100}{x^2}, & \text{for } x \geq 100. \end{cases}$$

Determine the probability that a component will survive 150 hours of operation.

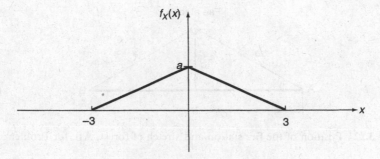

**Figure 3.20** The probability density function, $f_X(x)$, for Problem 3.5

3.7 Let $T$ denote the life (in months) of a light bulb and let

$$f_T(t) = \begin{cases} \dfrac{1}{15} - \dfrac{t}{450}, & \text{for } 0 \le t \le 30; \\ 0, & \text{elsewhere.} \end{cases}$$

(a) Plot $f_T(t)$ against $t$.
(b) Derive $F_T(t)$ and plot $F_T(t)$ against $t$.
(c) Determine using $f_T(t)$, the probability that the light bulb will last at least 15 months.
(d) Determine, using $F_T(t)$, the probability that the light bulb will last at least 15 months.
(e) A light bulb has already lasted 15 months. What is the probability that it will survive another month?

3.8 The time, in minutes, required for a student to travel from home to a morning class is uniformly distributed between 20 and 25. If the student leaves home promptly at 7:38 a.m., what is the probability that the student will not be late for class at 8:00 a.m.?

3.9 In constructing the bridge shown in Figure 3.21, an engineer is concerned with forces acting on the end supports caused by a randomly applied concentrated load $P$, the term 'randomly applied' meaning that the probability of the load lying in any region is proportional only to the length of that region. Suppose that the bridge has a span $2b$. Determine the PDF and pdf of random variable $X$, which is the distance from the load to the nearest edge support. Sketch these functions.

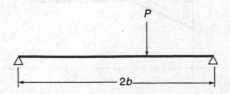

**Figure 3.21** Diagram of the bridge, for Problem 3.9

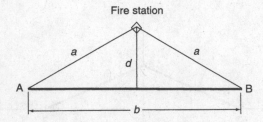

**Figure 3.22**   Position of the fire station and stretch of forest, AB, for Problem 3.10

3.10 Fire can erupt at random at any point along a stretch of forest AB. The fire station is located as shown in Figure 3.22. Determine the PDF and pdf of $X$, representing the distance between the fire and the fire station. Sketch these functions.

3.11 Pollutant concentrations caused by a pollution source can be modeled by the pdf $(a > 0)$:

$$f_R(r) = \begin{cases} 0, & \text{for } r < 0; \\ ae^{-ar}, & \text{for } r \geq 0; \end{cases}$$

where $R$ is the distance from the source. Determine the radius within which 95% of the pollutant is contained.

3.12 As an example of a mixed probability distribution, consider the following problem: a particle is at rest at the origin $(x = 0)$ at time $t = 0$. At a randomly selected time uniformly distributed over the interval $0 < t < 1$, the particle is suddenly given a velocity $v$ in the positive $x$ direction.
   (a) Show that $X$, the particle position at $t(0 < t < 1)$, has the PDF shown in Figure 3.23.
   (b) Calculate the probability that the particle is at least $v/3$ away from the origin at $t = 1/2$.

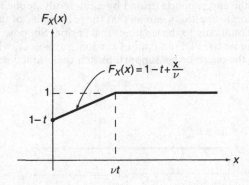

**Figure 3.23**   The probability distribution function, $F_X(x)$, for Problem 3.12

3.13 For each of the joint probability mass functions (jpmf), $p_{XY}(x, y)$, or joint probability density functions (jpdf), $f_{XY}(x, y)$, given below (cases 1–4), determine:
(a) the marginal mass or density functions,
(b) whether the random variables are independent.

(i)   Case 1

$$p_{XY}(x, y) = \begin{cases} 0.5, & \text{for } (x, y) = (1, 1); \\ 0.1, & \text{for } (x, y) = (1, 2); \\ 0.1, & \text{for } (x, y) = (2, 1); \\ 0.3, & \text{for } (x, y) = (2, 2). \end{cases}$$

(ii)  Case 2:

$$f_{XY}(x, y) = \begin{cases} a(x + y), & \text{for } 0 < x \leq 1, \text{and } 1 < y \leq 2; \\ 0, & \text{elsewhere.} \end{cases}$$

(iii) Case 3

$$f_{XY}(x, y) = \begin{cases} e^{-(x+y)}, & \text{for } (x, y) > (0, 0); \\ 0, & \text{elsewhere.} \end{cases}$$

(iv)  Case 4

$$f_{XY}(x, y) = \begin{cases} 4y(x - y)e^{-(x+y)}, & \text{for } 0 < x < \infty, \text{and } 0 < y \leq x; \\ 0, & \text{elsewhere.} \end{cases}$$

3.14 Suppose $X$ and $Y$ have jpmf

$$p_{XY}(x, y) = \begin{cases} 0.1, & \text{for } (x, y) = (1, 1); \\ 0.2, & \text{for } (x, y) = (1, 2); \\ 0.3, & \text{for } (x, y) = (2, 1); \\ 0.4, & \text{for } (x, y) = (2, 2). \end{cases}$$

(a) Determine marginal pmfs of $X$ and $Y$.
(b) Determine $P(X = 1)$.
(c) Determine $P(2X \leq Y)$.

3.15 Let $X_1$, $X_2$, and $X_3$ be independent random variables, each taking values $\pm 1$ with probabilities $1/2$. Define random variables $Y_1$, $Y_2$, and $Y_3$ by

$$Y_1 = X_1 X_2, \quad Y_2 = X_1 X_3, \quad Y_3 = X_2 X_3$$

Show that any two of these new random variables are independent but that $Y_1$, $Y_2$, and $Y_3$ are not independent.

3.16 The random variables $X$ and $Y$ are distributed according to the jpdf given by Case 2, in Problem 3.13(ii). Determine:
(a) $P(X \geq 0.5 \cap Y > 1.0)$.
(b) $P(XY < \frac{1}{2})$.

(c)  $P(X \le 0.5 | Y = 1.5)$.

(d)  $P(X \le 0.5 | Y \le 1.5)$.

3.17  Let random variable $X$ denote the time of failure in years of a system for which the PDF is $F_X(x)$. In terms of $F_X(x)$, determine the probability

$$P(X \le x | X \ge 100),$$

which is the conditional distribution function of $X$ given that the system did not fail up to 100 years.

3.18  The pdf of random variable $X$ is

$$f_X(x) = \begin{cases} 3x^2, & \text{for } -1 < x \le 0; \\ 0, & \text{elsewhere.} \end{cases}$$

Determine $P(X > b | X < b/2)$ with $-1 < b < 0$.

3.19  Using the joint probability distribution given in Example 3.5 for random variables $X$ and $Y$, determine:

(a)  $P(X > 3)$.

(b)  $P(0 \le Y < 3)$.

(c)  $P(X > 3 | Y \le 2)$.

3.20  Let

$$f_{XY}(x, y) = \begin{cases} ke^{-(x+y)}, & \text{for } 0 < x < 1, \text{and } 0 < y < 2; \\ 0, & \text{elsewhere.} \end{cases}$$

(a)  What must be the value of $k$?

(b)  Determine the marginal pdfs of $X$ and $Y$.

(c)  Are $X$ and $Y$ statistically independent? Why?

3.21  A commuter is accustomed to leaving home between 7:30 a.m and 8:00 a.m., the drive to the station taking between 20 and 30 minutes. It is assumed that departure time and travel time for the trip are independent random variables, uniformly distributed over their respective intervals. There are two trains the commuter can take; the first leaves at 8:05 a.m. and takes 30 minutes for the trip, and the second leaves at 8:25 a.m. and takes 35 minutes. What is the probability that the commuter misses both trains?

3.22  The distance $X$ (in miles) from a nuclear plant to the epicenter of potential earthquakes within 50 miles is distributed according to

$$f_X(x) = \begin{cases} \dfrac{2x}{2500}, & \text{for } 0 \le x \le 50; \\ 0, & \text{elsewhere;} \end{cases}$$

and the magnitude $Y$ of potential earthquakes of scales 5 to 9 is distributed according to

$$f_Y(y) = \begin{cases} \dfrac{3(9 - y)^2}{64}, & \text{for } 5 \le y \le 9; \\ 0, & \text{elsewhere.} \end{cases}$$

Assume that $X$ and $Y$ are independent. Determine $P(X \leq 25 \cap Y > 8)$, the probability that the next earthquake within 50 miles will have a magnitude greater than 8 and that its epicenter will lie within 25 miles of the nuclear plant.

3.23 Let random variables $X$ and $Y$ be independent and uniformly distributed in the square $(0,0) < (X, Y) < (1,1)$. Determine the probability that $XY < 1/2$.

3.24 In splashdown maneuvers, spacecrafts often miss the target because of guidance inaccuracies, atmospheric disturbances, and other error sources. Taking the origin of the coordinates as the designed point of impact, the $X$ and $Y$ coordinates of the actual impact point are random, with marginal density functions

$$f_X(x) = \frac{1}{\sigma(2\pi)^{1/2}} e^{-x^2/2\sigma^2}, \quad -\infty < x < \infty;$$

$$f_Y(y) = \frac{1}{\sigma(2\pi)^{1/2}} e^{-y^2/2\sigma^2}, \quad -\infty < y < \infty.$$

Assume that the random variables are independent. Show that the probability of a splashdown lying within a circle of radius $a$ centered at the origin is $1 - e^{-a^2/2\sigma^2}$.

3.25 Let $X_1, X_2, \ldots, X_n$ be independent and identically distributed random variables, each with PDF $F_X(x)$. Show that

$$P[\min(X_1, X_2, \ldots, X_n) \leq u] = 1 - [1 - F_X(u)]^n,$$
$$P[\max(X_1, X_2, \ldots, X_n) \leq u] = [F_X(u)]^n.$$

The above are examples of extreme-value distributions. They are of considerable practical importance and will be discussed in Section 7.6.

3.26 In studies of social mobility, assume that social classes can be ordered from 1 (professional) to 7 (unskilled). Let random variable $X_k$ denote the class order of the $k$th generation. Then, for a given region, the following information is given:
  (i)  The pmf of $X_0$ is described by $p_{X_0}(1) = 0.00$, $p_{X_0}(2) = 0.00$, $p_{X_0}(3) = 0.04$, $p_{X_0}(4) = 0.06$, $p_{X_0}(5) = 0.11$, $p_{X_0}(6) = 0.28$, and $p_{X_0}(7) = 0.51$.
  (ii) The conditional probabilities $P(X_{k+1} = i | X_k = j)$ for $i, j = 1, 2, \ldots, 7$ and for every $k$ are given in Table 3.2.

**Table 3.2**   $P(X_{k+1} = i | X_k = j)$ for Problem 3.26

| $i$ | $j$ | | | | | | |
| --- | --- | --- | --- | --- | --- | --- | --- |
|   | 1 | 2 | 3 | 4 | 5 | 6 | 7 |
| 1 | 0.388 | 0.107 | 0.035 | 0.021 | 0.009 | 0.000 | 0.000 |
| 2 | 0.146 | 0.267 | 0.101 | 0.039 | 0.024 | 0.013 | 0.008 |
| 3 | 0.202 | 0.227 | 0.188 | 0.112 | 0.075 | 0.041 | 0.036 |
| 4 | 0.062 | 0.120 | 0.191 | 0.212 | 0.123 | 0.088 | 0.083 |
| 5 | 0.140 | 0.206 | 0.357 | 0.430 | 0.473 | 0.391 | 0.364 |
| 6 | 0.047 | 0.053 | 0.067 | 0.124 | 0.171 | 0.312 | 0.235 |
| 7 | 0.015 | 0.020 | 0.061 | 0.062 | 0.125 | 0.155 | 0.274 |

(iii)  The outcome at the $(k+1)$th generation is dependent only on the class order at the $k$th generation and not on any generation prior to it; that is,

$$P(X_{k+1} = i | X_k = j \cap X_{k-1} = m \cap \ldots) = P(X_{k+1} = i | X_k = j)$$

Determine
(a)  The pmf of $X_3$.
(b)  The jpmf of $X_3$ and $X_4$.

# 4

# Expectations and Moments

While a probability distribution $[F_X(x), p_X(x),$ or $f_X(x)]$ contains a complete description of a random variable $X$, it is often of interest to seek a set of simple numbers that gives the random variable some of its dominant features. These numbers include moments of various orders associated with $X$. Let us first provide a general definition (Definition 4.1).

**Definition 4.1.** Let $g(X)$ be a real-valued function of a random variable $X$. The *mathematical expectation*, or simply *expectation*, of $g(X)$, denoted by $E\{g(X)\}$, is defined by

$$E\{g(X)\} = \sum_i g(x_i)p_X(x_i), \qquad (4.1)$$

if $X$ is discrete, where $x_1, x_2, \ldots$ are possible values assumed by $X$.

When the range of $i$ extends from 1 to infinity, the sum in Equation (4.1) exists if it converges absolutely; that is,

$$\sum_{i=1}^{\infty} |g(x_i)|p_X(x_i) < \infty.$$

The symbol $E\{\ \}$ is regarded here and in the sequel as the *expectation operator*. If random variable $X$ is continuous, the expectation $E\{g(X)\}$ is defined by

$$E\{g(X)\} = \int_{-\infty}^{\infty} g(x)f_X(x)\mathrm{d}x, \qquad (4.2)$$

if the improper integral is absolutely convergent, that is,

$$\int_{-\infty}^{\infty} |g(x)|f_X(x)\mathrm{d}x < \infty.$$

*Fundamentals of Probability and Statistics for Engineers* T.T. Soong © 2004 John Wiley & Sons, Ltd
ISBNs: 0-470-86813-9 (HB)  0-470-86814-7 (PB)

Let us note some basic properties associated with the expectation operator. For any constant $c$ and any functions $g(X)$ and $h(X)$ for which expectations exist, we have

$$
\left.
\begin{aligned}
&E\{c\} = c, \\
&E\{cg(X)\} = cE\{g(X)\}, \\
&E\{g(X) + h(X)\} = E\{g(X)\} + E\{h(X)\}, \\
&E\{g(X)\} \leq E\{h(X)\}, \quad \text{if } g(X) \leq h(X) \text{ for all values of } X.
\end{aligned}
\right\} \tag{4.3}
$$

These relations follow directly from the definition of $E\{g(X)\}$. For example,

$$
\begin{aligned}
E\{g(X) + h(X)\} &= \int_{-\infty}^{\infty} [g(x) + h(x)] f_X(x) \mathrm{d}x \\
&= \int_{-\infty}^{\infty} g(x) f_X(x) \mathrm{d}x + \int_{-\infty}^{\infty} h(x) f_X(x) \mathrm{d}x \\
&= E\{g(X)\} + E\{h(X)\},
\end{aligned}
$$

as given by the third of Equations (4.3). The proof is similar when $X$ is discrete.

## 4.1 MOMENTS OF A SINGLE RANDOM VARIABLE

Let $g(X) = X^n, n = 1, 2, \ldots$; the expectation $E\{X^n\}$, when it exists, is called the $n$th *moment* of $X$. It is denoted by $\alpha_n$ and is given by

$$
\alpha_n = E\{X^n\} = \sum_i x_i^n p_X(x_i), \quad \text{for } X \text{ discrete}; \tag{4.4}
$$

$$
\alpha_n = E\{X^n\} = \int_{-\infty}^{\infty} x^n f_X(x) \mathrm{d}x, \quad \text{for } X \text{ continuous}. \tag{4.5}
$$

### 4.1.1   MEAN, MEDIAN, AND MODE

One of the most important moments is $\alpha_1$, the first moment. Using the mass analogy for the probability distribution, the first moment may be regarded as the center of mass of its distribution. It is thus the average value of random variable $X$ and certainly reveals one of the most important characteristics of its distribution. The first moment of $X$ is synonymously called the *mean, expectation*, or *average value* of $X$. A common notation for it is $m_X$ or simply $m$.

**Example 4.1.** Problem: From Example 3.9 (page 64), determine the average number of cars turning west in a group of $n$ cars.

Answer: we wish to determine the mean of $Y, E\{Y\}$, for which the mass function is [from Equation (3.51)]

$$p_Y(k) = \binom{n}{k} q^k (1-q)^{n-k}, \quad k = 0, 1, 2, \ldots, n.$$

Equation (4.4) then gives

$$E\{Y\} = \sum_{k=0}^{n} k p_Y(k) = \sum_{k=0}^{n} k \binom{n}{k} q^k (1-q)^{n-k}$$

$$= \sum_{k=1}^{n} \frac{n!}{(k-1)!(n-k)!} q^k (1-q)^{n-k}.$$

Let $k - 1 = m$. We have

$$E\{Y\} = nq \sum_{m=0}^{n-1} \binom{n-1}{m} q^m (1-q)^{n-1-m}.$$

The sum in this expressions is simply the sum of binomial probabilities and hence equals one. Therefore,

$$E\{Y\} = nq,$$

which has a numerical value since $n$ and $q$ are known constants.

**Example 4.2.** Problem: the waiting time $X$ (in minutes) of a customer waiting to be served at a ticket counter has the density function

$$f_X(x) = \begin{cases} 2e^{-2x}, & \text{for } x \geq 0; \\ 0, & \text{elsewhere.} \end{cases}$$

Determine the average waiting time.

Answer: referring to Equation (4.5), we have, using integration by parts,

$$E\{X\} = \int_0^\infty x(2e^{-2x})dx = \frac{1}{2}\text{minute.}$$

**Example 4.3.** Problem: from Example 3.10 (pages 65), find the average resistance of the resistors after screening.

Answer: the average value required in this example is a *conditional mean* of $R$ given the event $A$. Although no formal definition is given, it should be clear that the desired average is obtained from

$$E\{R|A\} = \int_{48}^{52} rf_R(r|A)\mathrm{d}r = \int_{48}^{52} \frac{rf_R(r)}{c}\,\mathrm{d}r.$$

This integral can be evaluated when $f_R(r)$ is specified.

Two other quantities in common usage that also give a measure of centrality of a random variable are its *median* and *mode*.

A *median* of $X$ is any point that divides the mass of the distribution into two equal parts; that is, $x_0$ is a median of $X$ if

$$P(X \leq x_0) = \frac{1}{2}.$$

The mean of $X$ may not exist, but there exists at least one median.

In comparison with the mean, the median is sometimes preferred as a measure of central tendency when a distribution is skewed, particularly where there are a small number of extreme values in the distribution. For example, we speak of *median income* as a good central measure of personal income for a population. This is a better average because the median is not as sensitive to a small number of extremely high incomes or extremely low incomes as is the mean.

**Example 4.4.** Let $T$ be the time between emissions of particles by a radio-active atom. It is well established that $T$ is a random variable and that it obeys an exponential distribution; that is,

$$f_T(t) = \begin{cases} \lambda e^{-\lambda t}, & \text{for } t \geq 0; \\ 0, & \text{elsewhere;} \end{cases}$$

where $\lambda$ is a positive constant. The random variable $T$ is called the lifetime of the atom, and a common average measure of this lifetime is called the half-life, which is defined as the median of $T$. Thus, the half-life, $\tau$ is found from

$$\int_0^\tau f_T(t)\mathrm{d}t = \frac{1}{2},$$

or

$$\tau = \ln\left(\frac{2}{\lambda}\right).$$

Let us note that the mean life, $E\{T\}$, is given by

$$E\{T\} = \int_0^\infty t f_T(t)\mathrm{d}t = \frac{1}{\lambda}.$$

A point $x_i$ such that

$$p_X(x_i) > p_X(x_{i+1}) \quad \text{and} \quad p_X(x_i) > p_X(x_{i-1}), \quad X \text{ discrete,}$$
$$f_X(x_i) > f_X(x_i + \varepsilon) \quad \text{and} \quad f_X(x_i) > f_X(x_i - \varepsilon), \quad X \text{ continuous,}$$

where $\varepsilon$ is an arbitrarily small positive quantity, is called a *mode* of $X$. A mode is thus a value of $X$ corresponding to a peak in its mass function or density function. The term *unimodal distribution* refers to a probability distribution possessing a unique mode.

To give a comparison of these three measures of centrality of a random variable, Figure 4.1 shows their relative positions in three different situations. It is clear that the mean, the median, and the mode coincide when a unimodal distribution is symmetric.

## 4.1.2  CENTRAL MOMENTS, VARIANCE, AND STANDARD DEVIATION

Besides the mean, the next most important moment is the *variance*, which measures the dispersion or spread of random variable $X$ about its mean. Its definition will follow a general definition of central moments (see Definition 4.2).

**Definition 4.2.** The *central moments* of random variable $X$ are the moments of $X$ with respect to its mean. Hence, the $n$th *central moment* of $X$, $\mu_n$, is defined as

$$\mu_n = E\{(X-m)^n\} = \sum_i (x_i - m)^n p_X(x_i), \quad X \text{ discrete;} \tag{4.6}$$

$$\mu_n = E\{(X-m)^n\} = \int_{-\infty}^\infty (x-m)^n f_X(x)\mathrm{d}x, \quad X \text{ continuous.} \tag{4.7}$$

The *variance* of $X$ is the second central moment, $\mu_2$, commonly denoted by $\sigma_X^2$ or simply $\sigma^2$ or $\text{var}(X)$. It is the most common measure of dispersion of a distribution about its mean. Large values of $\sigma_X^2$ imply a large spread in the distribution of $X$ about its mean. Conversely, small values imply a sharp concentration of the mass of distribution in the neighborhood of the mean. This is illustrated in Figure 4.2 in which two density functions are shown with the same mean but different variances. When $\sigma_X^2 = 0$, the whole mass of the distribution is concentrated at the mean. In this extreme case, $X = m_X$ with probability 1.

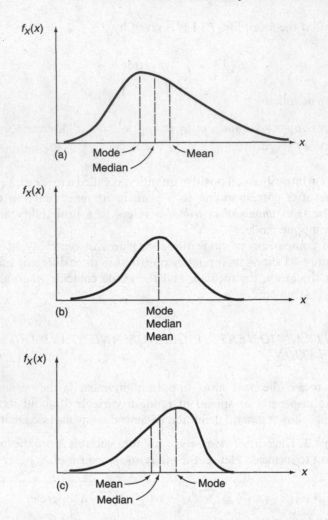

**Figure 4.1** Relative positions of the mean, median, and mode for three distributions: (a) positively shewed; (b) symmetrical; and (c) negatively shewed

An important relation between the variance and simple moments is

$$\sigma^2 = \alpha_2 - m^2. \tag{4.8}$$

This can be shown by making use of Equations (4.3). We get

$$\sigma^2 = E\{(X - m)^2\} = E\{X^2 - 2mX + m^2\} = E\{X^2\} - 2mE\{X\} + m^2$$
$$= \alpha_2 - 2m^2 + m^2 = \alpha_2 - m^2.$$

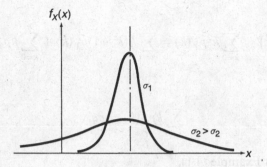

**Figure 4.2**   Density functions with different variances, $\sigma_1$, and $\sigma_2$

We note two other properties of the variance of a random variable $X$ which can be similarly verified. They are:

$$\left.\begin{array}{l} \mathrm{var}(X + c) = \mathrm{var}(X), \\ \mathrm{var}(cX) = c^2\mathrm{var}(X), \end{array}\right\} \tag{4.9}$$

where $c$ is any constant.

It is further noted from Equations (4.6) and (4.7) that, since each term in the sum in Equation (4.6) and the integrand in Equation (4.7) are nonnegative, the variance of a random variable is always nonnegative. The positive square root

$$\sigma_X = +[E\{(X - m)^2\}]^{1/2},$$

is called the *standard deviation* of $X$. An advantage of using $\sigma_X$ rather than $\sigma_X^2$ as a measure of dispersion is that it has the same unit as the mean. It can therefore be compared with the mean on the same scale to gain some measure of the degree of spread of the distribution. A dimensionless number that characterizes dispersion relative to the mean which also facilitates comparison among random variables of different units is the *coefficient of variation*, $v_X$, defined by

$$\boxed{v_X = \frac{\sigma_X}{m_X}.} \tag{4.10}$$

**Example 4.5.** Let us determine the variance of $Y$ defined in Example 4.1. Using Equation (4.8), we may write

$$\sigma_Y^2 = E\{Y^2\} - m_Y^2 = E\{Y^2\} - n^2q^2.$$

Now,

$$E\{Y^2\} = \sum_{k=0}^{n} k^2 p_Y(k) = \sum_{k=0}^{n} k(k-1)p_Y(k) + \sum_{k=0}^{n} kp_Y(k),$$

and

$$\sum_{k=0}^{n} kp_Y(k) = nq.$$

Proceeding as in Example (4.1),

$$\sum_{k=0}^{n} k(k-1)p_Y(k) = n(n-1)q^2 \sum_{k=2}^{n} \frac{(n-2)!}{(k-2)!(n-k)!} q^{k-2}(1-q)^{n-k}$$

$$= n(n-1)q^2 \sum_{j=0}^{m} \binom{m}{j} q^j(1-q)^{m-j}$$

$$= n(n-1)q^2.$$

Thus,

$$E\{Y^2\} = n(n-1)q^2 + nq,$$

and

$$\sigma_Y^2 = n(n-1)q^2 + nq - (nq)^2 = nq(1-q).$$

**Example 4.6.** We again use Equation (4.8) to determine the variance of $X$ defined in Example 4.2. The second moment of $X$ is, on integrating by parts,

$$E\{X^2\} = 2 \int_0^{\infty} x^2 e^{-2x} dx = \frac{1}{2}.$$

Hence,

$$\sigma_X^2 = E\{X^2\} - m_X^2 = \frac{1}{2} - \frac{1}{4} = \frac{1}{4}.$$

**Example 4.7.** Problem: owing to inherent manufacturing and scaling inaccuracies, the tape measures manufactured by a certain company have a standard deviation of 0.03 feet for a three-foot tape measure. What is a reasonable estimate of the standard deviation associated with three-yard tape measures made by the same manufacturer?

Answer: for this problem, it is reasonable to expect that errors introduced in the making of a three-foot tape measure again are accountable for inaccuracies in the three-yard tape measures. It is then reasonable to assume that the coefficient of variation $v = \sigma/m$ is constant for tape measures of all lengths manufactured by this company. Thus

$$v = \frac{0.03}{3} = 0.01,$$

and the standard deviation for a three-yard tape measures is $0.01 \times (9 \text{ feet}) = 0.09$ feet.

This example illustrates the fact that the coefficient of variation is often used as a measure of quality for products of different sizes or different weights. In the concrete industry, for example, the quality in terms of concrete strength is specified by a coefficient of variation, which is a constant for all mean strengths.

Central moments of higher order reveal additional features of a distribution. The *coefficient of skewness*, defined by

$$\gamma_1 = \frac{\mu_3}{\sigma^3} \tag{4.11}$$

gives a measure of the symmetry of a distribution. It is positive when a unimodal distribution has a dominant tail on the right. The opposite arrangement produces a negative $\gamma_1$. It is zero when a distribution is symmetrical about the mean. In fact, a symmetrical distribution about the mean implies that all odd-order central moments vanish.

The degree of flattening of a distribution near its peaks can be measured by the *coefficient of excess*, defined by

$$\gamma_2 = \frac{\mu_4}{\sigma^4} - 3. \tag{4.12}$$

A positive $\gamma_2$ implies a sharp peak in the neighborhood of a mode in a unimodal distribution, whereas a negative $\gamma_2$ implies, as a rule, a flattened peak. The significance of the number 3 in Equation (4.12) will be discussed in Section 7.2, when the normal distribution is introduced.

## 4.1.3   CONDITIONAL EXPECTATION

We conclude this section by introducing a useful relation involving conditional expectation. Let us denote by $E\{X|Y\}$ that function of random variable $Y$ for which the value at $Y = y_i$ is $E\{X|Y = y_i\}$. Hence, $E\{X|Y\}$ is itself a random variable, and one of its very useful properties is that

$$E\{X\} = E\{E\{X|Y\}\} \qquad (4.13)$$

If $Y$ is a discrete random variable taking on values $y_1, y_2, \ldots$, the above states that

$$E\{X\} = \sum_i E\{X|Y = y_i\}P(Y = y_i), \qquad (4.14)$$

and

$$E\{X\} = \int_{-\infty}^{\infty} E\{X|y\}f_Y(y)\mathrm{d}y, \qquad (4.15)$$

if $Y$ is continuous.

To establish the relation given by Equation (4.13), let us show that Equation (4.14) is true when both $X$ and $Y$ are discrete. Starting from the right-hand side of Equation (4.14), we have

$$\sum_i E\{X|Y = y_i\}P(Y = y_i) = \sum_i \sum_j x_j P(X = x_j | Y = y_i)P(Y = y_i).$$

Since, from Equation (2.24),

$$P(X = x_j | Y = y_i) = \frac{P(X = x_j \cap Y = y_i)}{P(Y = y_i)},$$

we have

$$\sum_i E\{X|Y = y_i\}P(Y = y_i) = \sum_i \sum_j x_j p_{XY}(x_j, y_i)$$

$$= \sum_j x_j \sum_i p_{XY}(x_j, y_i)$$

$$= \sum_j x_j p_X(x_j)$$

$$= E\{X\},$$

and the desired result is obtained.

The usefulness of Equation (4.13) is analogous to what we found in using the theorem of total probability discussed in Section 2.4 (see Theorem 2.1, page 23).

It states that, in order to determine $E\{X\}$, it can be found by taking a weighted average of the conditional expectation of $X$ given $Y = y_i$; each of these terms is weighted by probability $P(Y = y_i)$.

**Example 4.8.** Problem: the survival of a motorist stranded in a snowstorm depends on which of the three directions the motorist chooses to walk. The first road leads to safety after one hour of travel, the second leads to safety after three hours of travel, but the third will circle back to the original spot after two hours. Determine the average time to safety if the motorist is equally likely to choose any one of the roads.

Answer: let $Y = 1, 2$, and 3 be the events that the motorist chooses the first, second and third road, respectively. Then $P(Y = i) = 1/3$ for $i = 1, 2, 3$. Let $X$ be the time to safety, in hours. We have:

$$E\{X\} = \sum_{i=1}^{3} E\{X|Y = i\}P(Y = i)$$

$$= \frac{1}{3}\sum_{i=1}^{3} E\{X|Y = i\}.$$

Now,

$$\left.\begin{array}{l} E\{X|Y = 1\} = 1, \\[2mm] E\{X|Y = 2\} = 3, \\[2mm] E\{X|Y = 3\} = 2 + E\{X\}. \end{array}\right\} \tag{4.16}$$

Hence

$$E\{X\} = \frac{1}{3}(1 + 3 + 2 + E\{X\}),$$

or

$$E\{X\} = 3 \text{ hours.}$$

Let us remark that the third relation in Equations (4.16) is obtained by noting that, if the motorist chooses the third road, then it takes two hours to find that he or she is back to the starting point and the problem is as before. Hence, the motorist's expected additional time to safety is just $E\{X\}$. The result is thus $2 + E\{X\}$. We further remark that problems of this type would require much more work were other approaches to be used.

## 4.2   CHEBYSHEV INEQUALITY

In the discussion of expectations and moments, there are two aspects to be considered in applications. The first is that of calculating moments of various orders of a random variable knowing its distribution, and the second is concerned with making statements about the behavior of a random variable when only some of its moments are available. The second aspect arises in numerous practical situations in which available information leads only to estimates of some simple moments of a random variable.

The knowledge of mean and variance of a random variable, although very useful, is not sufficient to determine its distribution and therefore does not permit us to give answers to such questions as 'What is $P(X \leq 5)$?' However, as is shown in Theorem 4.1, it is possible to establish some probability bounds knowing only the mean and variance.

**Theorem 4.1:** the *Chebyshev inequality* states that

$$\boxed{P(|X - m_X| \geq k\sigma_X) \leq \frac{1}{k^2},}\qquad(4.17)$$

for any $k > 0$.

**Proof:** from the definition we have

$$\sigma_X^2 = \int_{-\infty}^{\infty} (x - m_X)^2 f_X(x)\mathrm{d}x \geq \int_{|x-m_X|\geq k\sigma_X} (x - m_X)^2 f_X(x)\mathrm{d}x$$

$$\geq k^2\sigma_X^2 \int_{|x-m_X|\geq k\sigma_X} f_X(x)\mathrm{d}x$$

$$= k^2\sigma_X^2 P(|X - m_X| \geq k\sigma_X).$$

Expression (4.17) follows. The proof is similar when $X$ is discrete

**Example 4.9.** In Example 4.7, for three-foot tape measures, we can write

$$P(|X - 3| \geq 0.03k) \leq \frac{1}{k^2}.$$

If $k = 2$,

$$P(|X - 3| \geq 0.06) \leq \frac{1}{4},$$

or

$$P(2.94 \leq X \leq 3.06) \geq \frac{3}{4}.$$

In words, the probability of a three-foot tape measure being in error less than or equal to $\pm 0.06$ feet is at least 0.75. Various probability bounds can be found by assigning different values to $k$.

The complete generality with which the Chebyshev inequality is derived suggests that the bounds given by Equation (4.17) can be quite conservative. This is indeed true. Sharper bounds can be achieved if more is known about the distribution.

## 4.3   MOMENTS OF TWO OR MORE RANDOM VARIABLES

Let $g(X, Y)$ be a real-valued function of two random variables $X$ and $Y$. Its expectation is defined by

$$E\{g(X, Y)\} = \sum_i \sum_j g(x_i, y_j) p_{XY}(x_i, y_j), \quad X \text{ and } Y \text{ discrete}, \tag{4.18}$$

$$E\{g(X, Y)\} = \int_{-\infty}^{\infty} \int_{-\infty}^{\infty} g(x, y) f_{XY}(x, y) \mathrm{d}x \mathrm{d}y, \quad X \text{ and } Y \text{ continuous}, \tag{4.19}$$

if the indicated sums or integrals exist.

In a completely analogous way, the *joint moments* $\alpha_{nm}$ of $X$ and $Y$ are given by, if they exist,

$$\alpha_{nm} = E\{X^n Y^m\}. \tag{4.20}$$

They are computed from Equation (4.18) or (4.19) by letting $g(X, Y) = X^n Y^m$.

Similarly, the *joint central moments* of $X$ and $Y$, when they exist, are given by

$$\mu_{nm} = E\{(X - m_X)^n (Y - m_Y)^m\}. \tag{4.21}$$

They are computed from Equation (4.18) or (4.19) by letting

$$g(X, Y) = (X - m_X)^n (Y - m_Y)^m.$$

Some of the most important moments in the two-random-variable case are clearly the individual means and variances of $X$ and $Y$. In the notation used here, the means of $X$ and $Y$ are, respectively, $\alpha_{10}$ and $\alpha_{01}$. Using Equation (4.19), for example, we obtain:

$$\alpha_{10} = E\{X\} = \int_{-\infty}^{\infty} \int_{-\infty}^{\infty} x f_{XY}(x, y) \mathrm{d}x \mathrm{d}y = \int_{-\infty}^{\infty} x \int_{-\infty}^{\infty} f_{XY}(x, y) \mathrm{d}y \mathrm{d}x$$

$$= \int_{-\infty}^{\infty} x f_X(x) \mathrm{d}x,$$

where $f_X(x)$ is the marginal density function of $X$. We thus see that the result is identical to that in the single-random-variable case.

This observation is, of course, also true for the individual variances. They are, respectively, $\mu_{20}$ and $\mu_{02}$, and can be found from Equation (4.21) with appropriate substitutions for $n$ and $m$. As in the single-random-variable case, we also have

or
$$\left.\begin{aligned} \mu_{20} = \alpha_{20} - \alpha_{10}^2 \qquad \sigma_X^2 = \alpha_{20} - m_X^2 \\ \mu_{02} = \alpha_{02} - \alpha_{01}^2 \qquad \sigma_Y^2 = \alpha_{02} - m_Y^2 \end{aligned}\right\} \tag{4.22}$$

### 4.3.1 COVARIANCE AND CORRELATION COEFFICIENT

The first and simplest joint moment of $X$ and $Y$ that gives some measure of their interdependence is $\mu_{11} = E\{(X - m_X)(Y - m_Y)\}$. It is called the *covariance* of $X$ and $Y$. Let us first note some of its properties.

**Property 4.1:**   the covariance is related to $\alpha_{nm}$ by

$$\mu_{11} = \alpha_{11} - \alpha_{10}\alpha_{01} = \alpha_{11} - m_X m_Y. \tag{4.23}$$

**Proof of Property 4.1:** Property 4.1 is obtained by expanding $(X - m_X)(Y - m_Y)$ and then taking the expectation of each term. We have:

$$\begin{aligned} \mu_{11} &= E\{(X - m_X)(Y - m_Y)\} = E\{XY - m_Y X - m_X Y + m_X m_Y\} \\ &= E\{XY\} - m_Y E\{X\} - m_X E\{Y\} + m_X m_Y \\ &= \alpha_{11} - \alpha_{10}\alpha_{01} - \alpha_{10}\alpha_{01} + \alpha_{10}\alpha_{01} \\ &= \alpha_{11} - \alpha_{10}\alpha_{01}. \end{aligned}$$

**Property 4.2:** let the *correlation coefficient* of $X$ and $Y$ be defined by

$$\rho = \frac{\mu_{11}}{(\mu_{20}\mu_{02})^{1/2}} = \frac{\mu_{11}}{\sigma_X \sigma_Y}. \tag{4.24}$$

Then, $|\rho| \le 1$.

**Proof of Property 4.2:** to show Property 4.2, let $t$ and $u$ be any real quantities and form

$$\phi(t, u) = E\{[t(X - m_x) + u(Y - m_Y)]^2\}$$
$$= \mu_{20}t^2 + 2\mu_{11}tu + \mu_{02}u^2.$$

Since the expectation of a nonnegative function of $X$ and $Y$ must be non-negative, $\phi(t, u)$ is a nonnegative quadratic form in $t$ and $u$, and we must have

$$\mu_{20}\mu_{02} - \mu_{11}^2 \geq 0, \tag{4.25}$$

which gives the desired result.

The normalization of the covariance through Equation (4.24) renders $\rho$ a useful substitute for $\mu_{11}$. Furthermore, the correlation coefficient is dimension-less and independent of the origin, that is, for any constants $a_1, a_2, b_1,$ and $b_2$ with $a_1 > 0$ and $a_2 > 0$, we can easily verify that

$$\rho(a_1 X + b_1, a_2 Y + b_2) = \rho(X, Y). \tag{4.26}$$

**Property 4.3.** If $X$ and $Y$ are independent, then

$$\mu_{11} = 0 \quad \text{and} \quad \rho = 0. \tag{4.27}$$

**Proof of Property 4.3:** let $X$ and $Y$ be continuous; their joint moment $\alpha_{11}$ is found from

$$\alpha_{11} = E\{XY\} = \int_{-\infty}^{\infty} \int_{-\infty}^{\infty} xy f_{XY}(x, y) \mathrm{d}x \mathrm{d}y.$$

If $X$ and $Y$ are independent, we see from Equation (3.45) that

$$f_{XY}(x, y) = f_X(x) f_Y(y),$$

and

$$\alpha_{11} = \int_{-\infty}^{\infty} \int_{-\infty}^{\infty} xy f_X(x) f_Y(y) \mathrm{d}x \mathrm{d}y = \int_{-\infty}^{\infty} x f_X(x) \mathrm{d}x \int_{-\infty}^{\infty} y f_Y(y) \mathrm{d}y$$
$$= m_X m_Y.$$

Equations (4.23) and (4.24) then show that $\mu_{11} = 0$ and $\rho = 0$. A similar result can be obtained for two independent discrete random variables.

This result leads immediately to an important generalization. Consider a function of $X$ and $Y$ in the form $g(X)h(Y)$ for which an expectation exists. Then, if $X$ and $Y$ are independent,

$$E\{g(X)h(Y)\} = E\{g(X)\}E\{h(Y)\}. \tag{4.28}$$

When the correlation coefficient of two random variables vanishes, we say they are *uncorrelated*. It should be carefully pointed out that what we have shown is that independence implies zero correlation. The converse, however, is not true. This point is more fully discussed in what follows.

The covariance or the correlation coefficient is of great importance in the analysis of two random variables. It is a measure of their *linear* interdependence in the sense that its value is a measure of accuracy with which one random variable can be approximated by a linear function of the other. In order to see this, let us consider the problem of approximating a random variable $X$ by a linear function of a second random variable $Y$, $aY + b$, where $a$ and $b$ are chosen so that the mean-square error $e$, defined by

$$e = E\{[X - (aY + b)]^2\}, \tag{4.29}$$

is minimized. Upon taking partial derivatives of $e$ with respect to $a$ and $b$ and setting them to zero, straightforward calculations show that this minimum is attained when

$$a = \frac{\sigma_X \rho}{\sigma_Y}$$

and

$$b = m_X - am_Y$$

Substituting these values into Equation (4.29) then gives $\sigma_X^2(1 - \rho^2)$ as the minimum mean-square error. We thus see that an exact fit in the mean-square sense is achieved when $|\rho| = 1$, and the linear approximation is the worst when $\rho = 0$. More specifically, when $\rho = +1$, the random variables $X$ and $Y$ are said to be *positively perfectly correlated* in the sense that the values they assume fall on a straight line with positive slope; they are *negatively perfectly correlated* when $\rho = -1$ and their values form a straight line with negative slope. These two extreme cases are illustrated in Figure 4.3. The value of $|\rho|$ decreases as scatter about these lines increases.

Let us again stress the fact that the correlation coefficient measures only the linear interdependence between two random variables. It is by no means a general measure of interdependence between $X$ and $Y$. Thus, $\rho = 0$ does not imply independence of the random variables. In fact, Example 4.10 shows, the

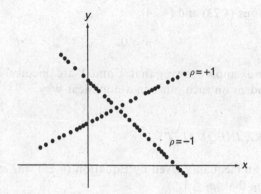

**Figure 4.3**   An illustration of perfect correlation, $\rho$

correlation coefficient can vanish when the values of one random variable are completely determined by the values of another.

**Example 4.10.** Problem: determine the correlation coefficient of random variables $X$ and $Y$ when $X$ takes values $\pm 1$ and $\pm 2$, each with probability $1/4$, and $Y = X^2$.

Answer: clearly, $Y$ assumes values 1 and 4, each with probability $1/2$, and their joint mass function is given by:

$$p_{XY}(x,y) = \begin{cases} \dfrac{1}{4}, & \text{for } (x,y) = (-1,1); \\[2mm] \dfrac{1}{4}, & \text{for } (x,y) = (1,1); \\[2mm] \dfrac{1}{4}, & \text{for } (x,y) = (-2,4); \\[2mm] \dfrac{1}{4}, & \text{for } (x,y) = (2,4). \end{cases}$$

The means and second moment $\alpha_{11}$ are given by

$$m_X = (-2)\left(\frac{1}{4}\right) + (-1)\left(\frac{1}{4}\right) + (1)\left(\frac{1}{4}\right) + (2)\left(\frac{1}{4}\right) = 0,$$

$$m_Y = (1)\left(\frac{1}{2}\right) + (4)\left(\frac{1}{2}\right) = 2.5,$$

$$\alpha_{11} = (-1)(1)\left(\frac{1}{4}\right) + (1)(1)\left(\frac{1}{4}\right) + (-2)(4)\left(\frac{1}{4}\right) + (2)(4)\left(\frac{1}{4}\right) = 0.$$

Hence,

$$\alpha_{11} - m_X m_Y = 0,$$

and, from Equations (4.23) and (4.24),

$$\rho = 0.$$

This is a simple example showing that $X$ and $Y$ are uncorrelated but they are completely dependent on each other in a nonlinear way.

### 4.3.2  SCHWARZ INEQUALITY

In Section 4.3.1, an inequality given by Equation (4.25) was established in the process of proving that $|\rho| \leq 1$:

$$\mu_{11}^2 = |\mu_{11}|^2 \leq \mu_{20}\mu_{02}. \tag{4.30}$$

We can also show, following a similar procedure, that

$$\boxed{E^2\{XY\} = |E\{XY\}|^2 \leq E\{X^2\}E\{Y^2\}.} \tag{4.31}$$

Equations (4.30) and (4.31) are referred to as the *Schwarz inequality*. We point them out here because they are useful in a number of situations involving moments in subsequent chapters.

### 4.3.3  THE CASE OF THREE OR MORE RANDOM VARIABLES

The expectation of a function $g(X_1, X_2, \ldots, X_n)$ of $n$ random variables $X_1, X_2, \ldots, X_n$ is defined in an analogous manner. Following Equations (4.18) and (4.19) for the two-random-variable case, we have

$$E\{g(X_1, \ldots, X_n)\} = \sum_{i_1} \cdots \sum_{i_n} g(x_{1i_1}, \ldots, x_{ni_n})p_{X_1\ldots X_n}(x_{1i_1}, \ldots, x_{ni_n}),$$

$$X_1, \ldots, X_n \text{ discrete}; \tag{4.32}$$

$$E\{g(X_1, \ldots, X_n)\} = \int_{-\infty}^{\infty} \cdots \int_{-\infty}^{\infty} g(x_1, \ldots, x_n)f_{X_1\ldots X_n}(x_1, \ldots, x_n)\,\mathrm{d}x_1 \ldots \mathrm{d}x_n,$$

$$X_1, \ldots, X_n \text{ continuous}; \tag{4.33}$$

where $p_{X_1\ldots X_n}$ and $f_{X_1\ldots X_n}$ are, respectively, the joint mass function and joint density function of the associated random variables.

The important moments associated with $n$ random variables are still the individual means, individual variances, and pairwise covariances. Let $X$ be the random column vector with components $X_1, \ldots, X_n$, and let the means of $X_1, \ldots, X_n$ be represented by the vector $\mathbf{m}_X$. A convenient representation of their variances and covariances is the *covariance matrix*, $\mathbf{\Lambda}$, defined by

$$\mathbf{\Lambda} = E\{(\mathbf{X} - \mathbf{m}_X)(\mathbf{X} - \mathbf{m}_X)^T\}, \tag{4.34}$$

where the superscript $T$ denotes the matrix transpose. The $n \times n$ matrix $\mathbf{\Lambda}$ has a structure in which the diagonal elements are the variances and in which the nondiagonal elements are covariances. Specifically, it is given by

$$\mathbf{\Lambda} = \begin{bmatrix} \text{var}(X_1) & \text{cov}(X_1, X_2) & \ldots & \text{cov}(X_1, X_n) \\ \text{cov}(X_2, X_1) & \text{var}(X_2) & \ldots & \text{cov}(X_2, X_n) \\ \vdots & \vdots & \ddots & \vdots \\ \text{cov}(X_n, X_1) & \text{cov}(X_n, X_2) & \ldots & \text{var}(X_n) \end{bmatrix}. \tag{4.35}$$

In the above 'var' reads 'variance of' and 'cov' reads 'covariance of'. Since $\text{cov}(X_i, X_j) = \text{cov}(X_j, X_i)$, the covariance matrix is always symmetrical.

In closing, let us state (in Theorem 4.2) without proof an important result which is a direct extension of Equation (4.28).

**Theorem 4.2:** if $X_1, X_2, \ldots, X_n$ are mutually independent, then

$$\boxed{E\{g_1(X_1)g_2(X_2)\ldots g_n(X_n)\} = E\{g_1(X_1)\}E\{g_2(X_2)\}\ldots E\{g_n(X_n)\},} \tag{4.36}$$

where $g_j(X_j)$ is an arbitrary function of $X_j$. It is assumed, of course, that all indicated expectations exist.

## 4.4  MOMENTS OF SUMS OF RANDOM VARIABLES

Let $X_1, X_2, \ldots, X_n$ be $n$ random variables. Their sum is also a random variable. In this section, we are interested in the moments of this sum in terms of those associated with $X_j, j = 1, 2, \ldots, n$. These relations find applications in a large number of derivations to follow and in a variety of physical situations.

Consider

$$Y = X_1 + X_2 + \cdots + X_n. \tag{4.37}$$

Let $m_j$ and $\sigma_j^2$ denote the respective mean and variance of $X_j$. Results 4.1–4.3 are some of the important results concerning the mean and variance of $Y$.

Verifications of these results are carried out for the case where $X_1, \ldots, X_n$ are continuous. The same procedures can be used when they are discrete.

**Result 4.1:** the mean of the sum is the sum of the means; that is,

$$\boxed{m_Y = m_1 + m_2 + \cdots + m_n.} \tag{4.38}$$

**Proof of Result 4.1:** to establish Result 4.1, consider

$$m_Y = E\{Y\} = E\{X_1 + X_2 + \cdots + X_n\}$$

$$= \int_{-\infty}^{\infty} \cdots \int_{-\infty}^{\infty} (x_1 + \cdots + x_n) f_{X_1 \ldots X_n}(x_1, \ldots, x_n) dx_1 \ldots dx_n$$

$$= \int_{-\infty}^{\infty} \cdots \int_{-\infty}^{\infty} x_1 f_{X_1 \ldots X_n}(x_1, \ldots, x_n) dx_1 \ldots dx_n$$

$$+ \int_{-\infty}^{\infty} \cdots \int_{-\infty}^{\infty} x_2 f_{X_1 \ldots X_n}(x_1, \ldots, x_n) dx_1 \ldots dx_n + \cdots$$

$$+ \int_{-\infty}^{\infty} \cdots \int_{-\infty}^{\infty} x_n f_{X_1 \ldots X_n}(x_1, \ldots, x_n) dx_1 \ldots dx_n.$$

The first integral in the final expression can be immediately integrated with respect to $x_2, x_3, \ldots, x_n$, yielding $f_{X_1}(x_1)$, the marginal density function of $X_1$. Similarly, the $(n-1)$-fold integration with respect to $x_1, x_3, \ldots, x_n$ in the second integral gives $f_{X_2}(x_2)$, and so on. Hence, the foregoing reduces to

$$m_Y = \int_{-\infty}^{\infty} x_1 f_{X_1}(x_1) dx_1 + \cdots + \int_{-\infty}^{\infty} x_n f_{X_n}(x_n) dx_n$$

$$= m_1 + m_2 + \cdots + m_n.$$

Combining Result 4.1 with some basic properties of the expectation we obtain some useful generalizations. For example, in view of the second of Equations (4.3), we obtain Result 4.2.

**Result 4.2:** if

$$Z = a_1 X_1 + a_2 X_2 + \cdots + a_n X_n, \tag{4.39}$$

where $a_1, a_2, \ldots, a_n$ are constants, then

$$m_Z = a_1 m_1 + a_2 m_2 + \cdots + a_n m_n \tag{4.40}$$

**Result 4.3:** let $X_1, \ldots, X_n$ be mutually independent random variables. Then the variance of the sum is the sum of the variances; that is,

$$\boxed{\sigma_Y^2 = \sigma_1^2 + \sigma_2^2 + \cdots + \sigma_n^2.}$$

(4.41)

Let us verify Result 4.3 for $n = 2$. The proof for the case of $n$ random variables follows at once by mathematical induction. Consider

$$Y = X_1 + X_2.$$

We know from Equation (4.38) that

$$m_Y = m_1 + m_2.$$

Subtracting $m_Y$ from $Y$, and $(m_1 + m_2)$ from $(X_1 + X_2)$ yields

$$Y - m_Y = (X_1 - m_1) + (X_2 - m_2)$$

and

$$\begin{aligned}
\sigma_Y^2 &= E\{(Y - m_Y)^2\} = E\{[(X_1 - m_1) + (X_2 - m_2)]^2\} \\
&= E\{(X_1 - m_1)^2 + 2(X_1 - m_1)(X_2 - m_2) + (X_2 - m_2)^2\} \\
&= E\{(X_1 - m_1)^2\} + 2E\{(X_1 - m_1)(X_2 - m_2)\} + E\{(X_2 - m_2)^2\} \\
&= \sigma_1^2 + 2\,\mathrm{cov}(X_1, X_2) + \sigma_2^2.
\end{aligned}$$

The covariance $\mathrm{cov}(X_1, X_2)$ vanishes, since $X_1$ and $X_2$ are independent [see Equation (4.27)], thus the desired result is obtained.

Again, many generalizations are possible. For example, if $Z$ is given by Equation (4.39), we have, following the second of Equations (4.9),

$$\sigma_Z^2 = a_1^2 \sigma_1^2 + \cdots + a_n^2 \sigma_n^2.$$

(4.42)

Let us again emphasize that, whereas Equation (4.38) is valid for any set of random variables $X_1, \ldots X_n$, Equation (4.41), pertaining to the variance, holds only under the independence assumption. Removal of the condition of independence would, as seen from the proof, add covariance terms to the right-hand side of Equation (4.41). It would then have the form

$$\sigma_Y^2 = \sigma_1^2 + \sigma_2^2 + \cdots + \sigma_n^2 + 2\,\mathrm{cov}(X_1, X_2) + 2\,\mathrm{cov}(X_1, X_3) + \cdots + 2\,\mathrm{cov}(X_{n-1}, X_n)$$

$$= \sum_{j=1}^{n} \sigma_j^2 + 2 \sum_{\substack{i=1 \\ i<j}}^{n-1} \sum_{j=2}^{n} \mathrm{cov}(X_i, X_j)$$

(4.43)

**Example 4.11.** Problem: an inspection is made of a group of $n$ television picture tubes. If each passes the inspection with probability $p$ and fails with probability $q$ ($p + q = 1$), calculate the average number of tubes in $n$ tubes that pass the inspection.

Answer: this problem may be easily solved if we introduce a random variable $X_j$ to represent the outcome of the $j$th inspection and define

$$X_j = \begin{cases} 1, & \text{if the } j\text{th tube passes inspection;} \\ 0, & \text{if the } j\text{th tube does not pass inspection.} \end{cases}$$

Then random variable $Y$, defined by

$$Y = X_1 + X_2 + \cdots + X_n,$$

has the desired property that its value is the total number of tubes passing the inspection. The mean of $X_j$ is

$$E\{X_j\} = 0(q) + 1(p) = p.$$

Therefore, as seen from Equation (4.38), the desired average number is given by

$$m_Y = E\{X_1\} + \cdots + E\{X_n\} = np.$$

We can also calculate the variance of $Y$. If $X_1, \ldots, X_n$ are assumed to be independent, the variance of $X_j$ is given by

$$\sigma_j^2 = E\{(X_j - p)^2\} = (0 - p)^2(q) + (1 - p)^2 p = pq.$$

Equation (4.41) then gives

$$\sigma_Y^2 = \sigma_1^2 + \cdots + \sigma_n^2 = npq.$$

**Example 4.12.** Problem: let $X_1, \ldots, X_n$ be a set of mutually independent random variables with a common distribution, each having mean $m$. Show that, for every $\varepsilon > 0$, and as $n \to \infty$,

$$P\left(\left|\frac{Y}{n} - m\right| \geq \varepsilon\right) \to 0, \quad \text{where } Y = X_1 + \cdots + X_n. \tag{4.44}$$

Note: this is a statement of the *law of large numbers*. The random variable $Y/n$ can be interpreted as an average of $n$ independently observed random variables from the same distribution. Equation (4.44) then states that the probability that this average will differ from the mean by greater than an arbitrarily prescribed

$\varepsilon$ tends to zero. In other words, random variable $Y/n$ approaches the true mean with probability 1.

Answer: to proceed with the proof of Equation (4.44), we first note that, if $\sigma^2$ is the variance of each $X_j$, it follows from Equation (4.41) that

$$\sigma_Y^2 = n\sigma^2.$$

According to the Chebyshev inequality, given by Expression (4.17), for every $k > 0$, we have

$$P(|Y - nm| \geq k) \leq \frac{n\sigma^2}{k^2}.$$

For $k = \varepsilon n$, the left-hand side is less than $\sigma^2/(\varepsilon^2 n)$, which tends to zero as $n \to \infty$. This establishes the proof.

Note that this proof requires the existence of $\sigma^2$. This is not necessary but more work is required without this restriction.

Among many of its uses, statistical sampling is an example in which the law of large numbers plays an important role. Suppose that in a group of $m$ families there are $m_j$ number of families with exactly $j$ children ($j = 0, 1, \ldots$, and $m_0 + m_1 + \ldots = m$). For a family chosen at random, the number of children is a random variable that assumes the value $r$ with probability $p_r = m_r/m$. A sample of $n$ families among this group represents $n$ observed independent random variables $X_1, \ldots, X_n$, with the same distribution. The quantity $(X_1 + \cdots + X_n)/n$ is the sample average, and the law of large numbers then states that, for sufficiently large samples, the sample average is likely to be close to

$$m = \sum_{r=0}^{\infty} r p_r = \sum_{r=0}^{\infty} r m_r/m,$$

the mean of the population.

**Example 4.13.** The random variable $Y/n$ in Example 4.12 is also called the *sample mean* associated with random variables $X_1, \ldots, X_n$ and is denoted by $\overline{X}$. In Example 4.12, if the coefficient of variation for each $X_i$ is $v$, the coefficient of variation $v_{\overline{X}}$ of $\overline{X}$ is easily derived from Equations (4.38) and (4.41) to be

$$v_{\overline{X}} = \frac{v}{n^{1/2}} \tag{4.45}$$

Equation (4.45) is the basis for the *law of $\sqrt{n}$* by Schrödinger, which states that the laws of physics are accurate within a probable relative error of the order of $n^{-1/2}$, where $n$ is the number of molecules that cooperate in a physical process. Basically, what Equation (4.45) suggests is that, if the action of each molecule exhibits a random variation measured by $v$, then a physical process resulting from additive actions of $n$ molecules will possess a random variation measured by $v/n^{1/2}$. It decreases as $n$ increases. Since $n$ is generally very large in the workings of physical processes, this result leads to the conjecture that the laws of physics can be exact laws despite local disorder.

## 4.5 CHARACTERISTIC FUNCTIONS

The expectation $E\{e^{jtX}\}$ of a random variable $X$ is defined as the *characteristic function* of $X$. Denoted by $\phi_X(t)$, it is given by

$$\phi_X(t) = E\{e^{jtX}\} = \sum_i e^{jtx_i} p_X(x_i), \quad X \text{ discrete;} \tag{4.46}$$

$$\phi_X(t) = E\{e^{jtX}\} = \int_{-\infty}^{\infty} e^{jtx} f_X(x)dx, \quad X \text{ continuous;} \tag{4.47}$$

where $t$ is an arbitrary real-valued parameter and $j = \sqrt{-1}$. The characteristic function is thus the expectation of a complex function and is generally complex valued. Since

$$|e^{jtX}| = |\cos tX + j\sin tX| = 1,$$

the sum and the integral in Equations (4.46) and (4.47) exist and therefore $\phi_X(t)$ always exists. Furthermore, we note

$$\left. \begin{array}{l} \phi_X(0) = 1, \\ \phi_X(-t) = \phi_X^*(t), \\ |\phi_X(t)| \le 1, \end{array} \right\} \tag{4.48}$$

where the asterisk denotes the complex conjugate. The first two properties are self-evident. The third relation follows from the observation that, since $f_X(x) \ge 0$,

$$|\phi_X(t)| = \left| \int_{-\infty}^{\infty} e^{jtx} f_X(x)dx \right| \le \int_{-\infty}^{\infty} f_X(x)dx = 1.$$

The proof is the same as that for discrete random variables.

We single this expectation out for discussion because it possesses a number of important properties that make it a powerful tool in random-variable analysis and probabilistic modeling.

### 4.5.1  GENERATION OF MOMENTS

One of the important uses of characteristic functions is in the determination of the moments of a random variable. Expanding $\phi_X(t)$ as a MacLaurin series, we see that (suppressing the subscript $X$ for convenience)

$$\phi(t) = \phi(0) + \phi'(0)t + \phi''(0)\frac{t^2}{2} + \cdots + \phi^{(n)}(0)\frac{t^n}{n!} + \ldots, \qquad (4.49)$$

where the primes denote derivatives. The coefficients of this power series are, from Equation (4.47),

$$\left.\begin{aligned}
\phi(0) &= \int_{-\infty}^{\infty} f_X(x)\mathrm{d}x = 1, \\[2mm]
\phi'(0) &= \frac{\mathrm{d}\phi(t)}{\mathrm{d}t}\bigg|_{t=0} = \int_{-\infty}^{\infty} \mathrm{j}x f_X(x)\mathrm{d}x = \mathrm{j}\alpha_1, \\[2mm]
&\ \ \vdots \\[2mm]
\phi^{(n)}(0) &= \frac{\mathrm{d}^n\phi(t)}{\mathrm{d}t^n}\bigg|_{t=0} = \int_{-\infty}^{\infty} \mathrm{j}^n x^n f_X(x)\mathrm{d}x = \mathrm{j}^n\alpha_n.
\end{aligned}\right\} \qquad (4.50)$$

Thus,

$$\phi(t) = 1 + \sum_{n=1}^{\infty} \frac{(\mathrm{j}t)^n \alpha_n}{n!}. \qquad (4.51)$$

The same results are obtained when $X$ is discrete.

Equation (4.51) shows that moments of all orders, if they exist, are contained in the expansion of $\phi(t)$, and these moments can be found from $\phi(t)$ through differentiation. Specifically, Equations (4.50) give

$$\boxed{\alpha_n = \mathrm{j}^{-n}\phi^{(n)}(0), \quad n = 1, 2, \ldots.} \qquad (4.52)$$

**Example 4.14.** Problem: determine $\phi(t)$, the mean, and the variance of a random variable $X$ if it has the binomial distribution

$$p_X(k) = \binom{n}{k}p^k(1-p)^{n-k}, \quad k = 0, 1, \ldots, n.$$

Answer: according to Equation (4.46),

$$\phi_X(t) = \sum_{k=0}^{n} e^{jtk} \binom{n}{k} p^k (1-p)^{n-k}$$

$$= \sum_{k=0}^{n} \binom{n}{k} (pe^{jt})^k (1-p)^{n-k} \tag{4.53}$$

$$= [pe^{jt} + (1-p)]^n.$$

Using Equation (4.52), we have

$$\alpha_1 = \frac{1}{j}\frac{d}{dt}[pe^{jt} + (1-p)]^n \Big|_{t=0} = n[pe^{jt} + (1-p)]^{n-1}(pe^{jt})\Big|_{t=0}$$

$$= np,$$

$$\alpha_2 = -\frac{d^2}{dt^2}[pe^{jt} + (1-p)]^n \Big|_{t=0} = np[(n-1)p + 1],$$

and

$$\sigma_X^2 = \alpha_2 - \alpha_1^2 = np[(n-1)p + 1] - n^2p^2 = np(1-p).$$

The results for the mean and variance are the same as those obtained in Examples 4.1 and 4.5.

**Example 4.15.** Problem: repeat the above when $X$ is exponentially distributed with density function

$$f_X(x) = \begin{cases} ae^{-ax}, & \text{for } x \geq 0; \\ 0, & \text{elsewhere.} \end{cases}$$

Answer: the characteristic function $\phi_X(t)$ in this case is

$$\phi_X(t) = \int_0^\infty e^{jtx}(ae^{-ax})dx = a\int_0^\infty e^{-(a-jt)x}dx = \frac{a}{a-jt}. \tag{4.54}$$

The moments are

$$\alpha_1 = \frac{1}{j}\frac{d}{dt}\left(\frac{a}{a-jt}\right)\Big|_{t=0} = \frac{1}{j}\left[\frac{ja}{(a-jt)^2}\right]\Big|_{t=0} = \frac{1}{a},$$

$$\alpha_2 = -\frac{d^2}{dt^2}\left(\frac{a}{a-jt}\right)\Big|_{t=0} = \frac{2}{a^2},$$

$$\sigma_X^2 = \alpha_2 - \alpha_1^2 = \frac{1}{a^2},$$

which agree with the moment calculations carried out in Examples 4.2 and 4.6.

Another useful expansion is the power series representation of the logarithm of the characteristic function; that is,

$$\log \phi_X(t) = \sum_{n=1}^{\infty} \frac{(jt)^n \lambda_n}{n!}, \qquad (4.55)$$

where coefficients $\lambda_n$ are again obtained from

$$\lambda_n = j^{-n} \frac{d^n}{dt^n} \log \phi_X(t) \bigg|_{t=0}. \qquad (4.56)$$

The relations between coefficients $\lambda_n$ and moments $\alpha_n$ can be established by forming the exponential of $\log \phi_X(t)$, expanding this in a power series of $jt$, and equating coefficients to those of corresponding powers in Equation (4.51). We obtain

$$\left. \begin{aligned} \lambda_1 &= \alpha_1, \\ \lambda_2 &= \alpha_2 - \alpha_1^2, \\ \lambda_3 &= \alpha_3 - 3\alpha_1\alpha_2 + 2\alpha_1^3, \\ \lambda_4 &= \alpha_4 - 3\alpha_2^2 - 4\alpha_1\alpha_3 + 12\alpha_1^2\alpha_2 - 6\alpha_1^4. \end{aligned} \right\} \qquad (4.57)$$

It is seen that $\lambda_1$ is the mean, $\lambda_2$ is the variance, and $\lambda_3$ is the third central moment. The higher order $\lambda_n$ are related to the moments of the same order or lower, but in a more complex way. Coefficients $\lambda_n$ are called *cumulants* of $X$ and, with a knowledge of these cumulants, we may obtain the moments and central moments.

### 4.5.2  INVERSION FORMULAE

Another important use of characteristic functions follows from the inversion formulae to be developed below.

Consider first a continuous random variable $X$. We observe that Equation (4.47) also defines $\phi_X(t)$ as the inverse Fourier transform of $f_X(x)$. The other half of the Fourier transform pair is

$$f_X(x) = \frac{1}{2\pi} \int_{-\infty}^{\infty} e^{-jtx} \phi_X(t) dt. \qquad (4.58)$$

This inversion formula shows that knowledge of the characteristic function specifies the distribution of $X$. Furthermore, it follows from the theory of

Fourier transforms that $f_X(x)$ is uniquely determined from Equation (4.58); that is, no two distinct density functions can have the same characteristic function.

This property of the characteristic function provides us with an alternative way of arriving at the distribution of a random variable. In many physical problems, it is often more convenient to determine the density function of a random variable by first determining its characteristic function and then performing the Fourier transform as indicated by Equation (4.58). Furthermore, we shall see that the characteristic function has properties that render it particularly useful for determining the distribution of a sum of independent random variables.

The inversion formula of Equation (4.58) follows immediately from the theory of Fourier transforms, but it is of interest to give a derivation of this equation from a probabilistic point of view.

**Proof of Equation (4.58):** an integration formula that can be found in any table of integrals is

$$\frac{1}{\pi} \int_{-\infty}^{\infty} \frac{\sin at}{t} \, dt = \begin{cases} -1, & \text{for } a < 0; \\ 0, & \text{for } a = 0; \\ 1, & \text{for } a > 0. \end{cases} \tag{4.59}$$

This leads to

$$\frac{1}{\pi} \int_{-\infty}^{\infty} \frac{\sin at + j(1 - \cos at)}{t} \, dt = \begin{cases} -1, & \text{for } a < 0; \\ 0, & \text{for } a = 0; \\ 1, & \text{for } a > 0; \end{cases} \tag{4.60}$$

because the function $(1 - \cos at)/t$ is an odd function of $t$ so that its integral over a symmetric range vanishes. Upon replacing $a$ by $X - x$ in Equation (4.60), we have

$$\frac{1}{2} - \frac{j}{2\pi} \int_{-\infty}^{\infty} \frac{1 - e^{j(X-x)t}}{t} \, dt = \begin{cases} 1, & \text{for } X < x; \\ \frac{1}{2}, & \text{for } X = x; \\ 0, & \text{for } X > x. \end{cases} \tag{4.61}$$

For a fixed value of $x$, Equation (4.61) is a function of random variable $X$, and it may be regarded as defining a new random variable $Y$. The random variable $Y$ is seen to be discrete, taking on values $1, \frac{1}{2}$, and $0$ with probabilities $P(X < x), P(X = x)$, and $P(X > x)$, respectively. The mean of $Y$ is thus equal to

$$E\{Y\} = (1)P(X < x) + \left(\frac{1}{2}\right)P(X = x) + (0)P(X > x).$$

However, notice that, since $X$ is continuous, $P(X = x) = 0$ if $x$ is a point of continuity in the distribution of $X$. Hence, using Equation (4.47),

$$E\{Y\} = P(X < x) = F_X(x)$$

$$= \frac{1}{2} - \frac{j}{2\pi} \int_{-\infty}^{\infty} \frac{1 - E\{e^{j(X-x)t}\}}{t} dt \qquad (4.62)$$

$$= \frac{1}{2} - \frac{j}{2\pi} \int_{-\infty}^{\infty} \frac{1 - e^{-jtx}\phi_X(t)}{t} dt.$$

The above defines the probability distribution function of $X$. Its derivative gives the inversion formula

$$f_X(x) = \frac{1}{2\pi} \int_{-\infty}^{\infty} e^{-jtx}\phi_X(t) dt, \qquad (4.63)$$

and we have Equation (4.58), as desired.

The inversion formula when $X$ is a discrete random variable is

$$p_X(x) = \lim_{u \to \infty} \frac{1}{2u} \int_{-u}^{u} e^{-jtx}\phi_X(t) dt. \qquad (4.64)$$

A proof of this relation can be constructed along the same lines as that given above for the continuous case.

**Proof of Equation (4.64):** first note the standard integration formula:

$$\frac{1}{2u} \int_{-u}^{u} e^{jat} dt = \begin{cases} \dfrac{\sin au}{au}, & \text{for } a \neq 0; \\ 1, & \text{for } a = 0. \end{cases} \qquad (4.65)$$

Replacing $a$ by $X - x$ and taking the limit as $u \to \infty$, we have a new random variable $Y$, defined by

$$Y = \lim_{u \to \infty} \frac{1}{2u} \int_{-u}^{u} e^{j(X-x)t} dt = \begin{cases} 0, & \text{for } X \neq x; \\ 1, & \text{for } X = x. \end{cases}$$

The mean of $Y$ is given by

$$E\{Y\} = (1)P(X = x) + (0)P(X \neq x) = P(X = x), \qquad (4.66)$$

and therefore

$$p_X(x) = \lim_{u \to \infty} \frac{1}{2u} \int_{-u}^{u} E\{e^{j(X-x)t}\} dt$$

(4.67)

$$= \lim_{u \to \infty} \frac{1}{2u} \int_{-u}^{u} e^{-jtx} \phi_X(t) dt,$$

which gives the desired inversion formula.

In summary, the transform pairs given by Equations (4.46), (4.47), (4.58), and (4.64) are collected and presented below for easy reference. For a continuous random variable $X$,

$$\left.\begin{aligned} \phi_X(t) &= \int_{-\infty}^{\infty} e^{jtx} f_X(x) dx, \\ f_X(x) &= \frac{1}{2\pi} \int_{-\infty}^{\infty} e^{-jtx} \phi_X(t) dt; \end{aligned}\right\}$$

(4.68)

and, for a discrete random variable $X$,

$$\left.\begin{aligned} \phi_X(t) &= \sum_{i} e^{jtx_i} p_X(x_i), \\ p_X(x) &= \lim_{u \to \infty} \frac{1}{2u} \int_{-u}^{u} e^{-jtx} \phi_X(t) dt. \end{aligned}\right\}$$

(4.69)

Of the two sets, Equations (4.68) for the continuous case are more important in terms of applicability. As we shall see in Chapter 5, probability mass functions for discrete random variables can be found directly without resorting to their characteristic functions.

As we have mentioned before, the characteristic function is particularly useful for the study of a sum of independent random variables. In this connection, let us state the following important theorem, (Theorem 4.3).

**Theorem 4.3:** The characteristic function of a sum of independent random variables is equal to the product of the characteristic functions of the individual random variables.

**Proof of Theorem 4.3:** Let

$$Y = X_1 + X_2 + \cdots + X_n.$$

(4.70)

Then, by definition,

$$\phi_Y(t) = E\{e^{jtY}\} = E\{e^{jt(X_1+X_2+\cdots+X_n)}\}$$
$$= E\{e^{jtX_1}e^{jtX_2}\ldots e^{jtX_n}\}.$$

Since $X_1, X_2, \ldots, X_n$ are mutually independent, Equation (4.36) leads to

$$E\{e^{jtX_1}e^{jtX_2}\ldots e^{jtX_n}\} = E\{e^{jtX_1}\}E\{e^{jtX_2}\}\ldots E\{e^{jtX_n}\}.$$

We thus have

$$\phi_Y(t) = \phi_{X_1}(t)\phi_{X_2}(t)\ldots\phi_{X_n}(t), \tag{4.71}$$

which was to be proved.

In Section (4.4), we obtained moments of a sum of random variables; Equation (4.71), coupled with the inversion formula in Equation (4.58) or Equation (4.64), enables us to determine the distribution of a sum of random variables from the knowledge of the distributions of $X_j, j = 1, 2, \ldots, n$, provided that they are mutually independent.

**Example 4.16.** Problem: let $X_1$ and $X_2$ be two independent random variables, both having an exponential distribution with parameter $a$, and let $Y = X_1 + X_2$. Determine the distribution of $Y$.

Answer: the characteristic function of an exponentially distributed random variable was obtained in Example 4.15. From Equation (4.54), we have

$$\phi_{X_1}(t) = \phi_{X_2}(t) = \frac{a}{a - jt}.$$

According to Equation (4.71), the characteristic function of $Y$ is simply

$$\phi_Y(t) = \phi_{X_1}(t)\phi_{X_2}(t) = \frac{a^2}{(a - jt)^2}.$$

Hence, the density function of $Y$ is, as seen from the inversion formula of Equations (4.68),

$$f_Y(y) = \frac{1}{2\pi}\int_{-\infty}^{\infty} e^{-jty}\phi_Y(t)dt$$
$$= \frac{a^2}{2\pi}\int_{-\infty}^{\infty}\frac{e^{-jty}}{(a - jt)^2}dt$$
$$= \begin{cases} a^2 y e^{-ay}, & \text{for } y \geq 0; \\ 0, & \text{elsewhere.} \end{cases} \tag{4.72}$$

The distribution given by Equation (4.72) is called a gamma distribution, which will be discussed extensively in Section 7.4.

**Example 4.17.** In 1827, Robert Brown, an English botanist, noticed that small particles of matter from plants undergo erratic movements when suspended in fluids. It was soon discovered that the erratic motion was caused by impacts on the particles by the molecules of the fluid in which they were suspended. This phenomenon, which can also be observed in gases, is called *Brownian motion*. The explanation of Brownian motion was one of the major successes of statistical mechanics. In this example, we study Brownian motion in an elementary way by using one-dimensional random walk as an adequate mathematical model.

Consider a particle taking steps on a straight line. It moves either one step to the right with probability $p$, or one step to the left with probability $q\,(p + q = 1)$. The steps are always of unit length, positive to the right and negative to the left, and they are taken independently. We wish to determine the probability mass function of its position after $n$ steps.

Let $X_i$ be the random variable associated with the $i$th step and define

$$X_i = \begin{cases} 1, & \text{if it is to the right;} \\ -1, & \text{if it is to the left.} \end{cases} \tag{4.73}$$

Then random variable $Y$, defined by

$$Y = X_1 + X_2 + \cdots + X_n,$$

gives the position of the particle after $n$ steps. It is clear that $Y$ takes integer values between $-n$ and $n$.

To determine $p_Y(k)$, $-n \le k \le n$, we first find its characteristic function. The characteristic function of each $X_i$ is

$$\phi_{X_i}(t) = E\{e^{jtX_i}\} = pe^{jt} + qe^{-jt}. \tag{4.74}$$

It then follows from Equation (4.71) that, in view of independence,

$$\phi_Y(t) = \phi_{X_1}(t)\phi_{X_2}(t)\dots\phi_{X_n}(t)$$
$$= (pe^{jt} + qe^{-jt})^n. \tag{4.75}$$

Let us rewrite it as

$$\phi_Y(t) = e^{-jnt}(pe^{2jt} + q)^n$$
$$= \sum_{i=0}^{n} \binom{n}{i} p^i q^{n-i} e^{j(2i-n)t}$$

Letting $k = 2i - n$, we get

$$\phi_Y(t) = \sum_{k=-n}^{n} \binom{n}{\frac{n+k}{2}} p^{(n+k)/2} q^{(n-k)/2} e^{jkt}. \tag{4.76}$$

Comparing Equation (4.76) with the definition in Equation (4.46) yields the mass function

$$p_Y(k) = \binom{n}{\frac{n+k}{2}} p^{(n+k)/2} q^{(n-k)/2}, \quad k = -n, -(n-2), \ldots, n. \tag{4.77}$$

Note that, if $n$ is even, $k$ must also be even, and, if $n$ is odd $k$ must be odd.

Considerable importance is attached to the symmetric case in which $k \ll n$, and $p = q = 1/2$. In order to consider this special case, we need to use *Stirling's formula*, which states that, for large $n$,

$$n! \cong (2\pi)^{1/2} e^{-n} n^{n+\frac{1}{2}} \tag{4.78}$$

Substituting this approximation into Equation (4.77) gives

$$p_Y(k) \cong \left(\frac{2}{n\pi}\right)^{1/2} e^{-k^2/2n}, \quad k = -n, \ldots, n. \tag{4.79}$$

A further simplification results when the length of each step is small. Assuming that $r$ steps occur in a unit time (i.e. $n = rt$) and letting $a$ be the length of each step, then, as $n$ becomes large, random variable $Y$ approaches a continuous random variable, and we can show that Equation (4.79) becomes

$$f_Y(y) = \frac{1}{(2\pi a^2 rt)^{1/2}} \exp\left(-\frac{y^2}{2a^2 rt}\right), \quad -\infty < y < \infty, \tag{4.80}$$

where $y = ka$. On letting

$$D = \frac{a^2 r}{2},$$

we have

$$f_Y(y) = \frac{1}{(4\pi Dt)^{1/2}} \exp\left(-\frac{y^2}{4Dt}\right), \quad -\infty < y < \infty. \tag{4.81}$$

The probability density function given above belongs to a *Gaussian* or *normal* random variable. This result is an illustration of the central limit theorem, to be discussed in Section 7.2.

Our derivation of Equation 4.81 has been purely analytical. In his theory of Brownian motion, Einstein also obtained this result with

$$D = \frac{2RT}{Nf}, \tag{4.82}$$

where $R$ is the universal gas constant, $T$ is the absolute temperature, $N$ is Avogadro's number, and $f$ is the coefficient of friction which, for liquid or gas at ordinary pressure, can be expressed in terms of its viscosity and particle size. Perrin, a French physicist, was awarded the Nobel Prize in 1926 for his success in determining, from experiment, Avogadro's number.

### 4.5.3  JOINT CHARACTERISTIC FUNCTIONS

The concept of characteristic functions also finds usefulness in the case of two or more random variables. The development below is concerned with continuous random variables only, but the principal results are equally valid in the case of discrete random variables. We also eliminate a bulk of the derivations involved since they follow closely those developed for the single-random-variable case.

The *joint characteristic function* of two random variables $X$ and $Y$, $\phi_{XY}(t, s)$, is defined by

$$\phi_{XY}(t, s) = E\{e^{j(tX+sY)}\} = \int_{-\infty}^{\infty} \int_{-\infty}^{\infty} e^{j(tx+sy)} f_{XY}(x, y) \mathrm{d}x\mathrm{d}y. \tag{4.83}$$

where $t$ and $s$ are two arbitrary real variables. This function always exists and some of its properties are noted below that are similar to those noted for Equations (4.48) corresponding to the single-random-variable case:

$$\left.\begin{aligned} \phi_{XY}(0,0) &= 1, \\ \phi_{XY}(-t, -s) &= \phi_{XY}^*(t, s), \\ |\phi_{XY}(t, s)| &\leq 1. \end{aligned}\right\} \tag{4.84}$$

Furthermore, it is easy to verify that joint characteristic function $\phi_{XY}(t, s)$ is related to marginal characteristic functions $\phi_X(t)$ and $\phi_Y(s)$ by

$$\left.\begin{aligned} \phi_X(t) &= \phi_{XY}(t, 0), \\ \phi_Y(s) &= \phi_{XY}(0, s). \end{aligned}\right\} \tag{4.85}$$

If random variables $X$ and $Y$ are independent, then we also have

$$\phi_{XY}(t,s) = \phi_X(t)\phi_Y(s). \tag{4.86}$$

To show the above, we simply substitute $f_X(x)f_Y(y)$ for $f_{XY}(x,y)$ in Equation (4.83). The double integral on the right-hand side separates, and we have

$$\phi_{XY}(t,s) = \int_{-\infty}^{\infty} e^{jtx}f_X(x)\mathrm{d}x \int_{-\infty}^{\infty} e^{jsy}f_Y(y)\mathrm{d}y$$

$$= \phi_X(t)\phi_Y(s),$$

and we have the desired result.

Analogous to the one-random-variable case, joint characteristic function $\phi_{XY}(t,s)$ is often called on to determine joint density function $f_{XY}(x,y)$ of $X$ and $Y$ and their joint moments. The density function $f_{XY}(x,y)$ is uniquely determined in terms of $\phi_{XY}(t,s)$ by the two-dimensional Fourier transform

$$f_{XY}(x,y) = \frac{1}{4\pi^2} \int_{-\infty}^{\infty} \int_{-\infty}^{\infty} e^{-j(tx+sy)}\phi_{XY}(t,s)\mathrm{d}t\mathrm{d}s; \tag{4.87}$$

and moments $E\{X^n Y^m\} = \alpha_{nm}$, if they exist, are related to $\phi_{XY}(t,s)$ by

$$\left. \frac{\partial^{n+m}}{\partial t^n \partial s^m}\phi_{XY}(t,s) \right|_{t,s=0} = j^{n+m} \int_{-\infty}^{\infty} \int_{-\infty}^{\infty} x^n y^m f_{XY}(x,y)\mathrm{d}x\mathrm{d}y$$

$$= j^{n+m}\alpha_{nm}. \tag{4.88}$$

The MacLaurin series expansion of $\phi_{XY}(t,s)$ thus takes the form

$$\phi_{XY}(t,s) = \sum_{i=0}^{\infty} \sum_{k=0}^{\infty} \frac{\alpha_{ik}}{i!k!}(jt)^i(js)^k. \tag{4.89}$$

The above development can be generalized to the case of more than two random variables in an obvious manner.

**Example 4.18.** Let us consider again the Brownian motion problem discussed in Example 4.17, and form two random variables $X'$ and $Y'$ as

$$\left. \begin{array}{l} X' = X_1 + X_2 + \cdots + X_{2n}, \\ Y' = X_{n+1} + X_{n+2} + \cdots + X_{3n}. \end{array} \right\} \tag{4.90}$$

They are, respectively, the position of the particle after $2n$ steps and its position after $3n$ steps relative to where it was after $n$ steps. We wish to determine the joint probability density function (jpdf) $f_{XY}(x, y)$ of random variables

$$X = \frac{X'}{n^{1/2}}$$

and

$$Y = \frac{Y'}{n^{1/2}}$$

for large values of $n$.

For the simple case of $p = q = \frac{1}{2}$, the characteristic function of each $X_k$ is [see Equation (4.74)]

$$\phi(t) = E\{e^{jtX_k}\} = \frac{1}{2}(e^{jt} + e^{-jt}) = \cos t, \tag{4.91}$$

and, following Equation (4.83), the joint characteristic function of $X$ and $Y$ is

$$\phi_{XY}(t, s) = E\{\exp[j(tX + sY)]\} = E\left\{\exp\left[j\left(\frac{tX'}{n^{1/2}} + \frac{sY'}{n^{1/2}}\right)\right]\right\}$$

$$= E\left\{\exp\left\langle\left(\frac{j}{n^{1/2}}\right)\left[t\sum_{k=1}^{n} X_k + (t+s)\sum_{k=n+1}^{2n} X_k + s\sum_{k=2n+1}^{3n} X_k\right]\right\rangle\right\}$$

$$= \left\{\phi\left(\frac{t}{n^{1/2}}\right)\phi\left[\frac{s+t}{n^{1/2}}\right]\phi\left(\frac{x}{n^{1/2}}\right)\right\}^n, \tag{4.92}$$

where $\phi(t)$ is given by Equation (4.91). The last expression in Equation (4.92) is obtained based on the fact that the $X_k, k = 1, 2, \ldots, 3n$, are mutually independent. It should be clear that $X$ and $Y$ are not independent, however.

We are now in the position to obtain $f_{XY}(x, y)$ from Equation (4.92) by using the inverse formula given by Equation (4.87). First, however, some simplifications are in order. As $n$ becomes large,

$$\left[\phi\left(\frac{t}{n^{1/2}}\right)\right]^n = \cos^n\left(\frac{t}{n^{1/2}}\right)$$

$$= \left(1 - \frac{t^2}{n2!} + \frac{t^4}{n^24!} - \cdots\right)^n \tag{4.93}$$

$$\cong e^{-t^2/2}.$$

Hence, as $n \to \infty$,

$$\phi_{XY}(t, s) \cong e^{-(t^2 + ts + s^2)}. \qquad (4.94)$$

Now, substituting Equation (4.94) into Equation (4.87) gives

$$f_{XY}(x, y) = \frac{1}{4\pi^2} \int_{-\infty}^{\infty} \int_{-\infty}^{\infty} e^{-j(tx + sy)} e^{-(t^2 + ts + s^2)} dt\, ds, \qquad (4.95)$$

which can be evaluated following a change of variables defined by

$$t = \frac{t' + s'}{\sqrt{2}}, \quad s = \frac{t' - s'}{\sqrt{2}}. \qquad (4.96)$$

The result is

$$f_{XY}(x, y) = \frac{1}{2\pi\sqrt{3}} \exp\left[ -\frac{x^2 - xy + y^2}{3} \right]. \qquad (4.97)$$

The above is an example of a *bivariate normal distribution*, to be discussed in Section 7.2.3.

Incidentally, the joint moments of $X$ and $Y$ can be readily found by means of Equation (4.88). For large $n$, the means of $X$ and $Y$, $\alpha_{10}$ and $\alpha_{01}$, are

$$\alpha_{10} = -j \frac{\partial \phi_{XY}(t, s)}{\partial t} \bigg|_{t,s=0} = -j(-2t - s)e^{-(t^2 + ts + s^2)} \bigg|_{t,s=0} = 0,$$

$$\alpha_{01} = -j \frac{\partial \phi_{XY}(t, s)}{\partial s} \bigg|_{t,s=0} = 0.$$

Similarly, the second moments are

$$\alpha_{20} = E\{X^2\} = -\frac{\partial^2 \phi_{XY}(t, s)}{\partial t^2} \bigg|_{t,s=0} = 2,$$

$$\alpha_{02} = E\{Y^2\} = -\frac{\partial^2 \phi_{XY}(t, s)}{\partial s^2} \bigg|_{t,s=0} = 2,$$

$$\alpha_{11} = E\{XY\} = -\frac{\partial^2 \phi_{XY}(t, s)}{\partial t \partial s} \bigg|_{t,s=0} = 1.$$

## FURTHER READING AND COMMENTS

As mentioned in Section 4.2, the Chebyshev inequality can be improved upon if some additional distribution features of a random variable are known beyond its first two moments. Some generalizations can be found in:

Mallows, C.L., 1956, 'Generalizations of Tchebycheff's Inequalities', *J. Royal Statistical Societies, Series B* **18** 139–176.

In many introductory texts, the discussion of characteristic functions of random variables is bypassed in favor of moment-generating functions. The moment-generating function $M_X(t)$ of a random variable $X$ is defined by

$$M_X(t) = E\{e^{tX}\}.$$

In comparison with characteristic functions, the use of $M_X(t)$ is simpler since it avoids computations involving complex numbers and it generates moments of $X$ in a similar fashion. However, there are two disadvantages in using $M_X(t)$. The first is that it may not exist for all values of $t$ whereas $\phi_X(t)$ always exists. In addition, powerful inversion formulae associated with characteristic functions no longer exist for moment-generating functions. For a discussion of the moment-generating function, see, for example:

Meyer, P.L., 1970, *Introductory Probability and Statistical Applications*, 2nd edn, Addison-Wesley, Reading, Mas, pp. 210–217.

## PROBLEMS

4.1 For each of the probability distribution functions (PDFs) given in Problem 3.1 (Page 67), determine the mean and variance, if they exist, of its associated random variable.

4.2 For each of the probability density functions (pdfs) given in Problem 3.4, determine the mean and variance, if they exist, of its associated random variable.

4.3 According to the PDF given in Example 3.4 (page 47), determine the average duration of a long-distance telephone call.

4.4 It is found that resistance of aircraft structural parts, $R$, in a nondimensionalized form, follows the distribution

$$f_R(r) = \begin{cases} \dfrac{2\sigma_R^3}{0.9996\pi[\sigma_R^2 + (r-1)^2]^2}, & \text{for } r \geq 0.33; \\ 0, & \text{elsewhere;} \end{cases}$$

where $\sigma_R = 0.0564$. Determine the mean of $R$.

4.5 A target is made of three concentric circles of radii $3^{-1/2}$, 1, and $3^{1/2}$ feet. Shots within the inner circle count 4 points, within the next ring 3 points, and within the third ring 2 points. Shots outside of the target count 0. Let $R$ be the random variable representing distance of the hit from the center. Suppose that the pdf of $R$ is

$$f_R(r) = \begin{cases} \dfrac{2}{\pi(1+r^2)}, & \text{for } r > 0; \\ 0, & \text{elsewhere.} \end{cases}$$

Compute the mean score of each shot.

4.6 A random variable $X$ has the exponential distribution

$$f_X(x) = \begin{cases} a\mathrm{e}^{-x/2}, & \text{for } x \geq 0; \\ 0, & \text{elsewhere.} \end{cases}$$

Determine:
(a) The value of $a$.
(b) The mean and variance of $X$.
(c) The mean and variance of $Y = (X/2) - 1$.

4.7 Let the mean and variance of $X$ be $m$ and $\sigma^2$, respectively. For what values of $a$ and $b$ does random variable $Y$, equal to $aX + b$, have mean 0 and variance 1?

4.8 Suppose that your waiting time (in minutes) for a bus in the morning is uniformly distributed over $(0, 5)$, whereas your waiting time in the evening is distributed as shown in Figure 4.4. These waiting times are assumed to be independent for any given day and from day to day.
(a) If you take the bus each morning and evening for five days, what is the mean of your total waiting time?
(b) What is the variance of your total five-day waiting time?
(c) What are the mean and variance of the difference between morning and evening waiting times on a given day?
(d) What are the mean and variance of the difference between total morning waiting time and total evening waiting time for five days?

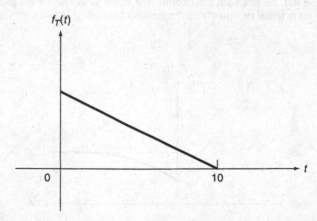

**Figure 4.4**   Density function of evening waiting times, for Problem 4.8

4.9  The diameter of an electronic cable, say $X$, is random, with pdf

$$f_X(x) = \begin{cases} 6x(1-x), & \text{for } 0 \le x \le 1; \\ 0, & \text{elsewhere.} \end{cases}$$

(a)  What is the mean value of the diameter?
(b)  What is the mean value of the cross-sectional area, $(\pi/4)X^2$?

4.10  Suppose that a random variable $X$ is distributed (arbitrarily) over the interval

$$a \le X \le b.$$

Show that:
(a)  $m_X$ is bounded by the same limits;
(b)  $\sigma_X^2 \le \dfrac{(b-a)^2}{4}$.

4.11  Show that, given a random variable $X$, $P(X = m_X) = 1$ if $\sigma_X^2 = 0$.

4.12  The waiting time $T$ of a customer at an airline ticket counter can be characterized by a mixed distribution function (see Figure 4.5):

$$F_T(t) = \begin{cases} 0, & \text{for } t < 0; \\ p + (1-p)(1 - e^{-\lambda t}) \end{cases}, \quad \text{for } t \ge 0.$$

Determine:
(a)  The average waiting time of an arrival, $E\{T\}$.
(b)  The average waiting time for an arrival given that a wait is required, $E\{T|T > 0\}$.

4.13  For the commuter described in Problem 3.21 (page 72), assuming that he or she makes one of the trains, what is the average arrival time at the destination?

4.14  A trapped miner has to choose one of two directions to find safety. If the miner goes to the right, then he will return to his original position after 3 minutes. If he goes to the left, he will with probability 1/3 reach safety and with probability 2/3 return to his original position after 5 minutes of traveling. Assuming that he is at all

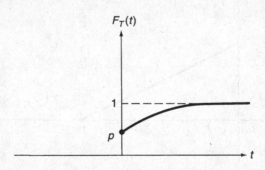

**Figure 4.5**  Distribution function, $F_T(t)$, of waiting times, for Problem 4.12

times equally likely to choose either direction, determine the average time interval (in minutes) that the miner will be trapped.

4.15 Show that:
   (a) $E\{X|Y = y\} = E\{X\}$ if $X$ and $Y$ are independent.
   (b) $E\{XY|Y = y\} = yE\{X|Y = y\}$.
   (c) $E\{XY\} = E\{YE[X|Y]\}$.

4.16 Let random variable $X$ be uniformly distributed over interval $0 \leq x \leq 2$. Determine a lower bound for $P(|X - 1| \leq 0.75)$ using the Chebyshev inequality and compare it with the exact value of this probability.

4.17 For random variable $X$ defined in Problem 4.16, plot $P(|X - m_X| \leq h)$ as a function of $h$ and compare it with its lower bound as determined by the Chebyshev inequality. Show that the lower bound becomes a better approximation of $P(|X - m_X| \leq h)$ as $h$ becomes large.

4.18 Let a random variable $X$ take only nonnegative values; show that, for any $a > 0$,

$$P(X \geq a) \leq \frac{m_X}{a}.$$

This is known as *Markov's inequality*.

4.19 The yearly snowfall of a given region is a random variable with mean equal to 70 inches.
   (a) What can be said about the probability that this year's snowfall will be between 55 and 85 inches?
   (b) Can your answer be improved if, in addition, the standard deviation is known to be 10 inches?

4.20 The number $X$ of airplanes arriving at an airport during a given period of time is distributed according to

$$p_X(k) = \frac{100^k e^{-100}}{k!}, \quad k = 0, 1, 2, \ldots.$$

Use the Chebyshev inequality to determine a lower bound for probability $P(80 \leq X \leq 120)$ during this period of time.

4.21 For each joint distribution given in Problem 3.13 (page 71), determine $m_X$, $m_Y$, $\sigma_X^2$, $\sigma_Y^2$, and $\rho_{XY}$ of random variables $X$ and $Y$.

4.22 In the circuit shown in Figure 4.6, the resistance $R$ is random and uniformly distributed between 900 and 1100 $\Omega$. The current $i = 0.01$ A and the resistance $r_0 = 1000 \Omega$ are constants.

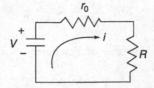

**Figure 4.6**  Circuit diagram for Problem 4.22

(a) Determine $m_V$ and $\sigma_V^2$ of voltage $V$, which is given by

$$V = (R + r_0)i.$$

(b) Determine the correlation coefficient of $R$ and $V$.

4.23 Let the jpdf of $X$ and $Y$ be given by

$$f_{XY}(x, y) = \begin{cases} xy, & \text{for } 0 < x < 1, \text{ and } 0 < y < 2; \\ 0, & \text{and elsewhere.} \end{cases}$$

Determine the mean of $Z$, equal to $(X^2 + Y^2)^{1/2}$.

4.24 The product of two random variables $X$ and $Y$ occurs frequently in applied problems. Let $Z = XY$ and assume that $X$ and $Y$ are independent. Determine the mean and variance of $Z$ in terms of $m_X, m_Y, \sigma_X^2$, and $\sigma_Y^2$.

4.25 Let $X = X_1 + X_2$, and $Y = X_2 + X_3$. Determine correlation coefficient $\rho_{XY}$ of $X$ and $Y$ in terms of $\sigma_{X_1}, \sigma_{X_2}$, and $\sigma_{X_3}$ when $X_1, X_2$, and $X_3$ are uncorrelated.

4.26 Let $X$ and $Y$ be discrete random variables with joint probability mass function (jpmf) given by Table 4.1. Show that $\rho_{XY} = 0$ but $X$ and $Y$ are not independent.

**Table 4.1** Joint probability mass function, $p_{XY}(x, y)$ for Problem 4.26

| $y$ | $x$ | | |
|---|---|---|---|
| | $-1$ | $0$ | $1$ |
| $-1$ | $a$ | $b$ | $a$ |
| $0$ | $b$ | $0$ | $b$ |
| $1$ | $a$ | $b$ | $a$ |

Note: $a + b = \dfrac{1}{4}$.

4.27 In a simple frame structure such as the one shown in Figure 4.7, the total horizontal displacement of top storey $Y$ is the sum of the displacements of individual storeys $X_1$ and $X_2$. Assume that $X_1$ and $X_2$ are independent and let $m_{X_1}, m_{X_2}, \sigma_{X_1}^2$, and $\sigma_{X_2}^2$ be their respective means and variances.
(a) Find the mean and variance of $Y$.
(b) Find the correlation coefficient between $X_2$ and $Y$. Discuss the result if $\sigma_{X_2}^2 \gg \sigma_{X_1}^2$.

4.28 Let $X_1, \ldots, X_n$ be a set of independent random variables, each of which has a probability density function (pdf) of the form

$$f_{X_j}(x_j) = \frac{1}{(2\pi)^{1/2}} e^{-x_j^2/2}, \quad j = 1, 2, \ldots, n, \quad -\infty < x_j < \infty.$$

Determine the mean and variance of $Y$, where

$$Y = \sum_{j=1}^{n} X_j^2.$$

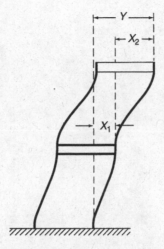

**Figure 4.7**   Frame structure, for Problem 4.27

4.29 Let $X_1, X_2, \ldots, X_n$ be independent random variables and let $\sigma_j^2$ and $\mu_j$ be the respective variance and third central moment of $X_j$. Let $\sigma^2$ and $\mu$ denote the corresponding quantities for $Y$, where $Y = X_1 + X_2 + \cdots + X_n$.
 (a) Show that $\sigma^2 = \sigma_1^2 + \sigma_2^2 + \cdots + \sigma_n^2$, and $\mu = \mu_1 + \mu_2 + \cdots + \mu_n$.
 (b) Show that this additive property does not apply to the fourth-order or higher-order central moments.

4.30 Determine the characteristic function corresponding to each of the PDFs given in Problem 3.1(a)–3.1(e) (page 67). Use it to generate the first two moments and compare them with results obtained in Problem 4.1. [Let $a = 2$ in part (e).]

4.31 We have shown that characteristic function $\phi_X(t)$ of random variable $X$ facilitates the determination of the moments of $X$. Another function $M_X(t)$, defined by

$$M_X(t) = E\{e^{tX}\},$$

and called the *moment-generating function* of $X$, can also be used to obtain moments of $X$. Derive the relationships between $M_X(t)$ and the moments of $X$.

4.32 Let

$$Y = a_1 X_1 + a_2 X_2 + \cdots + a_n X_n$$

where $X_1, X_2, \ldots, X_n$ are mutually independent. Show that

$$\phi_Y(t) = \phi_{X_1}(a_1 t)\phi_{X_2}(a_2 t)\ldots\phi_{X_n}(a_n t).$$

# 5

# Functions of Random Variables

The basic topic to be discussed in this chapter is one of determining the relationship between probability distributions of two random variables $X$ and $Y$ when
they are related by $Y = g(X)$. The functional form of $g(X)$ is given and deterministic. Generalizing to the case of many random variables, we are interested in the
determination of the joint probability distribution of $Y_j, j = 1, 2, \ldots, m$, which is
functionally dependent on $X_k, k = 1, 2, \ldots, n$, according to

$$Y_j = g_j(X_1, \ldots, X_n) , \quad j = 1, 2, \ldots, m, \ m \leq n, \tag{5.1}$$

when the joint probabilistic behavior of $X_k, k = 1, 2, \ldots, n$, is known.

Some problems of this type (i.e. transformations of random variables) have
been addressed in several places in Chapter 4. For example, Example 4.11 con-
siders transformation $Y = X_1 + \cdots + X_n$, and Example 4.18 deals with the transformation of $3n$ random variables $(X_1, X_2, \ldots, X_{3n})$ to two random variables
$(X', Y')$ defined by Equations (4.90). In science and engineering, most phenomena
are based on functional relationships in which one or more dependent variables
are expressed in terms of one or more independent variables. For example, force is
a function of cross-sectional area and stress, distance traveled over a time interval
is a function of the velocity, and so on. The techniques presented in this chapter
thus permit us to determine the probabilistic behavior of random variables that
are functionally dependent on some others with known probabilistic properties.

In what follows, transformations of random variables are treated in a systematic manner. In Equation (5.1), we are basically interested in the joint distributions
and joint moments of $Y_1, \ldots, Y_m$ given appropriate information on $X_1, \ldots, X_n$.

## 5.1 FUNCTIONS OF ONE RANDOM VARIABLE

Consider first a simple transformation involving only one random variable, and let

$$Y = g(X) \tag{5.2}$$

*Fundamentals of Probability and Statistics for Engineers* T.T. Soong © 2004 John Wiley & Sons, Ltd
ISBNs: 0-470-86813-9 (HB)   0-470-86814-7 (PB)

where $g(X)$ is assumed to be a continuous function of $X$. Given the probability distribution of $X$ in terms of its probability distribution function (PDF), probability mass function (pmf) or probability density function (pdf), we are interested in the corresponding distribution for $Y$ and its moment properties.

### 5.1.1  PROBABILITY DISTRIBUTION

Given the probability distribution of $X$, the quantity $Y$, being a function of $X$ as defined by Equation (5.2), is thus also a random variable. Let $R_X$ be the *range space* associated with random variable $X$, defined as the set of all possible values assumed by $X$, and let $R_Y$ be the corresponding range space associated with $Y$. A basic procedure of determining the probability distribution of $Y$ consists of the steps developed below.

For any outcome such as $X = x$, it follows from Equation (5.2) that $Y = y = g(x)$. As shown schematically in Figure 5.1, Equation (5.2) defines a mapping of values in range space $R_X$ into corresponding values in range space $R_Y$. Probabilities associated with each point (in the case of discrete random variable $X$) or with each region (in the case of continuous random variable $X$) in $R_X$ are carried over to the corresponding point or region in $R_Y$. The probability distribution of $Y$ is determined on completing this transfer process for every point or every region of nonzero probability in $R_X$. Note that many-to-one transformations are possible, as also shown in Figure 5.1. The procedure of determining the probability distribution of $Y$ is thus critically dependent on the functional form of $g$ in Equation (5.2).

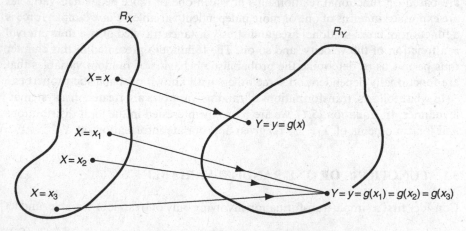

**Figure 5.1**  Transformation $y = g(x)$

### 5.1.1.1 Discrete Random Variables

Let us first dispose of the case when $X$ is a discrete random variable, since it requires only simple point-to-point mapping. Suppose that the possible values taken by $X$ can be enumerated as $x_1, x_2, \ldots$. Equation (5.2) shows that the corresponding possible values of $Y$ may be enumerated as $y_1 = g(x_1), y_2 = g(x_2), \ldots$. Let the pmf of $X$ be given by

$$p_X(x_i) = p_i, \quad i = 1, 2, \ldots. \tag{5.3}$$

The pmf of $y$ is simply determined as

$$p_Y(y_i) = p_Y[g(x_i)] = p_i, \quad i = 1, 2, \ldots. \tag{5.4}$$

**Example 5.1.** Problem: the pmf of a random variable $X$ is given as

$$p_X(x) = \begin{cases} \frac{1}{2}, & \text{for } x = -1; \\ \frac{1}{4}, & \text{for } x = 0; \\ \frac{1}{8}, & \text{for } x = 1; \\ \frac{1}{8}, & \text{for } x = 2; \end{cases}$$

Determine the pmf of $Y$ if $Y$ is related to $X$ by $Y = 2X + 1$.

Answer: the corresponding values of $Y$ are: $g(-1) = 2(-1) + 1 = -1$; $g(0) = 1$; $g(1) = 3$; and $g(2) = 5$. Hence, the pmf of $Y$ is given by

$$p_Y(y) = \begin{cases} \frac{1}{2}, & \text{for } y = -1; \\ \frac{1}{4}, & \text{for } y = 1; \\ \frac{1}{8}, & \text{for } y = 3; \\ \frac{1}{8}, & \text{for } y = 5. \end{cases}$$

**Example 5.2.** Problem: for the same $X$ as given in Example 5.1, determine the pmf of $Y$ if $Y = 2X^2 + 1$.

Answer: in this case, the corresponding values of $Y$ are: $g(-1) = 2(-1)^2 + 1 = 3$; $g(0) = 1$; $g(1) = 3$; and $g(2) = 9$, resulting in

$$p_Y(y) = \begin{cases} \frac{1}{4}, & \text{for } y = 1; \\ \frac{5}{8} \left(= \frac{1}{2} + \frac{1}{8}\right), & \text{for } y = 3; \\ \frac{1}{8}, & \text{for } y = 9. \end{cases}$$

### 5.1.1.2   Continuous Random Variables

A more frequently encountered case arises when $X$ is continuous with known PDF, $F_X(x)$, or pdf, $f_X(x)$. To carry out the mapping steps as outlined at the beginning of this section, care must be exercised in choosing appropriate corresponding regions in range spaces $R_X$ and $R_Y$, this mapping being governed by the transformation $Y = g(X)$. Thus, the degree of complexity in determining the probability distribution of $Y$ is a function of complexity in the transformation $g(X)$.

Let us start by considering a simple relationship

$$Y = g(X) = 2X + 1. \qquad (5.5)$$

The transformation $y = g(x)$ is presented graphically in Figure 5.2. Consider the PDF of $Y$, $F_Y(y)$; it is defined by

$$F_Y(y) = P(Y \leq y). \qquad (5.6)$$

The region defined by $Y \leq y$ in the range space $R_Y$ covers the heavier portion of the transformation curve, as shown in Figure 5.2, which, in the range space $R_X$, corresponds to the region $g(X) \leq y$, or $X \leq g^{-1}(y)$, where

$$g^{-1}(y) = \frac{y-1}{2}$$

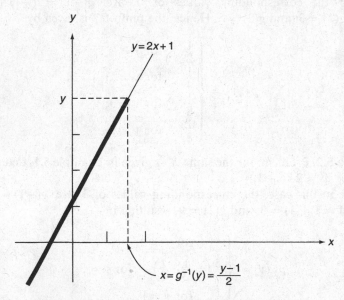

**Figure 5.2**   Transformation defined by Equation (5.5)

is the inverse function of $g(x)$, or the solution for $x$ in Equation (5.5) in terms of $y$. Hence,

$$F_Y(y) = P(Y \le y) = P[g(X) \le y] = P[X \le g^{-1}(y)] = F_X[g^{-1}(y)]. \quad (5.7)$$

Equation (5.7) gives the relationship between the PDF of $X$ and that of $Y$, our desired result.

The relationship between the pdfs of $X$ and $Y$ are obtained by differentiating both sides of Equation (5.7) with respect to $y$. We have:

$$f_Y(y) = \frac{dF_Y(y)}{dy} = \frac{d}{dy}\{F_X[g^{-1}(y)]\} = f_X[g^{-1}(y)]\frac{dg^{-1}(y)}{dy}. \quad (5.8)$$

It is clear that Equations (5.7) and (5.8) hold not only for the particular transformation given by Equation (5.5) but for all continuous $g(x)$ that are *strictly monotonic increasing functions* of $x$, that is, $g(x_2) > g(x_1)$ whenever $x_2 > x_1$.

Consider now a slightly different situation in which the transformation is given by

$$Y = g(X) = -2X + 1. \quad (5.9)$$

Starting again with $F_Y(y) = P(Y \le y)$, and reasoning as before, the region $Y \le y$ in the range space $R_Y$ is now mapped into the region $X > g^{-1}(y)$, as indicated in Figure 5.3. Hence, we have in this case

$$\begin{aligned}
F_Y(y) &= P(Y \le y) = P[X > g^{-1}(y)] \\
&= 1 - P[X \le g^{-1}(y)] = 1 - F_X[g^{-1}(y)].
\end{aligned} \quad (5.10)$$

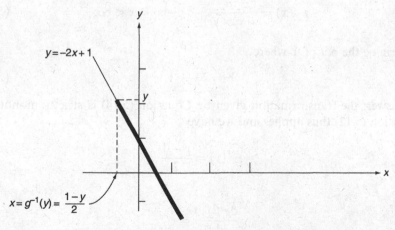

**Figure 5.3**   Transformation defined by Equation (5.9)

In comparison with Equation (5.7), Equation (5.10) yields a different relationship between the PDFs of $X$ and $Y$ owing to a different $g(X)$.

The relationship between the pdfs of $X$ and $Y$ for this case is again obtained by differentiating both sides of Equation (5.10) with respect to $y$, giving

$$\begin{aligned} f_Y(y) = \frac{dF_Y(y)}{dy} &= \frac{d}{dy}\{1 - F_X[g^{-1}(y)]\} \\ &= -f_X[g^{-1}(y)]\frac{dg^{-1}(y)}{dy}. \end{aligned} \tag{5.11}$$

Again, we observe that Equations (5.10) and (5.11) hold for all continuous $g(x)$ that are *strictly monotonic decreasing functions* of $x$, that is $g(x_2) < g(x_1)$ whenever $x_2 > x_1$.

Since the derivative $dg^{-1}(y)/dy$ in Equation (5.8) is always positive – as $g(x)$ is strictly monotonic increasing – and it is always negative in Equation (5.11) – as $g(x)$ is strictly monotonic decreasing – the results expressed by these two equations can be combined to arrive at Theorem 5.1.

**Theorem 5.1.** Let $X$ be a continuous random variable and $Y = g(X)$ where $g(X)$ is continuous in $X$ and strictly monotone. Then

$$\boxed{f_Y(y) = f_X[g^{-1}(y)]\left|\frac{dg^{-1}(y)}{dy}\right|,} \tag{5.12}$$

where $|u|$ denotes the absolute value of $u$.

**Example 5.3.** Problem: the pdf of $X$ is given by (Cauchy distribution):

$$f_X(x) = \frac{a}{\pi(x^2 + a^2)}, \quad -\infty < x < \infty. \tag{5.13}$$

Determine the pdf of $Y$ where

$$Y = 2X + 1. \tag{5.14}$$

Answer: the transformation given by Equation (5.14) is strictly monotone. Equation (5.12) thus applies and we have

$$g^{-1}(y) = \frac{y-1}{2},$$

and

$$\frac{dg^{-1}(y)}{dy} = \frac{1}{2}.$$

Following Equation (5.12), the result is

$$f_Y(y) = f_X\left[\frac{y-1}{2}\right]\left(\frac{1}{2}\right)$$

$$= \frac{a}{2\pi}\left[\frac{(y-1)^2}{4+a^2}\right]^{-1} \tag{5.15}$$

$$= \frac{2a}{\pi}\frac{1}{(y-1)^2 + 4a^2}, \quad -\infty < y < \infty.$$

It is valid over the entire range $-\infty < y < \infty$ as it is in correspondence with the range $-\infty < x < \infty$ defined in the range space $R_X$.

**Example 5.4.** Problem: the angle $\Phi$ of a pendulum as measured from the vertical is a random variable uniformly distributed over the interval $(-\pi/2 < \Phi < \pi/2)$. Determine the pdf of $Y$, the horizontal distance, as shown in Figure 5.4.

Answer: the transformation equation in this case is

$$Y = \tan\Phi, \tag{5.16}$$

where

$$f_\Phi(\phi) = \begin{cases} \dfrac{1}{\pi}, & \text{for } -\dfrac{\pi}{2} < \phi < \dfrac{\pi}{2}; \\ 0, & \text{elsewhere.} \end{cases} \tag{5.17}$$

As shown in Figure 5.5, Equation (5.16) is monotone within the range $-\pi/2 < \phi < \pi/2$. Hence, Equation (5.12) again applies and we have

$$g^{-1}(y) = \tan^{-1} y.$$

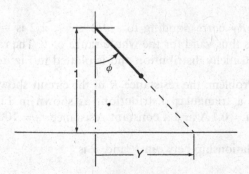

**Figure 5.4**   Pendulum, in Example 5.4

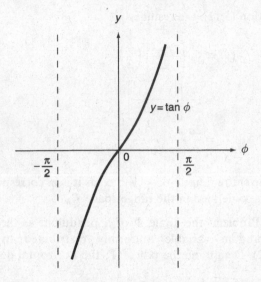

**Figure 5.5**  Transformation defined by Equation (5.16)

and

$$\frac{dg^{-1}(y)}{dy} = \frac{1}{1+y^2}.$$

The pdf of $Y$ is thus given by

$$f_Y(y) = \frac{f_\Phi(\tan^{-1} y)}{1+y^2}$$

$$= \frac{1}{\pi(1+y^2)}, \quad -\infty < y < \infty. \tag{5.18}$$

The range space $R_Y$ corresponding to $-\pi/2 < \phi < \pi/2$ is $-\infty < y < \infty$. The pdf given above is thus valid for the whole range of $y$. The random variable $Y$ has the so-called Cauchy distribution and is plotted in Figure 5.6.

**Example 5.5.** Problem: the resistance $R$ in the circuit shown in Figure 5.7 is random and has a triangular distribution, as shown in Figure 5.8. With a constant current $i = 0.1\,\text{A}$ and a constant resistance $r_0 = 100\,\Omega$; determine the pdf of voltage $V$.

Answer: the relationship between $V$ and $R$ is

$$V = i(R + r_0) = 0.1(R + 100), \tag{5.19}$$

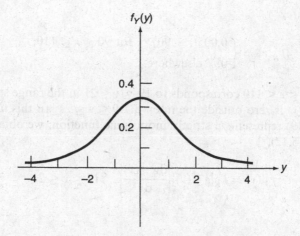

**Figure 5.6** Probability density function, $f_Y(y)$ in Example 5.4

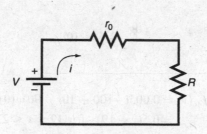

**Figure 5.7** Circuit for Example 5.5

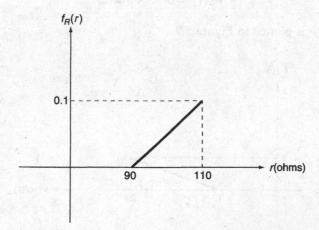

**Figure 5.8** Distribution, $f_R(r)$, in Example 5.5

and

$$f_R(r) = \begin{cases} 0.005(r - 90), & \text{for } 90 \leq r \leq 110; \\ 0, & \text{elsewhere.} \end{cases} \qquad (5.20)$$

The range $90 \leq r \leq 110$ corresponds to $19 \leq v \leq 21$ in the range space $R_V$. It is clear that $f_V(v)$ is zero outside the interval $19 \leq v \leq 21$. In this interval, since Equation (5.19) represents a strictly monotonic function, we obtain by means of Equation (5.12),

$$f_V(v) = f_R[g^{-1}(v)]\left|\frac{dg^{-1}(v)}{dv}\right|, \quad 19 \leq v \leq 21,$$

where

$$g^{-1}(v) = -100 + 10v,$$

and

$$\frac{dg^{-1}(v)}{dv} = 10.$$

We thus have

$$f_V(v) = 0.005(-100 + 10v - 90)(10)$$
$$= 0.5(v - 19), \text{ for } 19 \leq v \leq 21$$

and

$$f_V(v) = 0, \text{ elsewhere.}$$

The pdf of $V$ is plotted in Figure 5.9.

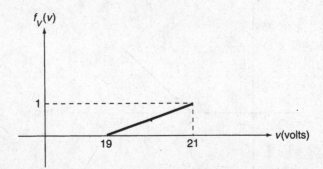

**Figure 5.9**   Density function $f_V(v)$, in Example 5.5

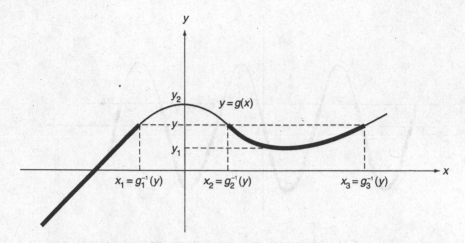

**Figure 5.10**   An example of nonmonotonic function $y = g(x)$

In the examples given above, it is easy to verify that all density functions obtained satisfy the required properties.

Let us now turn our attention to a more general case where function $Y = g(X)$ is not necessarily strictly monotonic. Two examples are given in Figures 5.10 and 5.11. In Figure 5.10, the monotonic property of the transformation holds for $y < y_1$, and $y > y_2$, and Equation (5.12) can be used to determine the pdf of $Y$ in these intervals of $y$. For $y_1 \leq y \leq y_2$, however, we must start from the beginning and consider $F_Y(y) = P(Y \leq y)$. The region defined by $Y \leq y$ in the range space $R_Y$ covers the heavier portions of the function $y = g(x)$, as shown in Figure 5.10. Thus:

$$F_Y(y) = P(Y \leq y) = P[X \leq g_1^{-1}(y)] + P[g_2^{-1}(y) < X \leq g_3^{-1}(y)]$$

$$= P[X \leq g_1^{-1}(y)] + P[X \leq g_3^{-1}(y)] - P[X \leq g_2^{-1}(y)] \qquad (5.21)$$

$$= F_X[g_1^{-1}(y)] + F_X[g_3^{-1}(y)] - F_X[g_2^{-1}(y)], \quad y_1 \leq y \leq y_2,$$

where $x_1 = g_1^{-1}(y)$, $x_2 = g_2^{-1}(y)$, and $x_3 = g_3^{-1}(y)$ are roots for $x$ of function $y = g(x)$ in terms of $y$.

As before, the relationship between the pdfs of $X$ and $Y$ is found by differentiating Equation (5.21) with respect to $y$. It is given by

$$f_Y(y) = f_X[g_1^{-1}(y)]\frac{dg_1^{-1}(y)}{dy} + f_X[g_3^{-1}(y)]\frac{dg_3^{-1}(y)}{dy} - f_X[g_2^{-1}(y)]\frac{dg_2^{-1}(y)}{dy}, \quad y_1 \leq y \leq y_2.$$

$$(5.22)$$

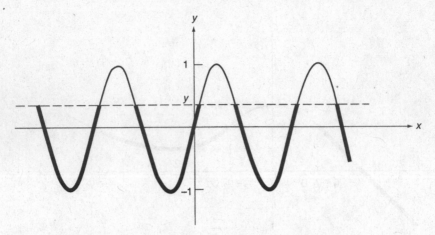

**Figure 5.11**   An example of nonmonotonic function $y = g(x)$

Since derivative $dg_2^{-1}(y)/dy$ is negative whereas the others are positive, Equation (5.22) takes the convenient form

$$f_Y(y) = \sum_{j=1}^{3} f_X[g_j^{-1}(y)] \left| \frac{dg_j^{-1}(y)}{dy} \right|, \qquad y_1 \le y \le y_2. \tag{5.23}$$

Figure 5.11 represents the transformation $y = \sin x$; this equation has an infinite (but countable) number of roots, $x_1 = g_1^{-1}(y)$, $x_2 = g_2^{-1}(y), \ldots,$ for any $y$ in the interval $-1 \le y \le 1$. Following the procedure outlined above, an equation similar to Equation (5.21) (but with an infinite number of terms) can be established for $F_Y(y)$ and, as seen from Equation (5.23), the pdf of $Y$ now has the form

$$f_Y(y) = \sum_{j=1}^{\infty} f_X[g_j^{-1}(y)] \left| \frac{dg_j^{-1}(y)}{dy} \right|, \qquad -1 \le y \le 1. \tag{5.24}$$

It is clear from Figure 5.11 that $f_Y(y) = 0$ elsewhere.

A general pattern now emerges when function $Y = g(X)$ is nonmonotonic. Equations (5.23) and (5.24) lead to Theorem 5.2.

**Theorem 5.2:** Let $X$ be a continuous random variable and $Y = g(X)$, where $g(X)$ is continuous in $X$, and $y = g(x)$ admits at most a countable number of roots $x_1 = g_1^{-1}(y)$, $x_2 = g_2^{-1}(y), \ldots.$ Then:

$$f_Y(y) = \sum_{j=1}^{r} f_X[g_j^{-1}(y)] \left| \frac{dg_j^{-1}(y)}{dy} \right|, \tag{5.25}$$

where $r$ is the number of roots for $x$ of equation $y = g(x)$. Clearly, Equation (5.12) is contained in this theorem as a special case ($r = 1$).

**Example 5.6.** Problem: in Example 5.4, let random variable $\Phi$ now be uniformly distributed over the interval $-\pi < \Phi < \pi$. Determine the pdf of $Y = \tan \Phi$.

Answer: the pdf of $\Phi$ is now

$$f_\Phi(\phi) = \begin{cases} \dfrac{1}{2\pi}, & \text{for } -\pi < \phi < \pi; \\ 0, & \text{elsewhere;} \end{cases}$$

and the relevant portion of the transformation equation is plotted in Figure 5.12. For each $y$, the two roots $\phi_1$ and $\phi_2$ of $y = \tan \phi$ are (see Figure 5.12)

$$\left.\begin{aligned} \phi_1 = g_1^{-1}(y) = \tan^{-1} y, \quad \text{for} -\frac{\pi}{2} < \phi_1 \leq 0 \\ \phi_2 = g_2^{-1}(y) = \tan^{-1} y, \quad \text{for } \frac{\pi}{2} < \phi_2 \leq \pi \end{aligned}\right\}, \quad y \leq 0;$$

$$\left.\begin{aligned} \phi_1 = \tan^{-1} y, \quad \text{for} -\pi < \phi_1 \leq -\frac{\pi}{2} \\ \phi_2 = \tan^{-1} y, \quad \text{for } 0 < \phi_2 \leq \frac{\pi}{2} \end{aligned}\right\}, \quad y > 0.$$

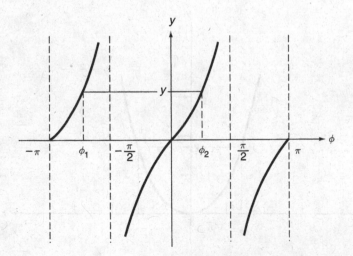

**Figure 5.12**   Transformation $y = \tan \phi$

For all $y$, Equation (5.25) yields

$$f_Y(y) = \sum_{j=1}^{2} f_\Phi[g_j^{-1}(y)] \left| \frac{dg_j^{-1}(y)}{dy} \right|$$

$$= \frac{1}{2\pi} \left( \frac{1}{1+y^2} \right) + \frac{1}{2\pi} \left( \frac{1}{1+y^2} \right) \qquad (5.26)$$

$$= \frac{1}{\pi(1+y^2)}, \quad -\infty < y < \infty,$$

a result identical to the solution for Example 5.4 [see Equation (5.18)].

**Example 5.7.** Problem: determine the pdf of $Y = X^2$ where $X$ is normally distributed according to

$$f_X(x) = \frac{1}{(2\pi)^{1/2}} e^{-x^2/2}, \quad -\infty < x < \infty. \qquad (5.27)$$

As shown in Figure 5.13, $f_Y(y) = 0$ for $y < 0$ since the transformation equation has no real roots in this range. For $y \geq 0$, the two roots of $y = x^2$ are

$$x_{1,2} = g_{1,2}^{-1}(y) = \pm y^{1/2}.$$

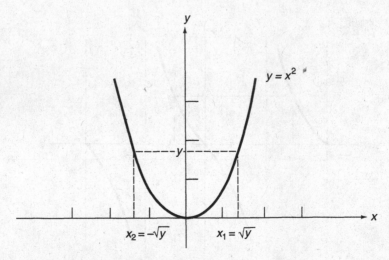

**Figure 5.13**   Transformation $y = x^2$

Hence, using Equation (5.25),

$$f_Y(y) = \sum_{j=1}^{2} f_X[g_j^{-1}(y)] \left| \frac{dg_j^{-1}(y)}{dy} \right|$$

$$= \frac{f_X(-y^{1/2})}{2y^{1/2}} + \frac{f_X(y^{1/2})}{2y^{1/2}}$$

$$= \frac{1}{(2\pi y)^{1/2}} e^{-y/2},$$

or

$$f_Y(y) = \begin{cases} \dfrac{1}{(2\pi y)^{1/2}} e^{-y/2}, & \text{for } y \geq 0; \\ 0, & \text{elsewhere.} \end{cases} \tag{5.28}$$

This is the so-called $\chi^2$ distribution, to be discussed in more detail in Section 7.4.2.

**Example 5.8.** Problem: a random voltage $V_1$ having a uniform distribution over interval $90\,\text{V} \leq V_1 \leq 110\,\text{V}$ is put into a nonlinear device (a limiter), as shown in Figure 5.14. Determine the probability distribution of the output voltage $V_2$.

Answer: the relationship between $V_1$ and $V_2$ is, as seen from Figure 5.14,

$$V_2 = g(V_1), \tag{5.29}$$

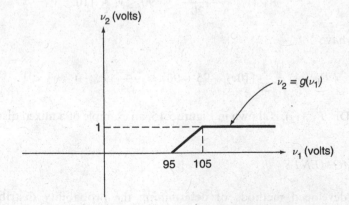

**Figure 5.14**  Transformation defined by Equation (5.29)

where

$$g(V_1) = 0, \quad V_1 < 95;$$

$$g(V_1) = \frac{V_1 - 95}{10}, \quad 95 \leq V_1 \leq 105;$$

$$g(V_1) = 1, \quad V_1 > 105.$$

The theorems stated in this section do not apply in this case to the portions $v_1 < 95\,\text{V}$ and $v_1 > 105\,\text{V}$ because infinite and noncountable number of roots for $v_1$ exist in these regions. However, we deduce immediately from Figure 5.14 that

$$P(V_2 = 0) = P(V_1 \leq 95) = F_{V_1}(95)$$

$$= \int_{90}^{95} f_{V_1}(v_1)\mathrm{d}v_1 = \frac{1}{4};$$

$$P(V_2 = 1) = P(V_1 > 105) = 1 - F_{V_1}(105)$$

$$= \frac{1}{4}.$$

For the middle portion, Equation (5.7) leads to

$$F_{V_2}(v_2) = F_{V_1}[g^{-1}(v_2)]$$
$$= F_{V_1}(10v_2 + 95), \quad 0 < v_2 < 1.$$

Now,

$$F_{V_1}(v_1) = \frac{v_1 - 90}{20}, \quad 90 \leq v_1 \leq 110.$$

We thus have

$$F_{V_2}(v_2) = \frac{1}{20}(10v_2 + 95 - 90) = \frac{2v_2 + 1}{4}, \quad 0 < v_2 < 1.$$

The PDF, $F_{V_2}(v_2)$, is shown in Figure 5.15, an example of a mixed distribution.

## 5.1.2  MOMENTS

Having developed methods of determining the probability distribution of $Y = g(X)$, it is a straightforward matter to calculate all the desired moments

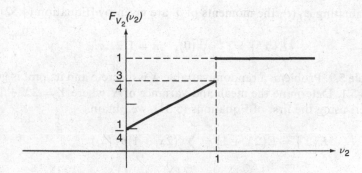

**Figure 5.15**   Distribution $F_{V_2}(v_2)$ in Example 5.8

of $Y$ if they exist. However, this procedure – the determination of moments of $Y$ on finding the probability law of $Y$ – is cumbersome and unnecessary if only the moments of $Y$ are of interest.

A more expedient and direct way of finding the moments of $Y = g(X)$, given the probability law of $X$, is to express moments of $Y$ as expectations of appropriate functions of $X$; they can then be evaluated directly within the probability domain of $X$. In fact, all the 'machinery' for proceeding along this line is contained in Equations (4.1) and (4.2).

Let $Y = g(X)$ and assume that all desired moments of $Y$ exist. The $n$th moment of $Y$ can be expressed as

$$E\{Y^n\} = E\{g^n(X)\}. \tag{5.30}$$

It follows from Equations (4.1) and (4.2) that, in terms of the pmf or pdf of $X$,

$$
\begin{aligned}
E\{Y^n\} = E\{g^n(X)\} &= \sum_i g^n(x_i)p_X(x_i), \quad X \text{ discrete;} \\
E\{Y^n\} = E\{g^n(X)\} &= \int_{-\infty}^{\infty} g^n(x)f_X(x)\mathrm{d}x, \quad X \text{ continuous.}
\end{aligned}
\tag{5.31}
$$

An alternative approach is to determine the characteristic function of $Y$ from which all moments of $Y$ can be generated through differentiation. As we see from the definition [Equations (4.46) and (4.47)], the characteristic function of $Y$ can be expressed by

$$
\left.
\begin{aligned}
\phi_Y(t) = E\{e^{jtY}\} = E\{e^{jtg(X)}\} &= \sum_i e^{jtg(x_i)}p_X(x_i), \quad X \text{ discrete;} \\
\phi_Y(t) = E\{e^{jtY}\} = E\{e^{jtg(X)}\} &= \int_{-\infty}^{\infty} e^{jtg(x)}f_X(x)\mathrm{d}x, \quad X \text{ continuous.}
\end{aligned}
\right\}
\tag{5.32}
$$

Upon evaluating $\phi_Y(t)$, the moments of $Y$ are given by [Equation (4.52)]:

$$E\{Y^n\} = j^{-n}\phi_Y^{(n)}(0), \quad n = 1, 2, \ldots. \tag{5.33}$$

**Example 5.9.** Problem: a random variable $X$ is discrete and its pmf is given in Example 5.1. Determine the mean and variance of Y where $Y = 2X + 1$.

Answer: using the first of Equations (5.31), we obtain

$$E\{Y\} = E\{2X + 1\} = \sum_i (2x_i + 1)p_X(x_i)$$

$$= (-1)\left(\frac{1}{2}\right) + (1)\left(\frac{1}{4}\right) + (3)\left(\frac{1}{8}\right) + (5)\left(\frac{1}{8}\right) \tag{5.34}$$

$$= \frac{3}{4};$$

$$E\{Y^2\} = E\{(2X + 1)^2\} = \sum_i (2x_i + 1)^2 p_X(x_i)$$

$$= (1)\left(\frac{1}{2}\right) + (1)\left(\frac{1}{4}\right) + (9)\left(\frac{1}{8}\right) + (25)\left(\frac{1}{8}\right) \tag{5.35}$$

$$= 5;$$

and

$$\sigma_Y^2 = E\{Y^2\} - E^2\{Y\} = 5 - \left(\frac{3}{4}\right)^2 = \frac{71}{16}. \tag{5.36}$$

Following the second approach, let us use the method of characteristic functions described by Equations (5.32) and (5.33). The characteristic function of $Y$ is

$$\phi_Y(t) = \sum_i e^{jt(2x_i+1)} p_X(x_i)$$

$$= e^{-jt}\left(\frac{1}{2}\right) + e^{jt}\left(\frac{1}{4}\right) + e^{3jt}\left(\frac{1}{8}\right) + e^{5jt}\left(\frac{1}{8}\right)$$

$$= \frac{1}{8}(4e^{-jt} + 2e^{jt} + e^{3jt} + e^{5jt}),$$

and we have

$$E\{Y\} = j^{-1}\phi_Y^{(1)}(0) = j^{-1}\left(\frac{j}{8}\right)(-4 + 2 + 3 + 5) = \frac{3}{4},$$

$$E\{Y^2\} = -\phi_Y^{(2)}(0) = \frac{1}{8}(4 + 2 + 9 + 25) = 5.$$

As expected, these answers agree with the results obtained earlier [Equations (5.34) and (5.35)].

Let us again remark that the procedures described above do not require knowledge of $f_Y(y)$. One can determine $f_Y(y)$ before moment calculations but it is less expedient when only moments of $Y$ are desired. Another remark to be made is that, since the transformation is linear ($Y = 2X + 1$) in this case, only the first two moments of $X$ are needed in finding the first two moments of $Y$, that is,

$$E\{Y\} = E\{2X + 1\} = 2E\{X\} + 1,$$
$$E\{Y^2\} = E\{(2X + 1)^2\} = 4E\{X^2\} + 4E\{X\} + 1,$$

as seen from Equations (5.34) and (5.35). When the transformation is nonlinear, however, moments of $X$ of different orders will be needed, as shown below.

**Example 5.10.** Problem: from Example 5.7, determine the mean and variance of $Y = X^2$. The mean of $Y$ is, in terms of $f_X(x)$,

$$E\{Y\} = E\{X^2\} = \frac{1}{(2\pi)^{1/2}} \int_{-\infty}^{\infty} x^2 e^{-x^2/2} dx = 1, \qquad (5.37)$$

and the second moment of $Y$ is given by

$$E\{Y^2\} = E\{X^4\} = \frac{1}{(2\pi)^{1/2}} \int_{-\infty}^{\infty} x^4 e^{-x^2/2} dx = 3. \qquad (5.38)$$

Thus,

$$\sigma_Y^2 = E\{Y^2\} - E^2\{Y\} = 3 - 1 = 2. \qquad (5.39)$$

In this case, complete knowledge of $f_X(x)$ is not needed but we to need to know the second and fourth moments of $X$.

## 5.2 FUNCTIONS OF TWO OR MORE RANDOM VARIABLES

In this section, we extend earlier results to a more general case. The random variable $Y$ is now a function of $n$ jointly distributed random variables, $X_1, X_2, \ldots, X_n$. Formulae will be developed for the corresponding distribution for $Y$.

As in the single random variable case, the case in which $X_1, X_2, \ldots$, and $X_n$ are discrete random variables presents no problem and we will demonstrate this by way of an example (Example 5.13). Our basic interest here lies in the determination of the distribution $Y$ when all $X_j, j = 1, 2, \ldots, n$, are continuous random variables. Consider the transformation

$$Y = g(X_1, \ldots, X_n) \tag{5.40}$$

where the joint distribution of $X_1, X_2, \ldots$, and $X_n$ is assumed to be specified in term of their joint probability density function (jpdf), $f_{X_1 \ldots X_n}(x_1, \ldots, x_n)$, or their joint probability distribution function (JPDF), $F_{X_1 \ldots X_n}(x_1, \ldots, x_n)$. In a more compact notation, they can be written as $f_X(x)$ and $F_X(x)$, respectively, where $X$ is an $n$-dimensional random vector with components $X_1, X_2, \ldots, X_n$.

The starting point of the derivation is the same as in the single-random-variable case; that is, we consider $F_Y(y) = P(Y \leq y)$. In terms of $X$, this probability is equal to $P[g(X) \leq y]$. Thus:

$$\begin{aligned} F_Y(y) = P(Y \leq y) &= P[g(X) \leq y] \\ &= F_X[x : g(x) \leq y]. \end{aligned} \tag{5.41}$$

The final expression in the above represents the JPDF of $X$ for which the argument $x$ satisfies $g(x) \leq y$. In terms of $f_X(x)$, it is given by

$$\boxed{ F_X[x : g(x) \leq y] = \int \cdots \int_{(R^n : g(x) \leq y)} f_X(x) dx } \tag{5.42}$$

where the limits of the integrals are determined by an $n$-dimensional region $R^n$ within which $g(x) \leq y$ is satisfied. In view of Equations (5.41) and (5.42), the PDF of $Y$, $F_Y(y)$, can be determined by evaluating the $n$-dimensional integral in Equation (5.42). The crucial step in this derivation is clearly the identification of $R^n$, which must be carried out on a problem-to-problem basis. As $n$ becomes large, this can present a formidable obstacle.

The procedure outlined above can be best demonstrated through examples.

**Example 5.11.** Problem: let $Y = X_1 X_2$. Determine the pdf of $Y$ in terms of $f_{X_1 X_2}(x_1, x_2)$.

Answer: from Equations (5.41) and (5.42), we have

$$F_Y(y) = \iint_{(R^2 : x_1 x_2 \leq y)} f_{X_1 X_2}(x_1, x_2) dx_1 dx_2. \tag{5.43}$$

The equation $x_1 x_2 = y$ is graphed in Figure 5.16 in which the shaded area represents $R^2$, or $x_1 x_2 \leq y$. The limits of the double integral can thus be determined and Equation (5.43) becomes

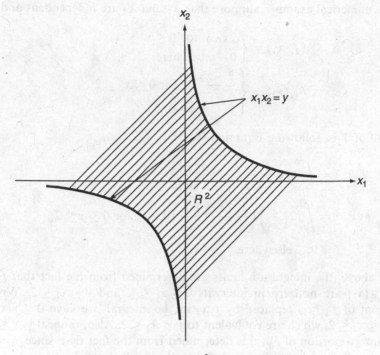

**Figure 5.16**  Region $R^2$, in Example 5.11

$$F_Y(y) = \int_0^\infty \int_{-\infty}^{y/x_2} f_{X_1 X_2}(x_1, x_2)\,\mathrm{d}x_1\,\mathrm{d}x_2 + \int_{-\infty}^0 \int_{y/x_2}^\infty f_{X_1 X_2}(x_1, x_2)\,\mathrm{d}x_1\,\mathrm{d}x_2. \quad (5.44)$$

Substituting $f_{X_1 X_2}(x_1, x_2)$ into Equation (5.44) enables us to determine $F_Y(y)$ and, on differentiating with respect to $y$, gives $f_Y(y)$.

For the special case where $X_1$ and $X_2$ are independent, we have $f_{X_1 X_2}(x_1, x_2) = f_{X_1}(x_1)f_{X_2}(x_2)$, and Equation (5.44) simplifies to

$$F_Y(y) = \int_0^\infty F_{X_1}\left(\frac{y}{x_2}\right)f_{X_2}(x_2)\,\mathrm{d}x_2 + \int_{-\infty}^0 \left[1 - F_{X_1}\left(\frac{y}{x_2}\right)\right]f_{X_2}(x_2)\,\mathrm{d}x_2,$$

and

$$f_Y(y) = \frac{\mathrm{d}F_Y(y)}{\mathrm{d}y} = \int_{-\infty}^\infty f_{X_1}\left(\frac{y}{x_2}\right)f_{X_2}(x_2)\left|\frac{1}{x_2}\right|\,\mathrm{d}x_2. \quad (5.45)$$

As a numerical example, suppose that $X_1$ and $X_2$ are independent and

$$f_{X_1}(x_1) = \begin{cases} 2x_1, & \text{for } 0 \le x_1 \le 1; \\ 0, & \text{elsewhere;} \end{cases}$$

$$f_{X_2}(x_2) = \begin{cases} \dfrac{2 - x_2}{2}, & \text{for } 0 \le x_2 \le 2; \\ 0, & \text{elsewhere.} \end{cases}$$

The pdf of $Y$ is, following Equation (5.45),

$$
\begin{aligned}
f_Y(y) &= \int_{-\infty}^{\infty} f_{X_1}\left(\frac{y}{x_2}\right) f_{X_2}(x_2) \left|\frac{1}{x_2}\right| dx_2; \\
&= \int_y^2 2\left(\frac{y}{x_2}\right)\left(\frac{2-x_2}{2}\right)\left(\frac{1}{x_2}\right) dx_2, \quad \text{for } 0 \le y \le 2; \\
&= 0, \quad \text{elsewhere.}
\end{aligned}
\tag{5.46}
$$

In the above, the integration limits are determined from the fact that $f_{X_1}(x_1)$ and $f_{X_2}(x_2)$ are nonzero in intervals $0 \le x_1 \le 1$, and $0 \le x_2 \le 2$. With the argument of $f_{X_1}(x_1)$ replaced by $y/x_2$ in the integral, we have $0 \le y/x_2 \le 1$, and $0 \le x_2 \le 2$, which are equivalent to $y \le x_2 \le 2$. Also, range $0 \le y \le 2$ for the nonzero portion of $f_Y(y)$ is determined from the fact that, since $y = x_1 x_2$, intervals $0 \le x_1 \le 1$, and $0 \le x_2 \le 2$ directly give $0 \le y \le 2$.

Finally, Equation (5.46) gives

$$f_Y(y) = \begin{cases} 2 + y(\ln y - 1 - \ln 2), & \text{for } 0 \le y \le 2; \\ 0, & \text{elsewhere.} \end{cases} \tag{5.47}$$

This is shown graphically in Figure 5.17. It is an easy exercise to show that

$$\int_0^2 f_Y(y)dy = 1.$$

**Example 5.12.** Problem: let $Y = X_1/X_2$ where $X_1$ and $X_2$ are independent and identically distributed according to

$$f_{X_1}(x_1) = \begin{cases} e^{-x_1}, & \text{for } x_1 > 0; \\ 0, & \text{elsewhere;} \end{cases} \tag{5.48}$$

and similarly for $X_2$. Determine $f_Y(y)$.

Answer: it follows from Equations (5.41) and (5.42) that

$$F_Y(y) = \iint\limits_{(R^2 \,:\, x_1/x_2 \le y)} f_{X_1 X_2}(x_1, x_2)dx_1 dx_2.$$

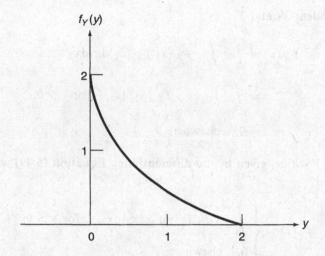

**Figure 5.17** Probability density function, $f_Y(y)$, in Example 5.11

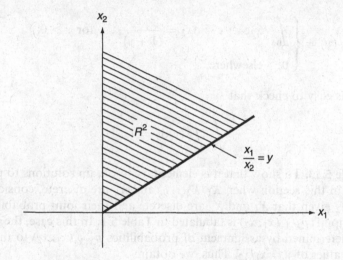

**Figure 5.18** Region $R^2$ in Example 5.12

The region $R^2$ for positive values of $x_1$ and $x_2$ is shown as the shaded area in Figure 5.18. Hence,

$$F_Y(y) = \begin{cases} \displaystyle\int_0^\infty \int_0^{x_2 y} f_{X_1 X_2}(x_1, x_2)\,dx_1\,dx_2, & \text{for } y > 0; \\ 0, & \text{elsewhere.} \end{cases}$$

For independent $X_1$ and $X_2$,

$$F_Y(y) = \int_0^\infty \int_0^{x_2 y} f_{X_1}(x_1) f_{X_2}(x_2) \mathrm{d}x_1 \mathrm{d}x_2;$$

$$= \int_0^\infty F_{X_1}(x_2 y) f_{X_2}(x_2) \mathrm{d}x_2, \quad \text{for } y > 0; \tag{5.49}$$

$$= 0, \quad \text{elsewhere.}$$

The pdf of $Y$ is thus given by, on differentiating Equation (5.49) with respect to $y$,

$$f_Y(y) = \begin{cases} \displaystyle\int_0^\infty x_2 f_{X_1}(x_2 y) f_{X_2}(x_2) \mathrm{d}x_2, & \text{for } y > 0; \\[2mm] 0, & \text{elsewhere;} \end{cases} \tag{5.50}$$

and, on substituting Equation (5.48) into Equation (5.50), it takes the form

$$f_Y(y) = \begin{cases} \displaystyle\int_0^\infty x_2 e^{-x_2 y} e^{-x_2} \mathrm{d}x_2 = \dfrac{1}{(1+y)^2}, & \text{for } y > 0; \\[2mm] 0, & \text{elsewhere.} \end{cases} \tag{5.51}$$

Again, it is easy to check that

$$\int_0^\infty f_Y(y) \mathrm{d}y = 1.$$

**Example 5.13.** To show that it is elementary to obtain solutions to problems discussed in this section when $X_1, X_2, \ldots,$ and $X_n$ are discrete, consider again $Y = X_1/X_2$ given that $X_1$ and $X_2$ are discrete and their joint probability mass function (jpmf) $p_{X_1 X_2}(x_1, x_2)$ is tabulated in Table 5.1. In this case, the pmf of $Y$ is easily determined by assignment of probabilities $p_{X_1 X_2}(x_1, x_2)$ to the corresponding values of $y = x_1/x_2$. Thus, we obtain:

$$p_Y(y) = \begin{cases} 0.5, & \text{for } y = \dfrac{1}{2}; \\[2mm] 0.24 + 0.04 = 0.28, & \text{for } y = 1; \\[2mm] 0.04, & \text{for } y = \dfrac{3}{2}; \\[2mm] 0.06, & \text{for } y = 2; \\[2mm] 0.12, & \text{for } y = 3. \end{cases}$$

**Table 5.1**  Joint probability mass function, $p_{X_1 X_2}(x_1, x_2)$, in Example 5.13

| $x_2$ | $x_1$ | | |
|---|---|---|---|
| | 1 | 2 | 3 |
| 1 | 0.04 | 0.06 | 0.12 |
| 2 | 0.5 | 0.24 | 0.04 |

**Example 5.14.** Problem: in structural reliability studies, the probability of failure $q$ is defined by

$$q = P(R \leq S),$$

where $R$ and $S$ represent, respectively, structural resistance and applied force. Let $R$ and $S$ be independent random variables taking only positive values. Determine $q$ in terms of the probability distributions associated with $R$ and $S$.

Answer: let $Y = R/S$. Probability $q$ can be expressed by

$$q = P\left(\frac{R}{S} \leq 1\right) = P(Y \leq 1) = F_Y(1).$$

Identifying $R$ and $S$ with $X_1$ and $X_2$, respectively, in Example 5.12, it follows from Equation (5.49) that

$$q = F_Y(1) = \int_0^\infty F_R(s) f_S(s) \mathrm{d}s.$$

**Example 5.15.** Problem: determine $F_Y(y)$ in terms of $f_{X_1 X_2}(x_1, x_2)$ when $Y = \min(X_1, X_2)$.

Answer: now,

$$F_Y(y) = \iint\limits_{(R^2 \,:\, \min(x_1, x_2) \leq y)} f_{X_1 X_2}(x_1, x_2) \mathrm{d}x_1 \mathrm{d}x_2,$$

where region $R^2$ is shown in Figure 5.19. Thus

$$F_Y(y) = \int_{-\infty}^y \int_{-\infty}^\infty f_{X_1 X_2}(x_1, x_2) \mathrm{d}x_1 \mathrm{d}x_2 + \int_y^\infty \int_{-\infty}^y f_{X_1 X_2}(x_1, x_2) \mathrm{d}x_1 \mathrm{d}x_2$$

$$= \int_{-\infty}^y \int_{-\infty}^\infty f_{X_1 X_2}(x_1, x_2) \mathrm{d}x_1 \mathrm{d}x_2 + \int_{-\infty}^\infty \int_{-\infty}^y f_{X_1 X_2}(x_1, x_2) \mathrm{d}x_1 \mathrm{d}x_2$$

$$- \int_{-\infty}^y \int_{-\infty}^y f_{X_1 X_2}(x_1, x_2) \mathrm{d}x_1 \mathrm{d}x_2$$

$$= F_{X_2}(y) + F_{X_1}(y) - F_{X_1 X_2}(y, y),$$

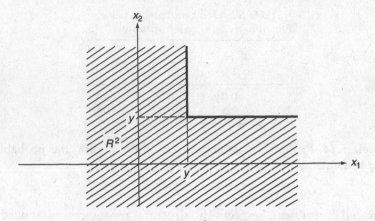

**Figure 5.19**   Region $R^2$ in Example 5.15

which is the desired solution. If random variables $X_1$ and $X_2$ are independent, straightforward differentiation shows that

$$f_Y(y) = f_{X_1}(y)[1 - F_{X_2}(y)] + f_{X_2}(y)[1 - F_{X_1}(y)].$$

Let us note here that the results given above can be obtained following a different, and more direct, procedure. Note that the event $[\min(X_1, X_2) \le y]$ is equivalent to the event $(X_1 \le y \cup X_2 \le y)$. Hence,

$$F_Y(y) = P(Y \le y) = P[\min(X_1, X_2) \le y]$$
$$= P(X_1 \le y \cup X_2 \le y).$$

Since

$$P(A \cup B) = P(A) + P(B) - P(AB),$$

we have

$$F_Y(y) = P(X_1 \le y) + P(X_2 \le y) - P(X_1 \le y \cap X_2 \le y)$$
$$= F_{X_1}(y) + F_{X_2}(y) - F_{X_1 X_2}(y, y).$$

If $X_1$ and $X_2$ are independent, we have

$$F_Y(y) = F_{X_1}(y) + F_{X_2}(y) - F_{X_1}(y)F_{X_2}(y),$$

and

$$f_Y(y) = \frac{dF_Y(y)}{dy} = f_{X_1}(y)[1 - F_{X_2}(y)] + f_{X_2}(y)[1 - F_{X_1}(y)].$$

We have not given examples in which functions of more than two random variables are involved. Although more complicated problems can be formulated in a similar fashion, it is in general more difficult to identify appropriate regions $R^n$ required by Equation (5.42), and the integrals are, of course, more difficult to carry out. In principle, however, no intrinsic difficulties present themselves in cases of functions of more than two random variables.

### 5.2.1  SUMS OF RANDOM VARIABLES

One of the most important transformations we encounter is a sum of random variables. It has been discussed in Chapter 4 in the context of characteristic functions. In fact, the technique of characteristic functions remains to be the most powerful technique for sums of *independent* random variables.

In this section, the procedure presented in the above is used to give an alternate method of attack.

Consider the sum

$$Y = g(X_1, \ldots, X_n) = X_1 + X_2 + \cdots + X_n. \tag{5.52}$$

It suffices to determine $f_Y(y)$ for $n = 2$. The result for this case can then be applied successively to give the probability distribution of a sum of any number of random variables. For $Y = X_1 + X_2$, Equations (5.41) and (5.42) give

$$F_Y(y) = \iint\limits_{(R^2 : x_1 + x_2 \leq y)} f_{X_1 X_2}(x_1, x_2) dx_1 dx_2,$$

and, as seen from Figure 5.20,

$$F_Y(y) = \int_{-\infty}^{\infty} \int_{-\infty}^{y - x_2} f_{X_1 X_2}(x_1, x_2) dx_1 dx_2. \tag{5.53}$$

Upon differentiating with respect to $y$ we obtain

$$f_Y(y) = \int_{-\infty}^{\infty} f_{X_1 X_2}(y - x_2, x_2) dx_2. \tag{5.54}$$

When $X_1$ and $X_2$ are independent, the above result further reduces to

$$f_Y(y) = \int_{-\infty}^{\infty} f_{X_1}(y - x_2) f_{X_2}(x_2) dx_2. \tag{5.55}$$

Integrals of the form given above arise often in practice. It is called *convolution* of the functions $f_{X_1}(x_1)$ and $f_{X_2}(x_2)$.

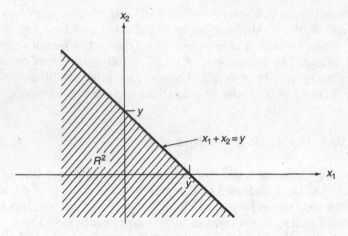

**Figure 5.20**   Region $R^2: x_1 + x_2 \leq y$

Considerable importance is attached to the results expressed by Equations (5.54) and (5.55) because sums of random variables occur frequently in practical situations. By way of recognizing this fact, Equation (5.55) is repeated now as Theorem 5.3.

**Theorem 5.3.** Let $Y = X_1 + X_2$, and let $X_1$ and $X_2$ be independent and continuous random variables. Then the pdf of $Y$ is the convolution of the pdfs associated with $X_1$ and $X_2$; that is,

$$f_Y(y) = \int_{-\infty}^{\infty} f_{X_1}(y - x_2)f_{X_2}(x_2)\mathrm{d}x_2 = \int_{-\infty}^{\infty} f_{X_2}(y - x_1)f_{X_1}(x_1)\mathrm{d}x_1. \qquad (5.56)$$

Repeated applications of this formula determine $f_Y(y)$ when $Y$ is a sum of any number of independent random variables.

**Example 5.16.** Problem: determine $f_Y(y)$ of $Y = X_1 + X_2$ when $X_1$ and $X_2$ are independent and identically distributed according to

$$f_{X_1}(x_1) = \begin{cases} ae^{-ax_1}, & \text{for } x_1 \geq 0; \\ 0, & \text{elsewhere}; \end{cases} \qquad (5.57)$$

and similarly for $X_2$.

Answer: Equation (5.56) in this case leads to

$$f_Y(y) = a^2 \int_0^y e^{-a(y-x_2)}e^{-ax_2}\mathrm{d}x_2, \quad y \geq 0, \qquad (5.58)$$

where the integration limits are determined from the requirements $y - x_2 > 0$, and $x_2 > 0$. The result is

$$f_Y(y) = \begin{cases} a^2 y e^{-ay}, & \text{for } y \geq 0; \\ 0, & \text{elsewhere.} \end{cases} \tag{5.59}$$

Let us note that this problem has also been solved in Example 4.16, by means of characteristic functions. It is to be stressed again that the method of characteristic functions is another powerful technique for dealing with sums of *independent* random variables. In fact, when the number of random variables involved in a sum is large, the method of characteristic function is preferred since there is no need to consider only two at a time as required by Equation (5.56).

**Example 5.17.** Problem: the random variables $X_1$ and $X_2$ are independent and uniformly distributed in intervals $0 \leq x_1 \leq 1$, and $0 \leq x_2 \leq 2$. Determine the pdf of $Y = X_1 + X_2$.

Answer: the convolution of $f_{X_1}(x_1) = 1$, $0 \leq x_1 \leq 1$, and $f_{X_2}(x_2) = 1/2$, $0 \leq x_2 \leq 2$, results in

$$f_Y(y) = \int_{-\infty}^{\infty} f_{X_1}(y - x_2) f_{X_2}(x_2) dx_2;$$

$$= \int_0^y (1)\left(\frac{1}{2}\right) dx_2 = \frac{y}{2}, \quad \text{for } 0 < y \leq 1;$$

$$= \int_{y-1}^y (1)\left(\frac{1}{2}\right) dx_2 = \frac{1}{2}, \quad \text{for } 1 < y \leq 2;$$

$$= \int_{y-1}^2 (1)\left(\frac{1}{2}\right) dx_2 = \frac{3-y}{2}, \quad \text{for } 2 < y \leq 3;$$

$$= 0, \quad \text{elsewhere.}$$

In the above, the limits of the integrals are determined from the requirements $0 \leq y - x_2 \leq 1$, and $0 \leq x_2 \leq 2$. The shape of $f_Y(y)$ is that of a trapezoid, as shown in Figure 5.21.

## 5.3   *m* FUNCTIONS OF *n* RANDOM VARIABLES

We now consider the general transformation given by Equation (5.1), that is,

$$Y_j = g_j(X_1, \ldots, X_n), \quad j = 1, 2, \ldots, m, \ m \leq n. \tag{5.60}$$

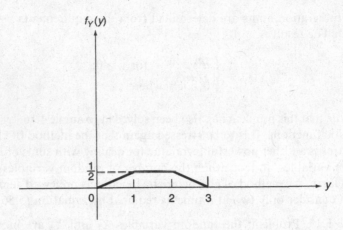

**Figure 5.21**  Probability density function, $f_Y(y)$, in Example 5.17

The problem is to obtain the joint probability distribution of random variables $Y_j$, $j = 1, 2, \ldots, m$, which arise as functions of $n$ jointly distributed random variables $X_k$, $k = 1, \ldots, n$. As before, we are primarily concerned with the case in which $X_1, \ldots, X_n$ are continuous random variables.

In order to develop pertinent formulae, the case of $m = n$ is first considered. We will see that the results obtained for this case encompass situations in which $m < n$.

Let $X$ and $Y$ be two $n$-dimensional random vectors with components $(X_1, \ldots, X_n)$ and $(Y_1, \ldots; Y_n)$, respectively. A vector equation representing Equation (5.60) is

$$Y = g(X) \tag{5.61}$$

where vector $g(X)$ has as components $g_1(X), g_2(X), \ldots g_n(X)$. We first consider the case in which functions $g_j$ in $g$ are continuous with respect to each of their arguments, have continuous partial derivatives, and define one-to-one mappings. It then follows that inverse functions $g_j^{-1}$ of $g^{-1}$, defined by

$$X = g^{-1}(Y), \tag{5.62}$$

exist and are unique. They also have continuous partial derivatives.

In order to determine $f_Y(y)$ in terms of $f_X(x)$, we observe that, if a closed region $R_X^n$ in the range space of $X$ is mapped into a closed region $R_Y^n$ in the range space of $Y$ under transformation $g$, the conservation of probability gives

$$\int_{R_Y^n} \cdots \int f_Y(y) \mathrm{d}y = \int_{R_X^n} \cdots \int f_X(x) \mathrm{d}x, \tag{5.63}$$

where the integrals represent $n$-fold integrals with respect to the components of $x$ and $y$, respectively. Following the usual rule of change of variables in multiple integrals, we can write (for example, see Courant, 1937):

$$\int \cdots \int_{R_X^n} f_X(x)dx = \int \cdots \int_{R_Y^n} f_X[g^{-1}(y)]|J|dy, \qquad (5.64)$$

where $J$ is the Jacobian of the transformation, defined as the determinant

$$J = \begin{vmatrix} \dfrac{\partial g_1^{-1}}{\partial y_1} & \dfrac{\partial g_1^{-1}}{\partial y_2} & \cdots & \dfrac{\partial g_1^{-1}}{\partial y_n} \\ \vdots & & & \vdots \\ \dfrac{\partial g_n^{-1}}{\partial y_1} & \dfrac{\partial g_n^{-1}}{\partial y_2} & \cdots & \dfrac{\partial g_n^{-1}}{\partial y_n} \end{vmatrix}. \qquad (5.65)$$

As a point of clarification, let us note that the vertical lines in Equation (5.65) denote determinant and those in Equation (5.64) represent absolute value.

Equations (5.63) and (5.64) then lead to the desired formula:

$$f_Y(y) = f_X[g^{-1}(y)]|J|. \qquad (5.66)$$

This result is stated as Theorem 5.4.

**Theorem 5.4.** For the transformation given by Equation (5.61) where $X$ is a continuous random vector and $g$ is continuous with continuous partial derivatives and defines a one-to-one mapping, the jpdf of $Y$, $f_Y(y)$, is given by

$$\boxed{f_Y(y) = f_X[g^{-1}(y)]|J|, \qquad (5.67)}$$

where $J$ is defined by Equation (5.65).

It is of interest to note that Equation (5.67) is an extension of Equation (5.12), which is for the special case of $n = 1$. Similarly, an extension is also possible of Equation (5.24) for the $n = 1$ case when the transformation admits more than one root. Reasoning as we have done in deriving Equation (5.24), we have Theorem 5.5.

**Theorem 5.5.** In Theorem 5.4, suppose transformation $y = g(x)$ admits at most a countable number of roots $x_1 = g_1^{-1}(y), x_2 = g_2^{-1}(y), \ldots$. Then

$$\boxed{f_Y(y) = \sum_{j=1}^{r} f_X[g_j^{-1}(y)]|J_j|, \qquad (5.68)}$$

where $r$ is the number of solutions for $\mathbf{x}$ of equation $y = g(\mathbf{x})$, and $J_j$ is defined by

$$
J_j =
\begin{vmatrix}
\dfrac{\partial g_{j1}^{-1}}{\partial y_1} & \dfrac{\partial g_{j1}^{-1}}{\partial y_2} & \cdots & \dfrac{\partial g_{j1}^{-1}}{\partial y_n} \\
\vdots & & & \vdots \\
\dfrac{\partial g_{jn}^{-1}}{\partial y_1} & \dfrac{\partial g_{jn}^{-1}}{\partial y_2} & \cdots & \dfrac{\partial g_{jn}^{-1}}{\partial y_n}
\end{vmatrix}
\tag{5.69}
$$

In the above, $g_{j1}, g_{j2}, \ldots$, and $g_{jn}$ are components of $\mathbf{g}_j$.

As we mentioned earlier, the results presented above can also be applied to the case in which the dimension of $\mathbf{Y}$ is smaller than that of $\mathbf{X}$. Consider the transformation represented in Equation (5.60) in which $m < n$. In order to use the formulae developed above, we first augment the $m$-dimensional random vector $\mathbf{Y}$ by another $(n - m)$ – dimensional random vector $\mathbf{Z}$. The vector $\mathbf{Z}$ can be constructed as a simple function of $\mathbf{X}$ in the form

$$
\mathbf{Z} = \mathbf{h}(\mathbf{X}),
\tag{5.70}
$$

where $\mathbf{h}$ satisfies conditions of continuity and continuity in partial derivatives. On combining Equations (5.60) and (5.70), we have now an $n$-random-variable to $n$-random-variable transformation, and the jpdf of $\mathbf{Y}$ and $\mathbf{Z}$ can be obtained by means of Equation (5.67) or Equation (5.68). The jpdf of $\mathbf{Y}$ alone is then found through integration with respect to the components of $\mathbf{Z}$.

**Example 5.18.** Problem: let random variables $X_1$ and $X_2$ be independent and identically and normally distributed according to

$$
f_{X_1}(x_1) = \frac{1}{(2\pi)^{1/2}} \exp\left(\frac{-x_1^2}{2}\right), \quad -\infty < x_1 < \infty,
$$

and similarly for $X_2$. Determine the jpdf of $Y_1 = X_1 + X_2$, and $Y_2 = X_1 - X_2$.

Answer: Equation (5.67) applies in this case. The solutions of $x_1$ and $x_2$ in terms of $y_1$ and $y_2$ are

$$
x_1 = g_1^{-1}(y) = \frac{y_1 + y_2}{2}, \quad x_2 = g_2^{-1}(y) = \frac{y_1 - y_2}{2}.
\tag{5.71}
$$

The Jacobian in this case takes the form

$$
J =
\begin{vmatrix}
\dfrac{\partial g_1^{-1}}{\partial y_1} & \dfrac{\partial g_1^{-1}}{\partial y_2} \\
\dfrac{\partial g_2^{-1}}{\partial y_1} & \dfrac{\partial g_2^{-1}}{\partial y_2}
\end{vmatrix}
=
\begin{vmatrix}
\dfrac{1}{2} & \dfrac{1}{2} \\
\dfrac{1}{2} & -\dfrac{1}{2}
\end{vmatrix}
= -\frac{1}{2}.
$$

Hence, Equation (5.67) leads to

$$f_{Y_1 Y_2}(y_1, y_2) = f_{X_1}[g_1^{-1}(y)] f_{X_2}[g_2^{-1}(y)] |J|$$

$$= \frac{1}{4\pi} \exp\left[\frac{-(y_1 + y_2)^2}{8}\right] \exp\left[\frac{-(y_1 - y_2)^2}{8}\right]$$

$$= \frac{1}{4\pi} \exp\left[\frac{-(y_1^2 + y_2^2)}{4}\right], \quad (-\infty, -\infty) < (y_1, y_2) < (\infty, \infty). \quad (5.72)$$

It is of interest to note that the result given by Equation (5.72) can be written as

$$f_{Y_1 Y_2}(y_1, y_2) = f_{Y_1}(y_1) f_{Y_2}(y_2), \tag{5.73}$$

where

$$f_{Y_1}(y_1) = \frac{1}{(4\pi)^{1/2}} \exp\left(\frac{-y_1^2}{4}\right), \quad -\infty < y_1 < \infty,$$

$$f_{Y_2}(y_2) = \frac{1}{(4\pi)^{1/2}} \exp\left(\frac{-y_2^2}{4}\right), \quad -\infty < y_2 < \infty,$$

implying that, although $Y_1$ and $Y_2$ are both functions of $X_1$ and $X_2$, they are independent and identically and normally distributed.

**Example 5.19.** Problem: for the same distributions assigned to $X_1$ and $X_2$ in Example 5.18, determine the jpdf of $Y_1 = (X_1^2 + X_2^2)^{1/2}$ and $Y_2 = X_1/X_2$.

Answer: let us first note that $Y_1$ takes values only in the positive range. Hence,

$$f_{Y_1 Y_2}(y_1, y_2) = 0, \quad y_1 < 0.$$

For $y_1 \geq 0$, the transformation $\mathbf{y} = \mathbf{g}(\mathbf{x})$ admits two solutions. They are:

$$x_{11} = g_{11}^{-1}(\mathbf{y}) = \frac{y_1 y_2}{(1 + y_2^2)^{1/2}},$$

$$x_{12} = g_{12}^{-1}(\mathbf{y}) = \frac{y_1}{(1 + y_2^2)^{1/2}},$$

and

$$x_{21} = g_{21}^{-1}(\mathbf{y}) = -x_{11},$$

$$x_{22} = g_{22}^{-1}(\mathbf{y}) = -x_{12}.$$

Equation (5.68) now applies and we have

$$f_{Y_1 Y_2}(y_1, y_2) = f_{\mathbf{X}}[g_1^{-1}(\mathbf{y})]|J_1| + f_{\mathbf{X}}[g_2^{-1}(\mathbf{y})]|J_2|, \qquad (5.74)$$

where

$$f_{\mathbf{X}}(\mathbf{x}) = f_{X_1}(x_1) f_{X_2}(x_2) = \frac{1}{2\pi} \exp\left[\frac{-x_1^2 + x_2^2}{2}\right], \qquad (5.75)$$

$$J_1 = J_2 = \begin{vmatrix} \dfrac{\partial g_{11}^{-1}}{\partial y_1} & \dfrac{\partial g_{11}^{-1}}{\partial y_2} \\ \dfrac{\partial g_{12}^{-1}}{\partial y_1} & \dfrac{\partial g_{12}^{-1}}{\partial y_2} \end{vmatrix} = -\frac{y_1}{1 + y_2^2}. \qquad (5.76)$$

On substituting Equations (5.75) and (5.76) into Equation (5.74), we have

$$f_{Y_1 Y_2}(y_1, y_2) = \left(\frac{y_1}{1 + y_2^2}\right)\left\{\frac{2}{2\pi}\exp\left[-\frac{y_1^2 y_2^2 + y_1^2}{2(1 + y_2^2)}\right]\right\};$$

$$= \frac{y_1}{(1 + y_2^2)\pi}\exp\left(\frac{-y_1^2}{2}\right), \quad \text{for } y_1 \geq 0, \text{and } -\infty < y_2 < \infty;$$

$$= 0, \quad \text{elsewhere.} \qquad (5.77)$$

We note that the result can again be expressed as the product of $f_{Y_1}(y_1)$ and $f_{Y_2}(y_2)$, with

$$f_{Y_1}(y_1) = \begin{cases} y_1 \exp\left(\dfrac{-y_1^2}{2}\right), & \text{for } y_1 \geq 0; \\ 0, & \text{elsewhere;} \end{cases}$$

$$f_{Y_2}(y_2) = \frac{1}{\pi(1 + y_2^2)}, \quad \text{for } -\infty < y_2 < \infty.$$

Thus random variables $Y_1$ and $Y_2$ are again independent in this case where $Y_1$ has the so-called Raleigh distribution and $Y_2$ is Cauchy distributed. We remark that the factor $(1/\pi)$ is assigned to $f_{Y_2}(y_2)$ to make the area under each pdf equal to 1.

**Example 5.20.** Let us determine the pdf of $Y$ considered in Example 5.11 by using the formulae developed in this section. The transformation is

$$Y = X_1 X_2. \qquad (5.78)$$

In order to conform with conditions stated in this section, we augment Equation (5.78) by some simple transformation such as

$$Z = X_2. \tag{5.79}$$

The random variables $Y$ and $Z$ now play the role of $Y_1$ and $Y_2$ in Equation (5.67) and we have

$$f_{YZ}(y,z) = f_{X_1 X_2}[g_1^{-1}(y,z), g_2^{-1}(y,z)]|J|, \tag{5.80}$$

where

$$g_1^{-1}(y,z) = \frac{y}{z},$$
$$g_2^{-1}(y,z) = z,$$
$$J = \begin{vmatrix} \dfrac{1}{z} & -\dfrac{y}{z^2} \\ 0 & 1 \end{vmatrix} = \frac{1}{z}.$$

Using specific forms of $f_{X_1}(x_1)$ and $f_{X_2}(x_2)$ given in Example (5.11), Equation (5.80) becomes

$$f_{YZ}(y,z) = f_{X_1}\left(\frac{y}{z}\right) f_{X_2}(z) \left|\frac{1}{z}\right| = \left(\frac{2y}{z}\right)\left(\frac{2-z}{2z}\right);$$
$$= \frac{y(2-z)}{z^2}, \quad \text{for } 0 \le y \le 2, \text{ and } y \le z \le 2; \tag{5.81}$$
$$= 0, \quad \text{elsewhere.}$$

Finally, pdf $f_Y(y)$ is found by performing integration of Equation (5.81) with respect to $z$:

$$f_Y(y) = \int_{-\infty}^{\infty} f_{YZ}(y,z)\,dz = \int_y^2 \left[\frac{y(2-z)}{z^2}\right] dz;$$
$$= 2 + y(\ln y - 1 - \ln 2), \quad \text{for } 0 \le y \le 2;$$
$$= 0, \quad \text{elsewhere.}$$

This result agrees with that given in Equation (5.47) in Example 5.11.

## REFERENCE

Courant, R., 1937, *Differential and Integral Calculus, Volume II*, Wiley-Interscience, New York.

## PROBLEMS

5.1 Determine the Probability distribution function (PDF) of $Y = 3X - 1$ if
   (a) Case 1:

$$F_X(x) = \begin{cases} 0, & \text{for } x < 3; \\ \dfrac{1}{3}, & \text{for } 3 \le x < 6; \\ 1, & \text{for } x \ge 6. \end{cases}$$

   (b) Case 2:

$$F_X(x) = \begin{cases} 0, & \text{for } x < 3; \\ \dfrac{x}{3} - 1, & \text{for } 3 \le x < 6; \\ 1, & \text{for } x \ge 6. \end{cases}$$

5.2 Temperature $C$ measured in degrees Celsius is related to temperature $X$ in degrees Fahrenheit by $C = 5(X - 32)/9$. Determine the probability density function (pdf) of $C$ if $X$ is random and is distributed uniformly in the interval (86, 95).

5.3 The random variable $X$ has a triangular distribution as shown in Figure 5.22. Determine the pdf of $Y = 3X + 2$.

5.4 Determine $F_Y(y)$ in terms of $F_X(x)$ if $Y = X^{1/2}$, where $F_X(x) = 0$, $x < 0$.

5.5 A random variable $Y$ has a 'log-normal' distribution if it is related to $X$ by $Y = e^X$, where $X$ is normally distributed according to

$$f_X(x) = \frac{1}{\sigma(2\pi)^{1/2}} \exp\left[\frac{-(x - m)^2}{2\sigma^2}\right], \quad -\infty < x < \infty$$

Determine the pdf of $Y$ for $m = 0$ and $\sigma = 1$.

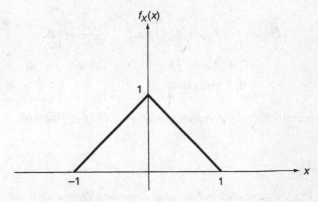

**Figure 5.22**  Distribution of $X$, for Problem 5.3

5.6 The following arises in the study of earthquake-resistant design. If $X$ is the magnitude of an earthquake and $Y$ is ground-motion intensity at distance $c$ from the earthquake, $X$ and $Y$ may be related by

$$Y = ce^X.$$

If $X$ has the distribution

$$f_X(x) = \begin{cases} \lambda e^{-\lambda x}, & \text{for } x \geq 0; \\ 0, & \text{elsewhere:} \end{cases}$$

(a) Show that the PDF of $Y$, $F_Y(y)$, is

$$F_Y(y) = \begin{cases} 1 - \left(\dfrac{y}{c}\right)^{-\lambda}, & \text{for } y \geq c; \\ 0, & \text{for } y < c. \end{cases}$$

(b) What is $f_Y(y)$?

5.7 The risk $R$ of an accident for a vehicle traveling at a 'constant' speed $V$ is given by

$$R = ae^{b(V-c)^2},$$

where $a$, $b$, and $c$ are positive constants. Suppose that speed $V$ of a class of vehicles is random and is uniformly distributed between $v_1$ and $v_2$. Determine the pdf of $R$ if (a) $(v_1, v_2) \geq c$, and (b) $v_1$ and $v_2$ are such that $c = (v_1 + v_2)/2$.

5.8 Let $Y = g(X)$, with $X$ uniformly distributed over the interval $a \leq x \leq b$. Suppose that the inverse function $X = g^{-1}(Y)$ is a single-valued function of $Y$ in the interval $g(a) \leq y \leq g(b)$. Show that the pdf of $Y$ is

$$f_Y(y) = \begin{cases} \dfrac{1}{b-a}\left[\dfrac{1}{g'[g^{-1}(y)]}\right], & \text{for } g(a) \leq y \leq g(b); \\ 0, & \text{elsewhere.} \end{cases}$$

where $g'(x) = dg(x)/dx$.

5.9 A rectangular plate of area $a$ is situated in a flow stream at an angle $\Theta$ measured from the streamline, as shown in Figure 5.23. Assuming that $\Theta$ is uniformly distributed from 0 to $\pi/2$, determine the pdf of the projected area perpendicular to the stream.

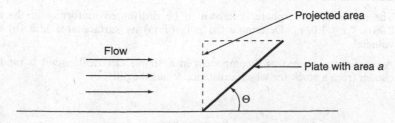

**Figure 5.23**   Plate in flow stream, for Problem 5.9

5.10 At a given location, the PDF of annual wind speed, $V$, in miles per hour is found to be

$$F_V(v) = \begin{cases} \exp\left[-\left(\dfrac{v}{36.6}\right)^{-6.96}\right], & \text{for } v > 0; \\ 0, & \text{elsewhere.} \end{cases}$$

The wind force $W$ exerted on structures is proportional to $V^2$. Let $W = aV^2$.
(a) Determine the pdf of $W$ and its mean and variance by using $f_W(w)$.
(b) Determine the mean and variance of $W$ directly from the knowledge of $F_V(v)$.

5.11 An electrical device called a full-wave rectifier transforms input $X$ to the device, to output $Y$ according to $Y = |X|$. If input $X$ has a pdf of the form

$$f_X(x) = \frac{1}{(2\pi)^{1/2}} \exp\left[\frac{-(x-1)^2}{2}\right], \quad -\infty < x < \infty,$$

(a) Determine the pdf of $Y$ and its mean and variance using $f_Y(y)$.
(b) Determine the mean and variance of $Y$ directly from the knowledge of $f_X(x)$.

5.12 An electrical device gives output $Y$ in terms of input $X$ according to

$$Y = g(X) = \begin{cases} 1, & \text{for } X > 0; \\ 0, & \text{for } X \le 0. \end{cases}$$

Is random variable $Y$ continuous or discrete? Determine its probability distribution in terms of the pdf of $X$.

5.13 The kinetic energy of a particle with mass $m$ and velocity $V$ is given by

$$X = \frac{mV^2}{2}$$

Suppose that $m$ is deterministic and $V$ is random with pdf given by

$$f_V(v) = \begin{cases} av^2 e^{-bv^2}, & \text{for } v > 0; \\ 0, & \text{elsewhere.} \end{cases}$$

Determine the pdf of $X$.

5.14 The radius $R$ of a sphere is known to be distributed uniformly in the range $0.99\,r_0 \le r \le 1.01\,r_0$. Determine the pdfs of (a) its surface area and (b) of its volume.

5.15 A resistor to be used as a component in a simple electrical circuit is randomly chosen from a stock for which resistance $R$ has the pdf

$$f_R(r) = \begin{cases} a^2 r e^{-ar}, & \text{for } r > 0; \\ 0, & \text{elsewhere.} \end{cases}$$

Suppose that voltage source $v$ in the circuit is a deterministic constant.
(a) Find the pdf of current $I$, where $I = v/R$, passing through the circuit.
(b) Find the pdf of power $W$, where $W = I^2 R$, dissipated in the resistor.

5.16 The independent random variables $X_1$ and $X_2$ are uniformly and identically distributed, with pdfs

$$f_{X_1}(x_1) = \begin{cases} \dfrac{1}{2}, & \text{for } -1 \le x_1 \le 1; \\ 0, & \text{elsewhere;} \end{cases}$$

and similarly for $X_2$. Let $Y = X_1 + X_2$.
(a) Determine the pdf of $Y$ by using Equation (5.56).
(b) Determine the pdf of $Y$ by using the method of characteristic functions developed in Section 4.5.

5.17 Two random variables, $T_1$ and $T_2$, are independent and exponentially distributed according to

$$f_{T_1}(t_1) = \begin{cases} 2e^{-2t_1}, & \text{for } t_1 \ge 0; \\ 0, & \text{elsewhere;} \end{cases}$$

$$f_{T_2}(t_2) = \begin{cases} 2e^{-2t_2}, & \text{for } t_2 \ge 0; \\ 0, & \text{elsewhere.} \end{cases}$$

(a) Determine the pdf of $T = T_1 - T_2$.
(b) Determine $m_T$ and $\sigma_T^2$.

5.18 A discrete random variable $X$ has a binomial distribution with parameters $(n, p)$. Its probability mass function (pmf) has the form

$$p_X(k) = \binom{n}{k} p^k (1 - p)^{n-k}, \quad k = 0, 1, 2, \ldots, n.$$

Show that, if $X_1$ and $X_2$ are independent and have binomial distributions with parameters $(n_1, p)$ and $(n_2, p)$, respectively, the sum $Y = X_1 + X_2$ has a binomial distribution with parameters $(n_1 + n_2, p)$.

5.19 Consider the sum of two independent random variables $X_1$ and $X_2$ where $X_1$ is discrete, taking values $a$ and $b$ with probabilities $P(X_1 = a) = p$, and $P(X_1 = b) = q$ $(p + q = 1)$, and $X_2$ is continuous with pdf $f_{X_2}(x_2)$.
(a) Show that $Y = X_1 + X_2$ is a continuous random variable with pdf

$$f_Y(y) = p f_{Y_1}(y) + q f_{Y_2}(y),$$

where $f_{Y_1}(y)$ and $f_{Y_2}(y)$ are, respectively, the pdfs of $Y_1 = a + X_2$, and $Y_2 = b + X_2$ at $y$.
(b) Plot $f_Y(y)$ by letting $a = 0$, $b = 1$, $p = \frac{1}{3}, q = \frac{2}{3}$, and

$$f_{X_2}(x_2) = \frac{1}{(2\pi)^{1/2}} \exp\left(\frac{-x_2^2}{2}\right), \quad -\infty < x_2 < \infty$$

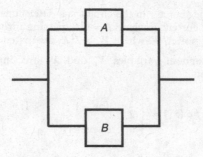

**Figure 5.24** Parallel arrangement of components A and B, for Problem 5.20

5.20 Consider a system with a parallel arrangement, as shown in Figure 5.24, and let $A$ be the primary component and $B$ its redundant mate (backup component). The operating lives of $A$ and $B$ are denoted by $T_1$ and $T_2$, respectively, and they follow the exponential distributions

$$f_{T_1}(t_1) = \begin{cases} a_1 e^{-a_1 t_1}, & \text{for } t_1 > 0; \\ 0, & \text{elsewhere}; \end{cases}$$

$$f_{T_2}(t_2) = \begin{cases} a_2 e^{-a_2 t_2}, & \text{for } t_2 > 0; \\ 0, & \text{elsewhere}. \end{cases}$$

Let the life of the system be denoted by $T$. Then $T = T_1 + T_2$ if the redundant part comes into operation only when the primary component fails (so-called 'cold redundancy') and $T = \max(T_1, T_2)$ if the redundant part is kept in a ready condition at all times so that delay is minimized in the event of changeover from the primary component to its redundant mate (so-called 'hot redundancy').
   (a) Let $T_C = T_1 + T_2$, and $T_H = \max(T_1, T_2)$. Determine their respective probability density functions.
   (b) Suppose that we wish to maximize the probability $P(T \geq t)$ for some $t$. Which type of redundancy is preferred?

5.21 Consider a system with components arranged in series, as shown in Figure 5.25, and let $T_1$ and $T_2$ be independent random variables, representing the operating lives of $A$ and $B$, for which the pdfs are given in Problem 5.20. Determine the pdf of system life $T = \min(T_1, T_2)$. Generalize to the case of $n$ components in series.

5.22 At a taxi stand, the number $X_1$ of taxis arriving during some time interval has a Poisson distribution with pmf given by

$$p_{X_1}(k) = \frac{\lambda^k e^{-\lambda}}{k!}, \quad k = 0, 1, 2, \ldots,$$

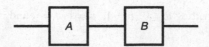

**Figure 5.25** Components A and B arranged in series, for Problem 5.21

where $\lambda$ is a constant. Suppose that demand $X_2$ at this location during the same time interval has the same distribution as $X_1$ and is independent of $X_1$. Determine the pdf of $Y = X_2 - X_1$ where $Y$ represents the excess of taxis in this time interval (positive and negative).

5.23 Determine the pdf of $Y = |X_1 - X_2|$ where $X_1$ and $X_2$ are independent random variables with respective pdfs $f_{X_1}(x_1)$ and $f_{X_2}(x_2)$.

5.24 The light intensity $I$ at a given point $X$ distance away from a light source is $I = C/X^2$ where $C$ is the source candlepower. Determine the pdf of $I$ if the pdfs of $C$ and $X$ are given by

$$f_C(c) = \begin{cases} \dfrac{1}{36}, & \text{for } 64 \le c \le 100; \\ 0, & \text{elsewhere}; \end{cases}$$

$$f_X(x) = \begin{cases} 1, & \text{for } 1 \le x \le 2; \\ 0, & \text{elsewhere}; \end{cases}$$

and $C$ and $X$ are independent.

5.25 Let $X_1$ and $X_2$ be independent and identically distributed according to

$$f_{X_1}(x_1) = \frac{1}{(2\pi)^{1/2}} \exp\left(\frac{-x_1^2}{2}\right), \quad -\infty < x_1 < \infty,$$

and similarly for $X_2$. By means of techniques developed in Section 5.2, determine the pdf of $Y$, where $Y = (X_1^2 + X_2^2)^{1/2}$. Check your answer with the result obtained in Example 5.19. (Hint: use polar coordinates to carry out integration.)

5.26 Extend the result of Problem 5.25 to the case of three independent and identically distributed random variables, that is, $Y = (X_1^2 + X_2^2 + X_3^2)^{1/2}$. (Hint: use spherical coordinates to carry out integration.)

5.27 The joint probability density function (jpdf) of random variables $X_1, X_2,$ and $X_3$ takes the form

$$f_{X_1 X_2 X_3}(x_1, x_2, x_3) = \begin{cases} \dfrac{6}{(1 + x_1 + x_2 + x_3)^4}, & \text{for } (x_1, x_2, x_3) > (0, 0, 0); \\ 0, & \text{elsewhere}. \end{cases}$$

Find the pdf of $Y = X_1 + X_2 + X_3$.

5.28 The pdfs of two independent random variables $X_1$ and $X_2$ are

$$f_{X_1}(x_1) = \begin{cases} e^{-x_1}, & \text{for } x_1 > 0; \\ 0, & \text{for } x_1 \le 0; \end{cases}$$

$$f_{X_2}(x_2) = \begin{cases} e^{-x_2}, & \text{for } x_2 > 0; \\ 0, & \text{for } x_2 \le 0. \end{cases}$$

Determine the jpdf of $Y_1$ and $Y_2$, defined by

$$Y_1 = X_1 + X_2, \quad Y_2 = \frac{X_1}{Y_1},$$

and show that they are independent.

5.29 The jpdf of $X$ and $Y$ is given by

$$f_{XY}(x, y) = \frac{1}{2\pi\sigma^2} \exp\left[\frac{-(x^2 + y^2)}{2\sigma^2}\right], \quad (-\infty, -\infty) < (x, y) < (\infty, \infty).$$

Determine the jpdf of $R$ and $\Phi$ and their respective marginal pdfs where $R = (X^2 + Y^2)^{1/2}$ is the vector length and $\Phi = \tan^{-1}(Y/X)$ is the phase angle. Are $R$ and $\Phi$ independent?

5.30 Show that an alternate formula for Equation (5.67) is

$$f_{\mathbf{Y}}(\mathbf{y}) = f_{\mathbf{X}}[g^{-1}(\mathbf{y})]|J'|^{-1},$$

where

$$J' = \begin{vmatrix} \dfrac{\partial g_1(\mathbf{x})}{\partial x_1} & \dfrac{\partial g_1(\mathbf{x})}{\partial x_2} & \cdots & \dfrac{\partial g_1(\mathbf{x})}{\partial x_n} \\ \vdots & & & \vdots \\ \dfrac{\partial g_n(\mathbf{x})}{\partial x_1} & \dfrac{\partial g_n(\mathbf{x})}{\partial x_2} & \cdots & \dfrac{\partial g_n(\mathbf{x})}{\partial x_n} \end{vmatrix}$$

is evaluated at $\mathbf{x} = g^{-1}(\mathbf{y})$. Similar alternate forms hold for Equations (5.12), (5.24) and (5.68).

# 6

# Some Important Discrete Distributions

This chapter deals with some distributions of discrete random variables that are important as models of scientific phenomena. The nature and applications of these distributions are discussed. An understanding of the situations in which these random variables arise enables us to choose an appropriate distribution for a scientific phenomenon under consideration. Thus, this chapter is also concerned with the induction step discussed in Chapter 1, by which a model is chosen on the basis of factual understanding of the physical phenomenon under study (step B to C in Figure 1.1).

Some important distributions of continuous random variables will be studied in Chapter 7.

## 6.1  BERNOULLI TRIALS

A large number of practical situations can be described by the repeated performance of a random experiment of the following basic nature: a sequence of trials is performed so that (a) for each trial, there are only two possible outcomes, say, success and failure; (b) the probabilities of the occurrence of these outcomes remain the same throughout the trials; and (c) the trials are carried out independently. Trials performed under these conditions are called *Bernoulli trials*. Despite of the simplicity of the situation, mathematical models arising from this basic random experiment have wide applicability. In fact, we have encountered Bernoulli trials in the random walk problems described in Examples 3.5 (page 52) and 4.17 (page 106) and also in the traffic problem examined in Example 3.9 (page 64). More examples will be given in the sections to follow.

Let us denote event 'success' by $S$, and event 'failure' by $F$. Also, let $P(S) = p$, and $P(F) = q$, where $p + q = 1$. Possible outcomes resulting from performing

*Fundamentals of Probability and Statistics for Engineers* T.T. Soong © 2004 John Wiley & Sons, Ltd
ISBNs: 0-470-86813-9 (HB)   0-470-86814-7 (PB)

a sequence of Bernoulli trials can be symbolically represented by

$$SSFFSFSSS \cdots FF$$
$$FSFSSFFFS \cdots SF$$

$$\vdots$$

and, owing to independence, the probabilities of these possible outcomes are easily computed. For example,

$$P(SSFFSF \cdots FF) = P(S)P(S)P(F)P(F)P(S)P(F) \cdots P(F)P(F)$$
$$= ppqqpq \cdots qq.$$

A number of these possible outcomes with their associated probabilities are of practical interest. We introduce three important distributions in this connection.

### 6.1.1  BINOMIAL DISTRIBUTION

The probability distribution of a random variable $X$ representing the number of successes in a sequence of $n$ Bernoulli trials, regardless of the order in which they occur, is frequently of considerable interest. It is clear that $X$ is a discrete random variable, assuming values $0, 1, 2, \ldots, n$. In order to determine its probability mass function, consider $p_X(k)$, the probability of having exactly $k$ successes in $n$ trials. This event can occur in as many ways as $k$ letters $S$ can be placed in $n$ boxes. Now, we have $n$ choices for the position of the first $S$, $n - 1$ choices for the second $S, \ldots$, and, finally, $n - k + 1$ choices for the position of the $k$th $S$. The total number of possible arrangements is thus $n(n - 1) \ldots (n - k + 1)$. However, as no distinction is made of the $S$s that are in the occupied positions, we must divide the number obtained above by the number of ways in which $k$ $S$s can be arranged in $k$ boxes, that is, $k(k - 1) \ldots 1 = k!$. Hence, the number of ways in which $k$ successes can happen in $n$ trials is

$$\frac{n(n - 1) \cdots (n - k + 1)}{k!} = \frac{n!}{k!(n - k)!}, \tag{6.1}$$

and the probability associated with each is $p^k q^{n-k}$. Hence, we have

$$p_X(k) = \binom{n}{k} p^k q^{n-k}, \quad k = 0, 1, 2, \ldots, n, \tag{6.2}$$

where

$$\binom{n}{k} = \frac{n!}{k!(n-k)!} \tag{6.3}$$

is the binomial coefficient in the binomial theorem

$$(a+b)^n = \sum_{k=0}^{n} \binom{n}{k} a^k b^{n-k}. \tag{6.4}$$

In view of its similarity in appearance to the terms of the binomial theorem, the distribution defined by Equation (6.2) is called the *binomial distribution*. It has two parameters, namely, $n$ and $p$. Owing to the popularity of this distribution, a random variable $X$ having a binomial distribution is often denoted by $B(n,p)$.

The shape of a binomial distribution is determined by the values assigned to its two parameters, $n$ and $p$. In general, $n$ is given as a part of the problem statement and $p$ must be estimated from observations.

A plot of probability mass function (pmf), $p_X(k)$, has been shown in Example 3.2 (page 43) for $n = 10$ and $p = 0.2$. The peak of the distribution will shift to the right as $p$ increases, reaching a symmetrical distribution when $p = 0.5$. More insight into the behavior of $p_X(k)$ can be gained by taking the ratio

$$\frac{p_X(k)}{p_X(k-1)} = \frac{(n-k+1)p}{kq} = 1 + \frac{(n+1)p - k}{kq}. \tag{6.5}$$

We see from Equation (6.5) that $p_X(k)$ is greater than $p_X(k-1)$ when $k < (n+1)p$ and is smaller when $k > (n+1)p$. Accordingly, if we define integer $k^*$ by

$$(n+1)p - 1 < k^* \le (n+1)p, \tag{6.6}$$

the value of $p_X(k)$ increases monotonically and attains its maximum value when $k = k^*$, then decreases monotonically. If $(n+1)p$ happens to be an integer, the maximum value takes place at both $p_X(k^* - 1)$ and $p_X(k^*)$. The integer $k^*$ is thus a mode of this distribution and is often referred to as 'the most probable number of successes'.

Because of its wide usage, pmf $p_X(k)$ is widely tabulated as a function of $n$ and $p$. Table A.1 in Appendix A gives its values for $n = 2, 3, \ldots, 10$, and $p = 0.01, 0.05, \ldots, 0.50$. Let us note that probability tables for the binomial and other commonly used distributions are now widely available in a number of computer software packages, and even on some calculators. For example, function BINOMDIST in Microsoft® Excel™ 2000 gives individual binomial probabilities given by Equation (6.2). Other statistical functions available in Excel™ 2000 are listed in Appendix B.

The calculation of $p_X(k)$ in Equation (6.2) is cumbersome as $n$ becomes large. An approximate way of determining $p_X(k)$ for large $n$ has been discussed in Example 4.17 (page 106) by means of Stirling's formula [Equation (4.78)]. Poisson approximation to the binomial distribution, to be discussed in Section 6.3.2, also facilitates probability calculations when $n$ becomes large.

The probability distribution function (PDF), $F_X(x)$, for a binomial distribution is also widely tabulated. It is given by

$$F_X(x) = \sum_{k=0}^{m \leq x} \binom{n}{k} p^k q^{n-k}, \qquad (6.7)$$

where $m$ is the largest integer less than or equal to $x$.

Other important properties of the binomial distribution have been derived in Example 4.1 (page 77), Example 4.5 (page 81), and Example 4.14 (page 99). Without giving details, we have, respectively, for the characteristic function, mean, and variance,

$$\phi_X(t) = (p e^{jt} + q)^n,$$
$$m_X = np, \qquad (6.8)$$
$$\sigma_X^2 = npq.$$

The fact that the mean of $X$ is $np$ suggests that parameter $p$ can be estimated based on the average value of the observed data. This procedure is used in Examples 6.2. We mention, however, that this parameter estimation problem needs to be examined much more rigorously, and its systematic treatment will be taken up in Part B.

Let us remark here that another formulation leading to the binomial distribution is to define random variable $X_j, j = 1, 2, \ldots, n$, to represent the outcome of the $j$th Bernoulli trial. If we let

$$X_j = \begin{cases} 0 & \text{if } j\text{th trial is a failure,} \\ 1 & \text{if } j\text{th trial is a success,} \end{cases} \qquad (6.9)$$

then the sum

$$X = X_1 + X_2 + \cdots + X_n \qquad (6.10)$$

gives the number of successes in $n$ trials. By definition, $X_1, \ldots,$ and $X_n$ are independent random variables.

The moments and distribution of $X$ can be easily found by using Equation (6.10). Since

$$E\{X_j\} = 0(q) + 1(p) = p, \; j = 1, 2, \ldots, n,$$

it follows from Equation (4.38) that

$$E\{X\} = p + p + \cdots + p = np, \tag{6.11}$$

which is in agreement with the corresponding expression in Equations (6.8). Similarly, its variance, characteristic function, and pmf are easily found following our discussion in Section 4.4 concerning sums of independent random variables.

We have seen binomial distributions in Example 3.5 (page 52), Example 3.9 (page 64), and Example 4.11 (page 96). Its applications in other areas are further illustrated by the following additional examples.

**Example 6.1.** Problem: a homeowner has just installed 20 light bulbs in a new home. Suppose that each has a probability 0.2 of functioning more than three months. What is the probability that at least five of these function more than three months? What is the average number of bulbs the homeowner has to replace in three months?

Answer: it is reasonable to assume that the light bulbs perform independently. If $X$ is the number of bulbs functioning more than three months (success), it has a binomial distribution with $n = 20$ and $p = 0.2$. The answer to the first question is thus given by

$$\sum_{k=5}^{20} p_X(k) = 1 - \sum_{k=0}^{4} p_X(k)$$

$$= 1 - \sum_{k=0}^{4} \binom{20}{k} (0.2)^k (0.8)^{20-k}$$

$$= 1 - (0.012 + 0.058 + 0.137 + 0.205 + 0.218) = 0.37.$$

The average number of replacements is

$$20 - E\{X\} = 20 - np = 20 - 20(0.2) = 16.$$

**Example 6.2.** Suppose that three telephone users use the same number and that we are interested in estimating the probability that more than one will use it at the same time. If independence of telephone habit is assumed, the probability of exactly $k$ persons requiring use of the telephone at the same time is given by the mass function $p_X(k)$ associated with the binomial distribution. Let it be given that, on average, a telephone user is on the phone 5 minutes per hour; an estimate of $p$ is

$$p = \frac{5}{60} = \frac{1}{12}.$$

The solution to this problem is given by

$$p_X(2) + p_X(3) = \binom{3}{2}\left(\frac{1}{12}\right)^2\left(\frac{11}{12}\right) + \binom{3}{3}\left(\frac{1}{12}\right)^3$$

$$= \frac{11}{864} = 0.0197.$$

**Example 6.3.** Problem: let $X_1$ and $X_2$ be two independent random variables, both having binomial distributions with parameters $(n_1, p)$ and $(n_2, p)$, respectively, and let $Y = X_1 + X_2$. Determine the distribution of random variable $Y$.

Answer: the characteristic functions of $X_1$ and $X_2$ are, according to the first of Equations (6.8),

$$\phi_{X_1}(t) = (pe^{jt} + q)^{n_1}, \phi_{X_2}(t) = (pe^{jt} + q)^{n_2}.$$

In view of Equation (4.71), the characteristic function of $Y$ is simply the product of $\phi_{X_1}(t)$ and $\phi_{X_2}(t)$. Thus,

$$\phi_Y(t) = \phi_{X_1}(t)\phi_{X_2}(t)$$
$$= (pe^{jt} + q)^{n_1+n_2}.$$

By inspection, it is the characteristic function corresponding to a binomial distribution with parameters $(n_1 + n_2, p)$. Hence, we have

$$p_Y(k) = \binom{n_1 + n_2}{k}p^k q^{n_1+n_2-k}, \quad k = 0, 1, \ldots, n_1 + n_2.$$

Generalizing the answer to Example 6.3, we have the following important result as stated in Theorem 6.1.

**Theorem 6.1:** The binomial distribution generates itself under addition of independent random variables with the same $p$.

**Example 6.4.** Problem: if random variables $X$ and $Y$ are independent binomial distributed random variables with parameters $(n_1, p)$ and $(n_2, p)$, determine the conditional probability mass function of $X$ given that

$$X + Y = m, \qquad 0 \leq m \leq n_1 + n_2.$$

Answer: for $k \leq \min(n_1, m)$, we have

$$P(X = k | X + Y = m) = \frac{P(X = k \cap X + Y = m)}{P(X + Y = m)}$$

$$= \frac{P(X = k \cap Y = m - k)}{P(X + Y = m)} = \frac{P(X = k)P(Y = m - k)}{P(X + Y = m)}$$

$$= \frac{\binom{n_1}{k} p^k (1 - p)^{n_1 - k} \binom{n_2}{m - k} p^{m - k} (1 - p)^{n_2 - m + k}}{\binom{n_1 + n_2}{m} p^m (1 - p)^{n_1 + n_2 - m}}$$

$$= \binom{n_1}{k} \binom{n_2}{m - k} \bigg/ \binom{n_1 + n_2}{m}, \quad k = 0, 1, \ldots, \min(n_1, m), \quad (6.12)$$

where we have used the result given in Example 6.3 that $X + Y$ is binomially distributed with parameters $(n_1 + n_2, p)$.

The distribution given by Equation (6.12) is known as the *hypergeometric distribution*. It arises as distributions in such cases as the number of black balls that are chosen when a sample of $m$ balls is randomly selected from a lot of $n$ items having $n_1$ black balls and $n_2$ white balls $(n_1 + n_2 = n)$. Let random variable $Z$ be this number. We have, from Equation (6.12), on replacing $n_2$ by $n - n_1$,

$$p_Z(k) = \binom{n_1}{k} \binom{n - n_1}{m - k} \bigg/ \binom{n}{m}, \quad k = 0, 1, \ldots, \min(n_1, m). \qquad (6.13)$$

### 6.1.2  GEOMETRIC DISTRIBUTION

Another event of interest arising from Bernoulli trials is the number of trials to (and including) the first occurrence of success. If $X$ is used to represent this number, it is a discrete random variable with possible integer values ranging from one to infinity. Its pmf is easily computed to be

$$p_X(k) = P(\underbrace{FF \ldots F}_{k-1} S) = P\underbrace{(F)P(F) \ldots P(F)}_{k-1} P(S)$$

$$= q^{k-1} p, \quad k = 1, 2, \ldots. \qquad (6.14)$$

This distribution is known as the *geometric distribution* with parameter $p$, where the name stems from its similarity to the familiar terms in geometric progression. A plot of $p_X(k)$ is given in Figure 6.1.

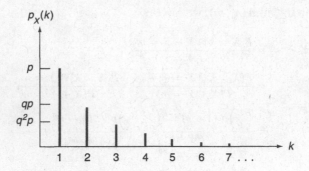

**Figure 6.1**   Geometric distribution $p_X(k)$

The corresponding probability distribution function is

$$F_X(x) = \sum_{k=1}^{m \leq x} p_X(k) = p + qp + \cdots + q^{m-1}p$$

$$= (1 - q)(1 + q + q^2 + \cdots + q^{m-1}) = 1 - q^m, \qquad (6.15)$$

where $m$ is the largest integer less than or equal to $x$. The mean and variance of $X$ can be found as follows:

$$E\{X\} = \sum_{k=1}^{\infty} kq^{k-1}p = p\sum_{k=1}^{\infty}\frac{\mathrm{d}}{\mathrm{d}q}q^k$$

$$= p\frac{\mathrm{d}}{\mathrm{d}q}\sum_{k=1}^{\infty}q^k = p\frac{\mathrm{d}}{\mathrm{d}q}\left(\frac{q}{1-q}\right) = \frac{1}{p}. \qquad (6.16)$$

In the above, the interchange of summation and differentiation is allowed because $|q| < 1$. Following the same procedure, the variance has the form

$$\sigma_X^2 = \frac{1 - p}{p^2}. \qquad (6.17)$$

**Example 6.5.** Problem: a driver is eagerly eyeing a precious parking space some distance down the street. There are five cars in front of the driver, each of which having a probability 0.2 of taking the space. What is the probability that the car immediately ahead will enter the parking space?

Answer: for this problem, we have a geometric distribution and need to evaluate $p_X(k)$ for $k = 5$ and $p = 0.2$. Thus,

$$p_X(5) = (0.8)^4(0.2) = 0.82,$$

which may seem much smaller than what we experience in similar situations.

**Example 6.6.** Problem: assume that the probability of a specimen failing during a given experiment is 0.1. What is the probability that it will take more than three specimens to have one surviving the experiment?

Answer: let $X$ denote the number of trials required for the first specimen to survive. It then has a geometric distribution with $p = 0.9$. The desired probability is

$$P(X > 3) = 1 - F_X(3) = 1 - (1 - q^3) = (0.1)^3 = 0.001.$$

**Example 6.7.** Problem: let the probability of occurrence of a flood of magnitude greater than a critical magnitude in any given year be 0.01. Assuming that floods occur independently, determine $E\{N\}$, the average return period. The *average return period*, or simply *return period*, is defined as the average number of years between floods for which the magnitude is greater than the critical magnitude.

Answer: it is clear that $N$ is a random variable with a geometric distribution and $p = 0.01$. The return period is then

$$E\{N\} = \frac{1}{p} = 100 \text{ years.}$$

The critical magnitude which gives rise to $E\{N\} = 100$ years is often referred to as the '100-year flood'.

## 6.1.3 NEGATIVE BINOMIAL DISTRIBUTION

A natural generalization of the geometric distribution is the distribution of random variable $X$ representing the number of Bernoulli trials necessary for the $r$th success to occur, where $r$ is a given positive integer.

In order to determine $p_X(k)$ for this case, let $A$ be the event that the first $k - 1$ trials yield exactly $r - 1$ successes, regardless of their order, and $B$ the event that a success turns up at the $k$th trial. Then, owing to independence,

$$p_X(k) = P(A \cap B) = P(A)P(B). \tag{6.18}$$

Now, $P(A)$ obeys a binomial distribution with parameters $k - 1$ and $r - 1$, or

$$P(A) = \binom{k-1}{r-1} p^{r-1} q^{k-r}, \quad k = r, r+1, \ldots, \tag{6.19}$$

and $P(B)$ is simply

$$P(B) = p. \tag{6.20}$$

Substituting Equations (6.19) and (6.20) into Equation (6.18) results in

$$p_X(k) = \binom{k-1}{r-1} p^r q^{k-r}, \quad k = r, r+1, \ldots. \tag{6.21}$$

We note that, as expected, it reduces to the geometric distribution when $r = 1$. The distribution defined by Equation (6.21) is known as the *negative binomial*, or *Pascal*, distribution with parameters $r$ and $p$. It is often denoted by NB$(r, p)$.

A useful variant of this distribution is obtained if we let $Y = X - r$. The random variable $Y$ is the number of Bernoulli trials *beyond* $r$ needed for the realization of the $r$th success, or it can be interpreted as the number of failures before the $r$th success.

The probability mass function of $Y, p_Y(m)$, is obtained from Equation (6.21) upon replacing $k$ by $m + r$. Thus,

$$\begin{aligned}
p_Y(m) &= \binom{m+r-1}{r-1} p^r q^m \\
&= \binom{m+r-1}{m} p^r q^m, \quad m = 0, 1, 2, \ldots.
\end{aligned} \tag{6.22}$$

We see that random variable $Y$ has the convenient property that the range of $m$ begins at zero rather than $r$ for values associated with $X$.

Recalling a more general definition of the binomial coefficient

$$\binom{a}{j} = \frac{a(a-1)\ldots(a-j+1)}{j!}, \tag{6.23}$$

for any real $a$ and any positive integer $j$, direct evaluation shows that the binomial coefficient in Equation (6.22) can be written in the form

$$\binom{m+r-1}{m} = (-1)^m \binom{-r}{m}. \tag{6.24}$$

Hence,

$$p_Y(m) = \binom{-r}{m} p^r (-q)^m, \quad m = 0, 1, 2, \ldots, \tag{6.25}$$

which is the reason for the name 'negative binomial distribution'.

The mean and variance of random variable $X$ can be determined either by following the standard procedure or by noting that $X$ can be represented by

$$X = X_1 + X_2 + \cdots + X_r, \tag{6.26}$$

where $X_j$ is the number of trials between the $(j-1)$th and (including) the $j$th successes. These random variables are mutually independent, each having the geometric distribution with mean $1/p$ and variance $(1-p)/p^2$. Therefore, the mean and variance of sum $X$ are, respectively, according to Equations (4.38) and (4.41),

$$m_X = \frac{r}{p}, \quad \sigma_X^2 = \frac{r(1-p)}{p^2}. \tag{6.27}$$

Since $Y = X - r$, the corresponding moments of $Y$ are

$$m_Y = \frac{r}{p} - r, \quad \sigma_Y^2 = \frac{r(1-p)}{p^2}. \tag{6.28}$$

**Example 6.8.** Problem: a curbside parking facility has a capacity for three cars. Determine the probability that it will be full within 10 minutes. It is estimated that 6 cars will pass this parking space within the timespan and, on average, 80% of all cars will want to park there.

Answer: the desired probability is simply the probability that the number of trials to the third success (taking the parking space) is less than or equal to 6. If $X$ is this number, it has a negative binomial distribution with $r = 3$ and $p = 0.8$. Using Equation (6.21), we have

$$P(X \le 6) = \sum_{k=3}^{6} p_X(k) = \sum_{k=3}^{6} \binom{k-1}{2}(0.8)^3(0.2)^{k-3}$$

$$= (0.8)^3[1 + (3)(0.2) + (6)(0.2)^2 + (10)(0.2)^3]$$

$$= 0.983.$$

Let us note that an alternative way of arriving at this answer is to sum the probabilities of having 3, 4, 5, and 6 successes in 6 Bernoulli trials using the binomial distribution. This observation leads to a general relationship between binomial and negative binomial distributions. Stated in general terms, if $X_1$ is $B(n, p)$ and $X_2$ is $NB(r, p)$, then

$$P(X_1 \ge r) = P(X_2 \le n),$$
$$P(X_1 < r) = P(X_2 > n). \tag{6.29}$$

**Example 6.9.** The negative binomial distribution is widely used in waiting-time problems. Consider, for example, a car waiting on a ramp to merge into freeway traffic. Suppose that it is 5th in line to merge and that the gaps between cars on the freeway are such that there is a probability of 0.4 that they are large enough for merging. Then, if $X$ is the waiting time before merging for this particular vehicle measured in terms of number of freeway gaps, it has a negative binomial distribution with $r = 5$ and $p = 0.4$. The mean waiting time is, as seen from Equation (6.27),

$$E\{X\} = \frac{5}{0.4} = 12.5 \text{ gaps.}$$

## 6.2 MULTINOMIAL DISTRIBUTION

Bernoulli trials can be generalized in several directions. A useful generalization is to relax the requirement that there be only two possible outcomes for each trial. Let there be $r$ possible outcomes for each trial, denoted by $E_1, E_2, \ldots, E_r$, and let $P(E_i) = p_i$, $i = 1, \ldots, r$, and $p_1 + p_2 + \cdots + p_r = 1$. A typical outcome of $n$ trials is a succession of symbols such as:

$$E_2 E_1 E_3 E_3 E_6 E_2 \ldots.$$

If we let random variable $X_i$, $i = 1, 2, \ldots, r$, represent the number of $E_i$ in a sequence of $n$ trials, the joint probability mass function (jpmf) of $X_1, X_2, \ldots, X_r$, is given by

$$p_{X_1 X_2 \ldots X_r}(k_1, k_2, \ldots, k_r) = \frac{n!}{k_1! k_2! \ldots k_r!} p_1^{k_1} p_2^{k_2} \ldots p_r^{k_r}, \tag{6.30}$$

where $k_j = 0, 1, 2, \ldots$, $j = 1, 2, \ldots, r$, and $k_1 + k_2 + \cdots + k_r = n$.

**Proof for Equation 6.30:** we want to show that the coefficient in Equation (6.30) is equal to the number of ways of placing $k_1$ letters $E_1, k_2$ letters $E_2, \ldots$, and $k_r$ letters $E_r$ in $n$ boxes. This can be easily verified by writing

$$\frac{n!}{k_1! k_2! \ldots k_r!} = \binom{n}{k_1} \binom{n - k_1}{k_2} \cdots \binom{n - k_1 - k_2 - \cdots - k_{r-1}}{k_r}.$$

The first binomial coefficient is the number of ways of placing $k_1$ letters $E_1$ in $n$ boxes; the second is the number of ways of placing $k_2$ letters $E_2$ in the remaining $n - k_1$ unoccupied boxes; and so on.

The formula given by Equation (6.30) is an important higher-dimensional joint probability distribution. It is called the *multinomial distribution* because it has the form of the general term in the multinomial expansion of $(p_1 + p_2 + \cdots + p_r)^n$. We note that Equation (6.30) reduces to the binomial distribution when $r = 2$ and with $p_1 = p$, $p_2 = q$, $k_1 = k$, and $k_2 = n - k$.

Since each $X_i$ defined above has a binomial distribution with parameters $n$ and $p_i$, we have

$$m_{X_i} = np_i, \quad \sigma^2_{X_i} = np_i(1 - p_i), \tag{6.31}$$

and it can be shown that the covariance is given by

$$\text{cov}(X_i, X_j) = -np_ip_j, \quad i,j = 1, 2, \ldots, r, \ i \neq j. \tag{6.32}$$

**Example 6.10.** Problem: income levels are classified as low, medium, and high in a study of incomes of a given population. If, on average, 10% of the population belongs to the low-income group and 20% belongs to the high-income group, what is the probability that, of the 10 persons studied, 3 will be in the low-income group and the remaining 7 will be in the medium-income group? What is the marginal distribution of the number of persons (out of 10) at the low-income level?

Answer: let $X_1$ be the number of low-income persons in the group of 10 persons, $X_2$ be the number of medium-income persons, and $X_3$ be the number of high-income persons. Then $X_1, X_2$, and $X_3$ have a multinomial distribution with $p_1 = 0.1$, $p_2 = 0.7$, and $p_3 = 0.2$; $n = 10$.
Thus

$$p_{X_1X_2X_3}(3,7,0) = \frac{10!}{3!7!0!}(0.1)^3(0.7)^7(0.2)^0 \cong 0.01.$$

The marginal distribution of $X_1$ is binomial with $n = 10$ and $p = 0.1$.

We remark that, while the single-random-variable marginal distributions are binomial, since $X_1, X_2, \ldots,$ and $X_r$ are not independent, the multinomial distribution is *not* a product of binomial distributions.


## 6.3  POISSON DISTRIBUTION

In this section we wish to consider a distribution that is used in a wide variety of physical situations. It is used in mathematical models for describing, in a specific interval of time, such events as the emission of $\alpha$ particles from a radioactive substance, passenger arrivals at an airline terminal, the distribution of dust particles reaching a certain space, car arrivals at an intersection, and many other similar phenomena.

To fix ideas in the following development, let us consider the problem of passenger arrivals at a bus terminal during a specified time interval. We shall use the notation $X(0, t)$ to represent the number of arrivals during time interval $[0, t)$, where the notation $[\ )$ denotes a left-closed and right-open interval; it is a discrete random variable taking possible values $0, 1, 2, \ldots$, whose distribution clearly depends on $t$. For clarity, its pmf is written as

$$p_k(0, t) = P[X(0, t) = k], \quad k = 0, 1, 2, \ldots, \tag{6.33}$$

to show its explicit dependence on $t$. Note that this is different from our standard notation for a pmf.

To study this problem, we make the following basic assumptions:

• Assumption 1: the random variables $X(t_1, t_2)$, $X(t_2, t_3), \ldots, X(t_n - 1, t_n)$, $t_1 < t_2 < \ldots < t_n$, are mutually independent, that is, the numbers of passenger arrivals in nonoverlapping time intervals are independent of each other.

• Assumption 2: for sufficiently small $\Delta t$,

$$p_1(t, t + \Delta t) = \lambda \Delta t + o(\Delta t) \tag{6.34}$$

where $o(\Delta t)$ stands for functions such that

$$\lim_{\Delta t \to 0} \frac{o(\Delta t)}{\Delta t} = 0. \tag{6.35}$$

This assumption says that, for a sufficiently small $\Delta t$, the probability of having exactly one arrival is proportional to the length of $\Delta t$. The parameter $\lambda$ in Equation (6.34) is called the *average density* or *mean rate of arrival* for reasons that will soon be made clear. For simplicity, it is assumed to be a constant in this discussion; however, there is no difficulty in allowing it to vary with time.

• Assumption 3: for sufficiently small $\Delta t$,

$$\sum_{k=2}^{\infty} p_k(t, t + \Delta t) = o(\Delta t) \tag{6.36}$$

This condition implies that the probability of having two or more arrivals during a sufficiently small interval is negligible.

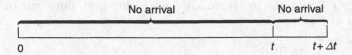

**Figure 6.2** Interval $[0, t + \Delta t)$

It follows from Equations (6.34) and (6.36) that

$$p_0(t, t + \Delta t) = 1 - \sum_{k=1}^{\infty} p_k(t, t + \Delta t)$$

$$= 1 - \lambda \Delta t + o(\Delta t). \qquad (6.37)$$

In order to determine probability mass function $p_k(0, t)$ based on the assumptions stated above, let us first consider $p_0(0, t)$. Figure 6.2 shows two nonoverlapping intervals, $[0, t)$ and $[t, t + \Delta t)$. In order that there are no arrivals in the total interval $[0, t + \Delta t)$, we must have no arrivals in both subintervals. Owing to the independence of arrivals in nonoverlapping intervals, we thus can write

$$p_0(0, t + \Delta t) = p_0(0, t) p_0(t, t + \Delta t)$$

$$= p_0(0, t)[1 - \lambda \Delta t + o(\Delta t)]. \qquad (6.38)$$

Rearranging Equation (6.38) and dividing both sides by $\Delta t$ gives

$$\frac{p_0(0, t + \Delta t) - p_0(0, t)}{\Delta t} = -p_0(0, t) \left[ \lambda - \frac{o(\Delta t)}{\Delta t} \right].$$

Upon letting $\Delta t \to 0$, we obtain the differential equation

$$\frac{dp_0(0, t)}{dt} = -\lambda p_0(0, t). \qquad (6.39)$$

Its solution satisfying the initial condition $p_0(0, 0) = 1$ is

$$p_0(0, t) = e^{-\lambda t}. \qquad (6.40)$$

The determination of $p_1(0, t)$ is similar. We first observe that one arrival in $[0, t + \Delta t)$ can be accomplished only by having no arrival in subinterval $[0, t)$ and one arrival in $[t, t + \Delta t)$, or one arrival in $[0, t)$ and no arrival in $[t, t + \Delta t)$. Hence we have

$$p_1(0, t + \Delta t) = p_0(0, t) p_1(t, t + \Delta t) + p_1(0, t) p_0(t, t + \Delta t). \qquad (6.41)$$

Substituting Equations (6.34), (6.37), and (6.40) into Equation (6.41) and letting $\Delta t \to 0$ we obtain

$$\frac{\mathrm{d}p_1(0, t)}{\mathrm{d}t} = -\lambda p_1(0, t) + \lambda e^{-\lambda t}, \quad p_1(0, 0) = 0, \tag{6.42}$$

which yields

$$p_1(0, t) = \lambda t e^{-\lambda t}. \tag{6.43}$$

Continuing in this way we find, for the general term,

$$\boxed{p_k(0, t) = \frac{(\lambda t)^k e^{-\lambda t}}{k!}, \quad k = 0, 1, 2, \ldots.} \tag{6.44}$$

Equation (6.44) gives the pmf of $X(0, t)$, the number of arrivals during time interval $[0, t)$ subject to the assumptions stated above. It is called the *Poisson distribution*, with parameters $\lambda$ and $t$. However, since $\lambda$ and $t$ appear in Equation (6.44) as a product, $\lambda t$, it can be replaced by a single parameter $\nu$, $\nu = \lambda t$, and so we can also write

$$\boxed{p_k(0, t) = \frac{\nu^k e^{-\nu}}{k!}, \quad k = 0, 1, 2, \ldots.} \tag{6.45}$$

The mean of $X(0, t)$ is given by

$$\begin{aligned}
E\{X(0, t)\} &= \sum_{k=0}^{\infty} k p_k(0, t) = e^{-\lambda t} \sum_{k=0}^{\infty} \frac{k(\lambda t)^k}{k!} \\
&= \lambda t e^{-\lambda t} \sum_{k=1}^{\infty} \frac{(\lambda t)^{k-1}}{(k-1)!} = \lambda t e^{-\lambda t} e^{\lambda t} = \lambda t.
\end{aligned} \tag{6.46}$$

Similarly, we can show that

$$\sigma_{X(0,t)}^2 = \lambda t. \tag{6.47}$$

It is seen from Equation (6.46) that parameter $\lambda$ is equal to the average number of arrivals per unit interval of time; the name 'mean rate of arrival' for $\lambda$, as mentioned earlier, is thus justified. In determining the value of this parameter in a given problem, it can be estimated from observations by $m/n$,

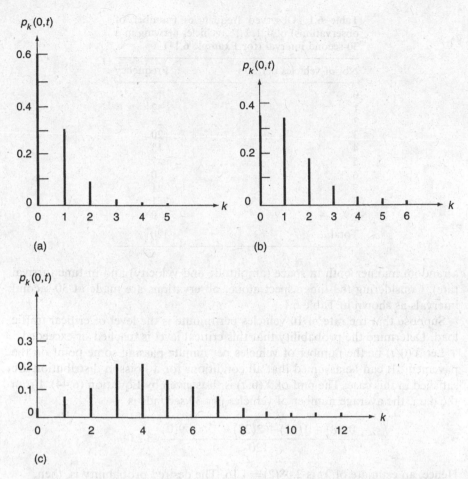

**Figure 6.3**    Poisson distribution $p_k(0, t)$, for several values of $\lambda t$: (a) $\lambda t = 0.5$; (b) $\lambda t = 1.0$; (c) $\lambda t = 4.0$

where $m$ is the observed number of arrivals in $n$ unit time intervals. Similarly, since $\nu = \lambda t$, $\nu$ represents the average number of arrivals in time interval $[0, t)$.

Also it is seen from Equation (6.47) that, as expected, the variance, as well as the mean, increases as the mean rate increases. The Poisson distribution for several values of $\lambda t$ is shown in Figure 6.3. In general, if we examine the ratio of $p_k(0, t)$ and $p_{k-1}(0, t)$, as we did for the binomial distribution, it shows that $p_k(0, t)$ increases monotonically and then decreases monotonically as $k$ increases, reaching its maximum when $k$ is the largest integer not exceeding $\lambda t$.

**Example 6.11.** Problem: traffic load in the design of a pavement system is an important consideration. Vehicles arrive at some point on the pavement in

**Table 6.1**  Observed frequencies (number of observations) of $0, 1, 2, \ldots$ vehicles arriving in a 30-second interval (for Example 6.11)

| No. of vehicles per 30 s | Frequency |
|---|---|
| 0 | 18 |
| 1 | 32 |
| 2 | 28 |
| 3 | 20 |
| 4 | 13 |
| 5 | 7 |
| 6 | 0 |
| 7 | 1 |
| 8 | 1 |
| $\geq 9$ | 0 |
| Total | 120 |

a random manner both in space (amplitude and velocity) and in time (arrival rate). Considering the time aspect alone, observations are made at 30-second intervals as shown in Table 6.1.

Suppose that the rate of 10 vehicles per minute is the level of critical traffic load. Determine the probability that this critical level is reached or exceeded.

Let $X(0, t)$ be the number of vehicles per minute passing some point on the pavement. It can be assumed that all conditions for a Poisson distribution are satisfied in this case. The pmf of $X(0, t)$ is thus given by Equation (6.44). From the data, the average number of vehicles per 30 seconds is

$$\frac{0(18) + 1(32) + 2(28) + \cdots + 9(0)}{120} \cong 2.08.$$

Hence, an estimate of $\lambda t$ is $2.08(2) = 4.16$. The desired probability is, then,

$$P[X(0, t) \geq 10] = \sum_{k=10}^{\infty} p_k(0, t) = 1 - \sum_{k=0}^{9} p_k(0, t)$$

$$= 1 - \sum_{k=0}^{9} \frac{(4.16)^k e^{-4.16}}{k!}$$

$$\cong 1 - 0.992 = 0.008.$$

The calculations involved in Example 6.11 are tedious. Because of its wide applicability, the Poisson distribution for different values of $\lambda t$ is tabulated in the literature. Table A.2 in Appendix A gives its mass function for values of $\lambda t$ ranging from 0.1 to 10. Figure 6.4 is also convenient for determining

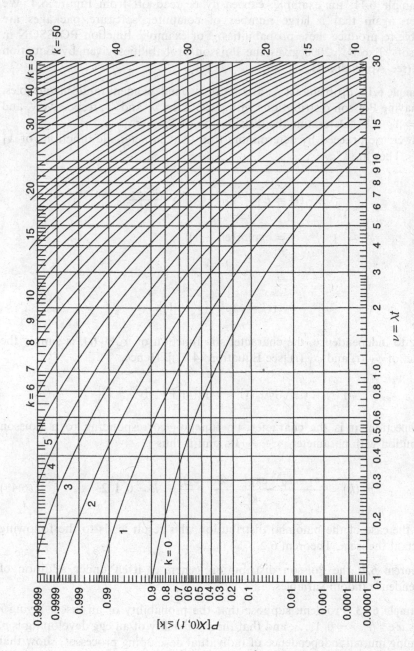

**Figure 6.4**  Probability distribution function of the Poisson distribution (Haight, 1967), Reproduced with permission of John Wiley

the PDF associated with a Poisson–distributed random variable. The answer to Example 6.11, for example, can easily be read off from Figure 6.4. We mention again that a large number of computer software packages are available to produce these probabilities. For example, function POISSON in Microsoft®Excel™ 2000 gives the Poisson probabilities given by Equation (6.44) (see Appendix B).

**Example 6.12.** Problem: let $X_1$ and $X_2$ be two independent random variables, both having Poisson distributions with parameters $\nu_1$ and $\nu_2$, respectively, and let $Y = X_1 + X_2$. Determine the distribution of $Y$.

Answer: we proceed by determining first the characteristic functions of $X_1$ and $X_2$. They are

$$\phi_{X_1}(t) = E\{e^{jtX_1}\} = e^{-\nu_1} \sum_{k=0}^{\infty} \frac{e^{jtk}\nu_1^k}{k!}$$

$$= \exp[\nu_1(e^{jt} - 1)]$$

and

$$\phi_{X_2}(t) = \exp[\nu_2(e^{jt} - 1)].$$

Owing to independence, the characteristic function of $Y$, $\phi_Y(t)$, is simply the product of $\phi_{X_1}(t)$ and $\phi_{X_2}(t)$ [see Equation (4.71)]. Hence,

$$\phi_Y(t) = \phi_{X_1}(t)\phi_{X_2}(t) = \exp[(\nu_1 + \nu_2)(e^{jt} - 1)].$$

By inspection, it is the characteristic function corresponding to a Poisson distribution with parameter $\nu_1 + \nu_2$. Its pmf is thus

$$p_Y(k) = \frac{(\nu_1 + \nu_2)^k \exp[-(\nu_1 + \nu_2)]}{k!}, \quad k = 0, 1, 2, \ldots. \tag{6.48}$$

As in the case of the binomial distribution, this result leads to the following important theorem, Theorem 6.2.

**Theorem 6.2:** the Poisson distribution generates itself under addition of independent random variables.

**Example 6.13.** Problem: suppose that the probability of an insect laying $r$ eggs is $\nu^r e^{-\nu}/r!$, $r = 0, 1, \ldots$, and that the probability of an egg developing is $p$. Assuming mutual independence of individual developing processes, show that the probability of a total of $k$ survivors is $(p\nu)^k e^{-p\nu}/k!$.

Answer: let $X$ be the number of eggs laid by the insect, and $Y$ be the number of eggs developed. Then, given $X = r$, the distribution of $Y$ is binomial with parameters $r$ and $p$. Thus,

$$P(Y = k|X = r) = \binom{r}{k}p^k(1 - p)^{r-k}, k = 0, 1, \ldots, r.$$

Now, using the total probability theorem, Theorem 2.1 [Equation (2.27)],

$$P(Y = k) = \sum_{r=k}^{\infty} P(Y = k|X = r)P(X = r)$$

$$= \sum_{r=k}^{\infty} \binom{r}{k}\frac{p^k(1 - p)^{r-k}\nu^r e^{-\nu}}{r!}. \tag{6.49}$$

If we let $r = k + n$, Equation (6.49) becomes

$$P(Y = k) = \sum_{n=0}^{\infty} \binom{n + k}{k}\frac{p^k(1 - p)^n \nu^{n+k}e^{-\nu}}{(n + k)!}$$

$$= \frac{(p\nu)^k e^{-\nu}}{k!} \sum_{n=0}^{\infty} \frac{(1 - p)^n \nu^n}{n!}$$

$$= \frac{(p\nu)^k e^{-\nu}e^{(1-p)\nu}}{k!} = \frac{(p\nu)^k e^{-p\nu}}{k!}, \quad k = 0, 1, 2, \ldots. \tag{6.50}$$

An important observation can be made based on this result. It implies that, if a random variable $X$ is Poisson distributed with parameter $\nu$, then a random variable $Y$, which is derived from $X$ by selecting only with probability $p$ each of the items counted by $X$, is also Poisson distributed with parameter $p\nu$. Other examples of the application of this result include situations in which $Y$ is the number of disaster-level hurricanes when $X$ is the total number of hurricanes occurring in a given year, or $Y$ is the number of passengers not being able to board a given flight, owing to overbooking, when $X$ is the number of passenger arrivals.

### 6.3.1 SPATIAL DISTRIBUTIONS

The Poisson distribution has been derived based on arrivals developing in time, but the same argument applies to distribution of points in space. Consider the distribution of flaws in a material. The number of flaws in a given volume has a Poisson distribution if Assumptions 1–3 are valid, with time intervals replaced by

**Table 6.2** Comparison of the observed and theoretical distributions of flying-bomb hits, for Example 6.14

| $n_k$ | $k$ | | | | | |
|---|---|---|---|---|---|---|
| | 0 | 1 | 2 | 3 | 4 | $\geq 5$ |
| $n_k^o$ | 229 | 211 | 93 | 35 | 7 | 1 |
| $n_k^p$ | 226.7 | 211.4 | 98.5 | 30.6 | 7.1 | 1.6 |

volumes, *and* if it is reasonable to assume that the probability of finding $k$ flaws in any region depends only on the volume and not on the shape of the region.

Other physical situations in which the Poisson distribution is used include bacteria counts on a Petri plate, the distribution of airplane-spread fertilizers in a field, and the distribution of industrial pollutants in a given region.

**Example 6.14.** A good example of this application is the study carried out by Clark (1946) concerning the distribution of flying-bomb hits in one part of London during World War 2. The area is divided into 576 small areas of $0.25\,\mathrm{km}^2$ each. In Table 6.2, the number $n_k$ of areas with exactly $k$ hits is recorded and is compared with the predicted number based on a Poisson distribution, with $\lambda t =$ number of total hits per number of areas $= 537/576 = 0.932$. We see an excellent agreement between the predicted and observed results.

### 6.3.2   THE POISSON APPROXIMATION TO THE BINOMIAL DISTRIBUTION

Let $X$ be a random variable having the binomial distribution with

$$p_X(k) = \binom{n}{k} p^k (1-p)^{n-k}, \quad k = 0, 1, \ldots, n. \tag{6.51}$$

Consider the case when $n \to \infty$, and $p \to 0$, in such a way that $np = \nu$ remains fixed. We note that $\nu$ is the mean of $X$, which is assumed to remain constant. Then,

$$p_X(k) = \binom{n}{k} \left(\frac{\nu}{n}\right)^k \left(1 - \frac{\nu}{n}\right)^{n-k}, \quad k = 0, 1, \ldots, n. \tag{6.52}$$

As $n \to \infty$, the factorials $n!$ and $(n-k)!$ appearing in the binomial coefficient can be approximated by using the Stirling's formula [Equation (4.78)]. We also note that

$$\lim_{n \to \infty} \left(1 + \frac{c}{n}\right)^n = e^c. \tag{6.53}$$

Using these relationships in Equation (6.52) then gives, after some manipulation,

$$p_X(k) = \frac{\nu^k e^{-\nu}}{k!}, \quad k = 0, 1, \ldots. \tag{6.54}$$

This Poisson approximation to the binomial distribution can be used to advantage from the point of view of computational labor. It also establishes the fact that a close relationship exists between these two important distributions.

**Example 6.15.** Problem: suppose that the probability of a transistor manufactured by a certain firm being defective is 0.015. What is the probability that there is no defective transistor in a batch of 100?

Answer: let $X$ be the number of defective transistors in 100. The desired probability is

$$p_X(0) = \binom{100}{0}(0.015)^0(0.985)^{100-0} = (0.985)^{100} = 0.2206.$$

Since $n$ is large and $p$ is small in this case, the Poisson approximation is appropriate and we obtain

$$p_X(0) = \frac{(1.5)^0 e^{-1.5}}{0!} = e^{-1.5} = 0.223,$$

which is very close to the exact answer. In practice, the Poisson approximation is frequently used when $n > 10$, and $p < 0.1$.

**Example 6.16.** Problem: in oil exploration, the probability of an oil strike in the North Sea is 1 in 500 drillings. What is the probability of having exactly 3 oil-producing wells in 1000 explorations?

Answer: in this case, $n = 1000$, and $p = 1/500 = 0.002$, and the Poisson approximation is appropriate. Using Equation (6.54), we have $\nu = np = 2$, and the desired probability is

$$p_X(3) = \frac{2^3 e^{-2}}{3!} = 0.18.$$

The examples above demonstrate that the Poisson distribution finds applications in problems where the probability of an event occurring is small. For this reason, it is often referred to as the *distribution of rare events*.

## 6.4  SUMMARY

We have introduced in this chapter several discrete distributions that are used extensively in science and engineering. Table 6.3 summarizes some of the important properties associated with these distributions, for easy reference.

**Table 6.3**   Summary of discrete distributions

| Distribution | Probability mass function | Parameters | Mean | Variance |
|---|---|---|---|---|
| Binomial | $\binom{n}{k}p^k(1-p)^{n-k},$ <br> $k = 0, 1, \ldots, n$ | $n, p$ | $np$ | $np(1-p)$ |
| Hypergeometric | $\binom{n_1}{k}\binom{n-n_1}{m-k}\Big/\binom{n}{m},$ <br> $k = 0, 1, \ldots, \min(n_1, m)$ | $n, n_1, m$ | $\dfrac{mn_1}{n}$ | $\dfrac{mn_1(n-n_1)(n-m)}{n^2(n-1)}$ |
| Geometric | $(1-p)^{k-1}p,$ <br> $k = 1, 2, \ldots$ | $p$ | $\dfrac{1}{p}$ | $\dfrac{1-p}{p^2}$ |
| Negative binomial (Pascal) | $\binom{k-1}{r-1}p^r(1-p)^{k-r},$ <br> $k = r, r+1, \ldots$ | $r, p$ | $\dfrac{r}{p}$ | $\dfrac{r(1-p)}{p^2}$ |
| Multinomial | $\dfrac{n!}{k_1!k_2!\cdots k_r!}p_1^{k_1}p_2^{k_2}\cdots p_r^{k_r},$ <br> $k_1, \ldots, k_r = 0, 1, 2, \ldots,$ <br> $\sum_{i=1}^{r}k_i = n, \ \sum_{i=1}^{r}p_i = 1,$ <br> $i = 1, \ldots, r$ | $n, p_i, i = 1, \cdots, r$ | $np_i$ | $np_i(1-p_i)$ |
| Poisson | $\dfrac{(\lambda t)^k e^{-\lambda t}}{k!}, k = 0, 1, 2, \ldots$ | $\lambda t$ | $\lambda t$ | $\lambda t$ |

# FURTHER READING

Clark, R.D., 1946, "An Application of the Poisson Distribution", *J. Inst. Actuaries* **72** 48–52.
Solloway, C.B., 1993, "A Simplified Statistical Model for Missile Launching: III", TM 312–287, Jet Propulsion Laboratory, Pasadena, CA.

Binomial and Poisson distributions are widely tabulated in the literature. Additional references in which these tables can be found are:

Arkin, H., and Colton, R. 1963, *Tables for Statisticians*, 2nd eds, Barnes and Noble, New York.
Beyer, W.H., 1966, *Handbook of Tables for Probability and Statistics*, Chemical Rubber Co., Cleveland, OH.
Grant, E.C., 1964, *Statistical Quality Control*, 3rd eds, McGraw-Hill, New York.
Haight, F.A., 1967, *Handbook of the Poisson Distribution*, John Wiley & Sons Inc., New York.
Hald, A., 1952, *Statistical Tables and Formulas*, John Wiley & Sons Inc., New York.
Molina, E.C., 1949, *Poisson's Exponential Binomial Limit*, Von Nostrand, New York.
National Bureau of Standards, 1949, *Tables of the Binomial Probability Distributions: Applied Mathematics Series 6*, US Government Printing Office, Washington, DC.
Owen, D., 1962, *Handbook of Statistical Tables*, Addison-Wesley, Reading, MA.
Pearson, E.S., and Harley, H.O. (eds), 1954, *Biometrika Tables for Statisticians, Volume 1*, Cambridge University Press, Cambridge, England.

## PROBLEMS

6.1 The random variable $X$ has a binomial distribution with parameters $(n, p)$. Using the formulation given by Equation (6.10), derive its probability mass function (pmf), mean, and variance and compare them with results given in Equations (6.2) and (6.8). (Hint: see Example 4.18, page 109).

6.2 Let $X$ be the number of defective parts produced on a certain production line. It is known that, for a given lot, $X$ is binomial, with mean equal to 240, and variance 48. Determine the pmf of $X$ and the probability that none of the parts is defective in this lot.

6.3 An experiment is repeated 5 times. Assuming that the probability of an experiment being successful is 0.75 and assuming independence of experimental outcomes:
   (a) What is the probability that all five experiments will be successful?
   (b) How many experiments are expected to succeed on average?

6.4 Suppose that the probability is 0.2 that the air pollution level in a given region will be in the unsafe range. What is the probability that the level will be unsafe 7 days in a 30-day month? What is the average number of 'unsafe' days in a 30-day month?

6.5 An airline estimates that 5% of the people making reservations on a certain flight will not show up. Consequently, their policy is to sell 84 tickets for a flight that can only hold 80 passengers. What is the probability that there will be a seat available for every passenger that shows up? What is the average number of no-shows?

6.6 Assuming that each child has probability 0.51 of being a boy:
   (a) Find the probability that a family of four children will have (i) exactly one boy, (ii) exactly one girl, (iii) at least one boy, and (iv) at least one girl.
   (b) Find the number of children a couple should have in order that the probability of their having at least two boys will be greater than 0.75.

6.7  Suppose there are five customers served by a telephone exchange and that each customer may demand one line or none in any given minute. The probability of demanding one line is 0.25 for each customer, and the demands are independent.
   (a) What is the probability distribution function of $X$, a random variable representing the number of lines required in any given minute?
   (b) If the exchange has three lines, what is the probability that the customers will all be satisfied?

6.8  A park-by-permit-only facility has $m$ parking spaces. A total of $n$ $(n \geq m)$ parking permits are issued, and each permit holder has a probability $p$ of using the facility in a given period.
   (a) Determine the probability that a permit holder will be denied a parking space in the given time period.
   (b) Determine the expected number of people turned away in the given time period.

6.9  For the hypergeometric distribution given by Equation (6.13), show that as $n \to \infty$ it approaches the binomial distribution with parameters $m$ and $n_1/n$; that is,

$$p_Z(k) = \binom{m}{k}\left(\frac{n_1}{n}\right)^k\left(1 - \frac{n_1}{n}\right)^{m-k}, \quad k = 0, 1, \ldots, m.$$

and thus that the hypergeometric distribution can be approximated by a binomial distribution as $n \to \infty$.

6.10 A manufacturing firm receives a lot of 100 parts, of which 5 are defective. Suppose that the firm accepts all 100 parts if and only if no defective ones are found in a sample of 10 parts randomly selected for inspection. Determine the probability that this lot will be accepted.

6.11 A shipment of 10 boxes of meat contains 2 boxes of contaminated goods. An inspector randomly selects 4 boxes; let $Z$ be the number of boxes of contaminated meat among the selected 4 boxes.
(a) What is the pmf of $Z$?
(b) What is the probability that at least one of the four boxes is contaminated?
(c) How many boxes must be selected so that the probability of having at last one contaminated box is larger than 0.75?

6.12 In a sequence of Bernoulli trials with probability $p$ of success, determine the probability that $r$ successes will occur before $s$ failures.

6.13 Cars arrive independently at an intersection. Assuming that, on average, 25% of the cars turn left and that the left-turn lane has a capacity for 5 cars, what is the probability that capacity is reached in the left-turn lane when 10 cars are delayed by a red signal?

6.14 Suppose that $n$ independent steps must be taken in the sterilization procedure for a biological experiment, each of which has a probability $p$ of success. If a failure in any of the $n$ steps would cause contamination, what is the probability of contamination if $n = 10$ and $p = 0.99$?

6.15 An experiment is repeated in a civil engineering laboratory. The outcomes of these experiments are considered independent, and the probability of an experiment being successful is 0.7.
(a) What is the probability that no more than 6 attempts are necessary to produce 3 successful experiments?
(b) What is the average number of failures before 3 successful experiments occur?
(c) Suppose one needs 3 *consecutive* successful experiments. What is the probability that exactly 6 attempts are necessary?

6.16 The definition of the 100-year flood is given in Example 6.7.
(a) Determine the probability that exactly one flood equaling or exceeding the 100-year flood will occur in a 100-year period.
(b) Determine the probability that one or more floods equaling or exceeding the 100-year flood will occur in a 100-year period.

6.17 A shipment of electronic parts is sampled by testing items sequentially until the first defective part is found. If 10 or more parts are tested before the first defective part is found, the shipment is accepted as meeting specifications.
(a) Determine the probability that the shipment will be accepted if it contains 10% defective parts.
(b) How many items need to be sampled if it is desired that a shipment with 25% defective parts be rejected with probability of at least 0.75?

6.18 Cars enter an interchange from the south. On average, 40% want to go west, 10% east, and 50% straight on (north). Of 8 cars entering the interchange:
(a) Determine the joint probability mass function (jpmf) of $X_1$ (cars westbound), $X_2$ (cars eastbound), and $X_3$ (cars going straight on).
(b) Determine the probability that half will go west and half will go east.
(c) Determine the probability that more than half will go west.

6.19 For Example 6.10, determine the jpmf of $X_1$ and $X_2$. Determine the probability that, of the 10 persons studied, fewer than 2 persons will be in the low-income group and fewer than 3 persons will be in the middle-income group.

6.20 The following describes a simplified countdown procedure for launching 3 space vehicles from 2 pads:

- Two vehicles are erected simultaneously on two pads and the countdown proceeds on one vehicle.
- When the countdown has been successfully completed on the first vehicle, the countdown is initiated on the second vehicle, the following day.
- Simultaneously, the vacated pad is immediately cleaned and prepared for the third vehicle. There is a (fixed) period of $r$ days delay after the launching before the same pad may be utilized for a second launch attempt (the turnaround time).
- After the third vehicle is erected on the vacated pad, the countdown procedure is *not* initiated until the day after the second vehicle is launched.
- Each vehicle is independent of, and identical to, the others. On any single countdown attempt there is a probability $p$ of a successful completion and a probability $q$ ($q = 1 - p$) of failure. Any failure results in the termination of that countdown attempt and a new attempt is made the *following day*. That is, any failure leads to a one-day delay. It is assumed that a successful countdown attempt can be completed in one day.
- The failure to complete a countdown does not affect subsequent attempts in any way; that is, the trials are independent from day to day as well as from vehicle to vehicle.

Let $X$ be the number of days until the third successful countdown. Show that the pmf of $X$ is given by:

$$p_X(k) = (k - r - 1)p^2 q^{k-r-2}(1 - q^{r-1}) + \frac{(k - r)!}{2(k - r - 2)!}p^3 q^{k-3}, k = r + 2, r + 3, \ldots.$$

6.21 Derive the variance of a Poisson-distributed random variable $X$ as given by Equation (6.47).

6.22 Show that, for the Poisson distribution, $p_k(0, t)$ increases monotonically and then decreases monotonically as $k$ increases, reaching its maximum when $k$ is the largest integer not exceeding $\lambda t$.

6.23 At a certain plant, accidents have been occurring at an average rate of 1 every 2 months. Assume that the accidents occur independently.
(a) What is the average number of accidents per year?
(b) What is the probability of there being no accidents in a given month?

6.24 Assume that the number of traffic accidents in New York State during a 4-day memorial day weekend is Poisson-distributed with parameter $\lambda = 3.25$ per day. Determine the probability that the number of accidents is less than 10 in this 4-day period.

6.25 A radioactive source is observed during 7 time intervals, each interval being 10 seconds in duration. The number of particles emitted during each period is counted. Suppose that the number of particles emitted, say $X$, during each observed period has an average rate of 0.5 particles per second.
(a) What is the probability that 4 or more particles are emitted in each interval?
(b) What is the probability that in at least 1 of 7 time intervals, 4 or more particles are emitted?

6.26 Each air traffic controller at an airport is given the responsibility of monitoring at most 20 takeoffs and landings per hour. During a given period, the average rate of takeoffs and landings is 1 every 2 minutes. Assuming Poisson arrivals and departures, determine the probability that 2 controllers will be needed in this time period.

6.27 The number of vehicles crossing a certain point on a highway during a unit time period has a Poisson distribution with parameter $\lambda$. A traffic counter is used to record this number but, owing to limited capacity, it registers the maximum number of 30 whenever the count equals or exceeds 30. Determine the pmf of $Y$ if $Y$ is the number of vehicles recorded by the counter.

6.28 As an application of the Poisson approximation to the binomial distribution, estimate the probability that in a class of 200 students exactly 20 will have birthdays on any given day.

6.29 A book of 500 pages contains on average 1 misprint per page. Estimate the probability that:
(a) A given page contains at least 1 misprint.
(b) At least 3 pages will contain at least 1 misprint.

6.30 Earthquakes are registered at an average frequency of 250 per year in a given region. Suppose that the probability is 0.09 that any earthquake will have a magnitude greater than 5 on the Richter scale. Assuming independent occurrences of earthquakes, determine the pmf of $X$, the number of earthquakes greater than 5 on the Richter scale per year.

6.31 Let $X$ be the number of accidents in which a driver is involved in $t$ years. In proposing a distribution for $X$, the 'accident likelihood' $\Lambda$ varies from driver to driver and is considered as a random variable. Suppose that the conditional pmf $p_{X\Lambda}(x|\lambda)$ is given by the Poisson distribution,

$$p_{X\Lambda}(k|\lambda) = \frac{(\lambda t)^k e^{-\lambda t}}{k!}, \quad k = 0, 1, 2, \ldots,$$

and suppose that the probability density function (pdf) of $\Lambda$ is of the form $(a, b > 0)$

$$f_\Lambda(\lambda) = \begin{cases} \dfrac{a}{b\Gamma(a)} \left(\dfrac{a\lambda}{b}\right)^{a-1} e^{-a\lambda/b}, & \text{for } \lambda \geq 0, \\ 0, & \text{elsewhere,} \end{cases}$$

where $\Gamma(a)$ is the gamma function, defined by

$$\Gamma(a) = \int_0^\infty x^{a-1} e^{-x} dx.$$

Show that the pmf of $X$ has a negative binomial distribution in the form

$$p_X(k) = \frac{\Gamma(a+k)}{k!\Gamma(a)} \left(\frac{a}{a+bt}\right)^a \left(\frac{bt}{a+bt}\right)^k, \quad k = 0, 1, 2, \ldots.$$

6.32 Suppose that $\lambda$, the mean rate of arrival, in the Poisson distribution is time-dependent and is given by

$$\lambda = \frac{vt^{v-1}}{w}.$$

Determine pmf $p_k(0, t)$, the probability of exactly $k$ arrivals in the time interval $[0, t)$. [Note that differential equations such as Equations (6.39) and (6.42) now have time-dependent coefficients.]

6.33 Derive the jpmf of two Poisson random variables $X_1$ and $X_2$, where $X_1 = X(0, t_1)$, and $X_2 = X(0, t_2), t_2 > t_1$, with the same mean rate of arrival $\lambda$. Determine probability $P(X_1 \le \lambda t_1 \cap X_2 \le \lambda t_2)$. This is the probability that the numbers of arrivals in intervals $[0, t_1)$ and $[0, t_2)$ are both equal to or less than the average arrivals in their respective intervals.

# 7

# Some Important Continuous Distributions

Let us turn our attention to some important continuous probability distributions. Physical quantities such as time, length, area, temperature, pressure, load, intensity, etc., when they need to be described probabilistically, are modeled by continuous random variables. A number of important continuous distributions are introduced in this chapter and, as in Chapter 6, we are also concerned with the nature and applications of these distributions in science and engineering.

## 7.1 UNIFORM DISTRIBUTION

A continuous random variable $X$ has a *uniform distribution* over an interval $a$ to $b(b > a)$ if it is equally likely to take on any value in this interval. The probability density function (pdf) of $X$ is constant over interval $(a, b)$ and has the form

$$
f_X(x) = \begin{cases} \dfrac{1}{b-a}, & \text{for } a \leq x \leq b; \\ 0, & \text{elsewhere.} \end{cases} \tag{7.1}
$$

As we see from Figure 7.1(a), it is constant over $(a, b)$, and the height must be $1/(b - a)$ in order that the area under the density function is unity.

The probability distribution function (PDF) is, on integrating Equation (7.1),

$$
F_X(x) = \begin{cases} 0, & \text{for } x < a; \\ \dfrac{x-a}{b-a}, & \text{for } a \leq x \leq b; \\ 1, & \text{for } x > b; \end{cases} \tag{7.2}
$$

*Fundamentals of Probability and Statistics for Engineers* T.T. Soong © 2004 John Wiley & Sons, Ltd
ISBNs: 0-470-86813-9 (HB)    0-470-86814-7 (PB)

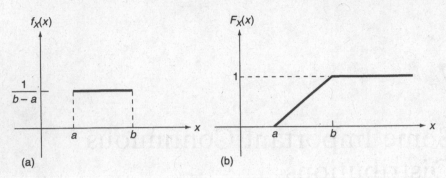

**Figure 7.1**  (a) The probability density function, $f_X(x)$, and (b) the probability distribution function, $F_X(x)$, of $X$

which is graphically presented in Figure 7.1(b).

The mean, $m_X$, and variance, $\sigma_X^2$, of $X$ are easily found to be

$$m_X = \int_a^b x f_X(x) \mathrm{d}x = \frac{1}{b-a} \int_a^b x \mathrm{d}x = \frac{a+b}{2};$$

$$\sigma_X^2 = \frac{1}{b-a} \int_a^b \left( x - \frac{a+b}{2} \right)^2 \mathrm{d}x = \frac{(b-a)^2}{12}.$$

(7.3)

The uniform distribution is one of the simplest distributions and is commonly used in situations where there is no reason to give unequal likelihoods to possible ranges assumed by the random variable over a given interval. For example, the arrival time of a flight might be considered uniformly distributed over a certain time interval, or the distribution of the distance from the location of live loads on a bridge to an end support might be adequately represented by a uniform distribution over the bridge span. Let us also comment that one often assigns a uniform distribution to a specific random variable simply because of a lack of information, beyond knowing the range of values it spans.

**Example 7.1.** Problem: owing to unpredictable traffic situations, the time required by a certain student to travel from her home to her morning class is uniformly distributed between 22 and 30 minutes. If she leaves home at precisely 7.35 a.m., what is the probability that she will not be late for class, which begins promptly at 8:00 a.m.?

Answer: let $X$ be the class arrival time of the student in minutes after 8:00 a.m. It then has a uniform distribution given by

$$f_X(x) = \begin{cases} \dfrac{1}{8}, & \text{for } -3 \leq x \leq 5; \\ 0, & \text{elsewhere}. \end{cases}$$

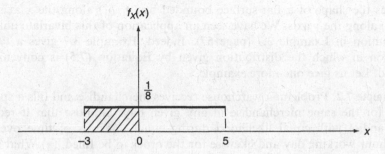

**Figure 7.2** Probability density function, $f_X(x)$, of $X$, in Example 7.1

We are interested in the probability $P(-3 \leq X \leq 0)$. As seen from Figure 7.2, it is clear that this probability is equal to the ratio of the shaded area and the unit total area. Hence,

$$P(-3 \leq X \leq 0) = 3\left(\frac{1}{8}\right) = \frac{3}{8}.$$

It is also clear that, owing to uniformity in the distribution, the solution can be found simply by taking the ratio of the length from $-3$ to $0$ to the total length of the distribution interval. Stated in general terms, if a random variable $X$ is uniformly distributed over an interval $A$, then the probability of $X$ taking values in a subinterval $B$ is given by

$$P(X \text{ in } B) = \frac{\text{length of } B}{\text{length of } A}. \tag{7.4}$$

### 7.1.1   BIVARIATE UNIFORM DISTRIBUTION

Let random variable $X$ be uniformly distributed over an interval $(a_1, b_1)$, and let random variable $Y$ be uniformly distributed over an interval $(a_2, b_2)$. Furthermore, let us assume that they are independent. Then, the joint probability density function of $X$ and $Y$ is simply

$$f_{XY}(x, y) = f_X(x) f_Y(y) = \begin{cases} \dfrac{1}{(b_1 - a_1)(b_2 - a_2)}, & \text{for } a_1 \leq x \leq b_1, \text{ and } a_2 \leq y \leq b_2; \\ 0, & \text{elsewhere.} \end{cases}$$

$$\tag{7.5}$$

It takes the shape of a flat surface bounded by $(a_1, b_1)$ along the $x$ axis and $(a_2, b_2)$ along the $y$ axis. We have seen an application of this bivariate uniform distribution in Example 3.7 (page 57). Indeed, Example 3.7 gives a typical situation in which the distribution given by Equation (7.5) is conveniently applied. Let us give one more example.

**Example 7.2.** Problem: a warehouse receives merchandise and fills a specific order for the same merchandise in any given day. Suppose that it receives merchandise with equal likelihood during equal intervals of time over the eight-hour working day and likewise for the order to be filled. (a) What is the probability that the order will arrive after the merchandise is received and (b) what is the probability that the order will arrive within two hours after the receipt of merchandise?

Answer: let $X$ be the time of receipt of merchandise expressed as a fraction of an eight-hour working day, and let $Y$ be the time of receipt of the order similarly expressed. Then

$$f_X(x) = \begin{cases} 1, & \text{for } 0 \leq x \leq 1; \\ 0, & \text{elsewhere;} \end{cases} \tag{7.6}$$

and similarly for $f_Y(y)$. The joint probability density function (jpdf) of $X$ and $Y$ is, assuming independence,

$$f_{XY}(x, y) = \begin{cases} 1, & \text{for } 0 \leq x \leq 1, \text{ and } 0 \leq y \leq 1; \\ 0, & \text{elsewhere;} \end{cases}$$

and is shown in Figure 7.3.

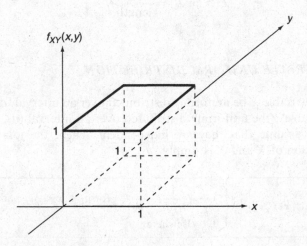

**Figure 7.3**   Joint probability density function, $f_{XY}(x,y)$, of $X$ and $Y$ in Example 7.2

To answer the first question, in part (a), we integrate $f_{XY}(x, y)$ over an appropriate region in the $(x, y)$ plane satisfying $y \geq x$. Since $f_{XY}(x, y)$ is a constant over $(0, 0) \leq (x, y) \leq (1, 1)$, this is the same as taking the ratio of the area satisfying $y \geq x$ to the total area bounded by $(0, 0) \leq (x, y) \leq (1, 1)$, which is unity. As seen from Figure 7.4(a), we have

$$P(Y \geq X) = \text{shaded area } A = \frac{1}{2}.$$

We proceed the same way in answering the second question, in part (b). It is easy to see that the appropriate region for this part is the shaded area $B$, as shown in Figure 7.4(b). The desired probability is, after dividing area $B$ into the two subregions as shown,

$$P\left(X \leq Y \leq X + \frac{1}{4}\right) = \text{shaded area } B$$

$$= \frac{1}{4}\left(\frac{3}{4}\right) + \frac{1}{2}\left(\frac{1}{4}\right)\left(\frac{1}{4}\right) = \frac{7}{32}.$$

We see from Example 7.2 that calculations of various probabilities of interest in this situation involve taking ratios of appropriate areas. If random variables $X$ and $Y$ are independent and uniformly distributed over a region $A$, then the probability of $X$ and $Y$ taking values in a subregion $B$ is given by

$$P[(X, Y) \text{in } B] = \frac{\text{area of } B}{\text{area of } A}. \tag{7.7}$$

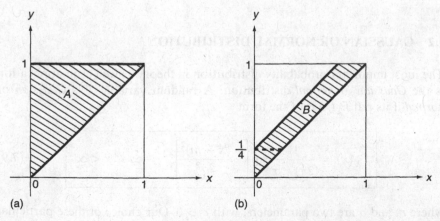

**Figure 7.4**   (a) Region $A$ and (b) region $B$ in the $(x, y)$ plane in Example 7.2

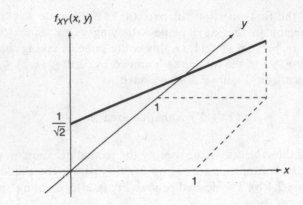

**Figure 7.5** Joint probability density function, $f_{XY}(x,y)$, of $X$ and $Y$, given by Equation (7.8)

It is noteworthy that, if the independence assumption is removed, the jpdf of two uniformly distributed random variables will not take the simple form as given by Equation 7.5. In the extreme case when $X$ and $Y$ are perfectly correlated, the jpdf of $X$ and $Y$ degenerates from a surface into a line over the $(x, y)$ plane. For example, let $X$ and $Y$ be uniformly and identically distributed over the interval $(0, 1)$ and let $X = Y$. Then the jpdf and $X$ and $Y$ has the form

$$f_{XY}(x,y) = \frac{1}{\sqrt{2}}, \quad x = y, \text{ and } (0,0) \leq (x,y) \leq (1,1), \qquad (7.8)$$

which is graphically presented in Figure 7.5. More detailed discussions on correlated and uniformly distributed random variables can be found in Kramer (1940).

## 7.2 GAUSSIAN OR NORMAL DISTRIBUTION

The most important probability distribution in theory as well as in application is the *Gaussian* or *normal* distribution. A random variable $X$ is *Gaussian* or *normal* if its pdf $f_X(x)$ is of the form

$$f_X(x) = \frac{1}{(2\pi)^{1/2}\sigma} \exp\left[-\frac{(x-m)^2}{2\sigma^2}\right], \quad -\infty < x < \infty \qquad (7.9)$$

where $m$ and $\sigma$ are two parameters, with $\sigma > 0$. Our choice of these particular symbols for the parameters will become clear presently.

Its corresponding PDF is

$$F_X(x) = \frac{1}{(2\pi)^{1/2}\sigma} \int_{-\infty}^{x} \exp\left[-\frac{(u-m)^2}{2\sigma^2}\right] du, \quad -\infty < x < \infty, \qquad (7.10)$$

which cannot be expressed in closed form analytically but can be numerically evaluated for any $x$.

The pdf and PDF expressed by Equations (7.9) and (7.10), respectively, are graphed in Figures 7.6(a) and 7.6(b), respectively, for $m = 0$ and $\sigma = 1$. The

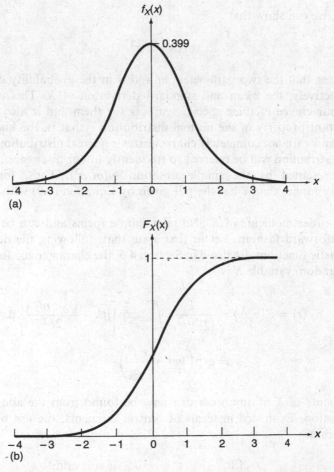

**Figure 7.6** (a) Probability density function, $f_X(x)$, and (b) probability distribution function, $F_X(x)$, of $X$ for $m = 0$ and $\sigma = 1$

graph of $f_X(x)$ in this particular case is the well-known bell-shaped curve, symmetrical about the origin [Figure 7.6(a)].

Let us determine the mean and variance of $X$. By definition, the mean of $X$, $E\{X\}$, is given by

$$E\{X\} = \int_{-\infty}^{\infty} x f_X(x) dx = \frac{1}{(2\pi)^{1/2}\sigma} \int_{-\infty}^{\infty} x \exp\left[-\frac{(x-m)^2}{2\sigma^2}\right] dx,$$

which yields

$$E\{X\} = m.$$

Similarly, we can show that

$$\mathrm{var}(X) = \sigma^2. \tag{7.11}$$

We thus see that the two parameters $m$ and $\sigma$ in the probability distribution are, respectively, the mean and standard derivation of $X$. This observation justifies our choice of these special symbols for them and it also points out an important property of the normal distribution – that is, the knowledge of its mean and variance completely characterizes a normal distribution. Since the normal distribution will be referred to frequently in our discussion, it is sometimes represented by the simple notation $N(m, \sigma^2)$. Thus, for example, $X$: $N(0, 9)$ implies that $X$ has the pdf given by Equation (7.9) with $m = 0$ and $\sigma = 3$.

Higher-order moments of $X$ also take simple forms and can be derived in a straightforward fashion. Let us first state that, following the definition of characteristic functions discussed in Section 4.5, the characteristic function of a normal random variable $X$ is

$$\phi_X(t) = E\{e^{jtX}\} = \frac{1}{(2\pi)^{1/2}\sigma} \int_{-\infty}^{\infty} \exp\left[jtx - \frac{(x-m)^2]}{2\sigma^2}\right] dx$$

$$= \exp\left(jmt - \frac{\sigma^2 t^2}{2}\right), \tag{7.12}$$

The moments of $X$ of any order can now be found from the above through differentiation. Expressed in terms of central moments, the use of Equation (4.52) gives us

$$\mu_n = \begin{cases} 0, & \text{if } n \text{ is odd;} \\ 1(3)\cdots(n-1)\sigma^n, & \text{if } n \text{ is even.} \end{cases} \tag{7.13}$$

Let us note in passing that $\gamma_2$, the coefficient of excess, defined by Equation (4.12), for a normal distribution is zero. Hence, it is used as the reference distribution for $\gamma_2$.

## 7.2.1  THE CENTRAL LIMIT THEOREM

The great practical importance associated with the normal distribution stems from the powerful central limit theorem stated below (Theorem 7.1). Instead of giving the theorem in its entire generality, it serves our purposes quite well by stating a more restricted version attributable to Lindberg (1922).

**Theorem 7.1: the central limit theorem.** Let $\{X_n\}$ be a sequence of mutually independent and identically distributed random variables with means $m$ and variances $\sigma^2$. Let

$$Y = \sum_{j=1}^{n} X_j, \qquad (7.14)$$

and let the normalized random variable $Z$ be defined as

$$Z = \frac{(Y - nm)}{n^{1/2}\sigma}. \qquad (7.15)$$

Then the probability distribution function of $Z$, $F_Z(z)$, converges to $N(0, 1)$ as $n \to \infty$ for every fixed $z$.

**Proof of Theorem 7.1:** We first remark that, following our discussion in Section 4.4 on moments of sums of random variables, random variable $Y$ defined by Equation (7.14) has mean $nm$ and standard deviation $n^{1/2}\sigma$. Hence, $Z$ is simply the standardized random variable $Y$ with zero mean and unit standard deviation. In terms of characteristic functions $\phi_X(t)$ of random variables $X_j$, the characteristic function of $Y$ is simply

$$\phi_Y(t) = [\phi_X(t)]^n. \qquad (7.16)$$

Consequently, $Z$ possesses the characteristic function

$$\phi_Z(t) = \left[ \exp\left( -\frac{jmt}{n^{1/2}\sigma} \right) \phi_X\left( \frac{t}{n^{1/2}\sigma} \right) \right]^n. \qquad (7.17)$$

Expanding $\phi_X(t)$ in a MacLaurin series as indicated by Equation (4.49), we can write

$$
\begin{aligned}
\phi_Z(t) &= \left\{ \exp\left(-\frac{jmt}{n^{1/2}\sigma}\right) \left[ 1 + \frac{mjt}{n^{1/2}\sigma} + \frac{(\sigma^2 + m^2)}{2}\left(\frac{jt}{n^{1/2}\sigma}\right)^2 + \cdots \right] \right\}^n \\
&= \left[ 1 - \frac{t^2}{2n} + \mathrm{o}\left(\frac{t^2}{n}\right) \right]^n \hspace{3cm} (7.18) \\
&\to \exp\left(\frac{-t^2}{2}\right), \quad \text{as } n \to \infty.
\end{aligned}
$$

In the last step we have used the elementary identity

$$
\lim_{n\to\infty} \left(1 + \frac{c}{n}\right)^n = \mathrm{e}^c \hspace{3cm} (7.19)
$$

for any real $c$.

Comparing the result given by Equation (7.18) with the form of the characteristic function of a normal random variable given by Equation (7.12), we see that $\phi_Z(t)$ approaches the characteristic function of the zero-mean, unit-variance normal distribution. The proof is thus complete.

As we mentioned earlier, this result is a somewhat restrictive version of the central limit theorem. It can be extended in several directions, including cases in which $Y$ is a sum of dependent as well as nonidentically distributed random variables.

The central limit theorem describes a very general class of random phenomena for which distributions can be approximated by the normal distribution. In words, when the randomness in a physical phenomenon is the cumulation of many small additive random effects, it tends to a normal distribution irrespective of the distributions of individual effects. For example, the gasoline consumption of all automobiles of a particular brand, supposedly manufactured under identical processes, differs from one automobile to another. This randomness stems from a wide variety of sources, including, among other things: inherent inaccuracies in manufacturing processes, nonuniformities in materials used, differences in weight and other specifications, difference in gasoline quality, and different driver behavior. If one accepts the fact that each of these differences contribute to the randomness in gasoline consumption, the central limit theorem tells us that it tends to a normal distribution. By the same reasoning, temperature variations in a room, readout errors associated with an instrument, target errors of a certain weapon, and so on can also be reasonably approximated by normal distributions.

Let us also mention that, in view of the central limit theorem, our result in Example 4.17 (page 106) concerning a one-dimensional random walk should be of no surprise. As the number of steps increases, it is expected that position of the particle becomes normally distributed in the limit.

### 7.2.2   PROBABILITY TABULATIONS

Owing to its importance, we are often called upon to evaluate probabilities associated with a normal random variable $X: N(m, \sigma^2)$, such as

$$P(a < X \le b) = \frac{1}{(2\pi)^{1/2}\sigma} \int_a^b \exp\left[-\frac{(x-m)^2}{2\sigma^2}\right] dx. \qquad (7.20)$$

However, as we commented earlier, the integral given above cannot be evaluated by analytical means and is generally performed numerically. For convenience, tables are provided that enable us to determine probabilities such as the one expressed by Equation (7.20).

The tabulation of the PDF for the normal distribution with $m = 0$ and $\sigma = 1$ is given in Appendix A, Table A.3. A random variable with distribution $N(0, 1)$ is called a *standardized* normal random variable, and we shall denote it by $U$. Table A.3 gives $F_U(u)$ for points in the right half of the distribution only (i.e. for $u \ge 0$). The corresponding values for $u < 0$ are obtained from the symmetry property of the standardized normal distribution [see Figure 7.6(a)] by the relationship

$$F_U(-u) = 1 - F_U(u). \qquad (7.21)$$

First, Table A.3 in conjunction with Equation (7.21) can be used to determine $P(a < U \le b)$ for any $a$ and $b$. Consider, for example, $P(-1.5 < U \le 2.5)$. It is given by

$$P(-1.5 < U \le 2.5) = F_U(2.5) - F_U(-1.5).$$

The value of $F_U(2.5)$ is found from Table A.3 to be 0.9938; $F_U(-1.5)$ is equal to $1 - F_U(1.5)$, with $F_U(1.5) = 0.9332$, as seen from Table A.3. Thus

$$P(-1.5 < U \le 2.5) = F_U(2.5) - [1 - F_U(1.5)]$$
$$= 0.994 - 1 + 0.933 = 0.927.$$

More importantly, Table A.3 and Equation (7.21) are also sufficient for determining probabilities associated with normal random variables with arbitrary means and variances. To do this, let us first state Theorem 7.2.

**Theorem 7.2:** Let $X$ be a normal random variable with distribution $N(m, \sigma^2)$. Then $(X - m)/\sigma$ is the standardized normal random variable with distribution $N(0, 1)$, or

$$U = \frac{X - m}{\sigma}. \tag{7.22}$$

**Proof of Theorem 7.2:** the characteristic function of random variable $(X - m)/\sigma$ is

$$E\left\{\exp\left[\frac{jt(X - m)}{\sigma}\right]\right\} = \exp\left(\frac{-jtm}{\sigma}\right) E\left\{\exp\left(\frac{jtX}{\sigma}\right)\right\} = \exp\left(\frac{-jtm}{\sigma}\right) \phi_X(t/\sigma).$$

From Equation (7.12) we have

$$\phi_X(t) = \exp\left(jmt - \frac{\sigma^2 t^2}{2}\right). \tag{7.23}$$

Hence,

$$E\left\{\exp\left[\frac{jt(X - m)}{\sigma}\right]\right\} = \exp\left[-\frac{jmt}{\sigma} + \frac{jmt}{\sigma} - \frac{\sigma^2}{2}\left(\frac{t}{\sigma}\right)^2\right]$$

$$= \exp\left(\frac{-t^2}{2}\right). \tag{7.24}$$

The result given above takes the form of $\phi_X(t)$ with $m = 0$ and $\sigma = 1$, and the proof is complete.

Theorem 7.2 implies that

$$P(a < X \le b) = P[a < (U\sigma + m) \le b] = P\left(\frac{a - m}{\sigma} < U \le \frac{b - m}{\sigma}\right). \tag{7.25}$$

The value of the right-hand side can now be found from Table A.3, with the aid of Equation (7.21) if necessary.

As has been noted, probabilities provided by Table A.3 can also be obtained from a number of computer software packages such as Microsoft® Excel™ 2000 (see Appendix B).

**Example 7.3.** Problem: owing to many independent error sources, the length of a manufactured machine part is normally distributed with $m = 11$ cm and $\sigma = 0.2$ cm. If specifications require that the length be between 10.6 cm and 11.2 cm, what proportion of the manufactured parts will be rejected on average?

Answer: If $X$ is used to denote the part length in centimeters, it is reasonable to assume that it is distributed according to N(11, 0.04). Thus, on average, the proportion of acceptable parts is $P(10.6 < X \leq 11.2)$. From Equation (7.25), and using Table A.3, we have

$$P(10.6 < X \leq 11.2) = P\left(\frac{10.6 - 11}{0.2} < U \leq \frac{11.2 - 11}{0.2}\right)$$

$$= P(-2 < U \leq 1) = F_U(1) - [1 - F_U(2)]$$

$$= 0.8413 - (1 - 0.9772) = 0.8185.$$

The desired answer is then $1 - 0.8185$, which gives 0.1815.

The use of the normal distribution in Example 7.3 raises an immediate concern. Normal random variables assume values in positive and negative ranges, whereas the length of a machine part as well as many other physical quantities cannot take negative values. However, from a modeling point of view, it is a commonly accepted practice that normal random variables are valid representations for nonnegative quantities in as much as probability $P(X < 0)$ is sufficiently small. In Example 7.3, for example, this probability is

$$P(X < 0) = P\left(U < -\frac{11}{0.2}\right) = P(U < -55) \cong 0$$

**Example 7.4.** Let us compute $P(m - k\sigma < X \leq m + k\sigma)$ where $X$ is distributed N$(m, \sigma^2)$. It follows from Equations (7.21) and (7.25) that

$$P(m - k\sigma < X \leq m + k\sigma) = P(-k < U \leq k)$$

$$= F_U(k) - F_U(-k) = 2F_U(k) - 1. \qquad (7.26)$$

We note that the result in Example 7.4 is independent of $m$ and $\sigma$ and is a function only of $k$. Thus, the probability that $X$ takes values within $k$ standard deviations about its expected value depends only on $k$ and is given by Equation (7.26). It is seen from Table A.3 that 68.3%, 95.5%, and 99.7% of the area under a normal density function are located, respectively, in the ranges $m \pm \sigma$, $m \pm 2\sigma$, and $m \pm 3\sigma$. This is illustrated in Figures 7.7(a)–7.7(c). For example, the chances are about 99.7% that a randomly selected sample from a normal distribution is within the range of $m \pm 3\sigma$ [Figure 7.7(c)].

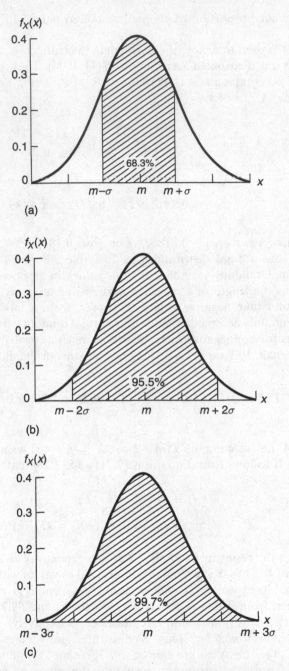

**Figure 7.7**   The Area under the normal density function within the range (a) $m \pm \sigma$, (b) $m \pm 2\sigma$, and (c) $m \pm 3\sigma$

### 7.2.3   MULTIVARIATE NORMAL DISTRIBUTION

Consider two random variables $X$ and $Y$. They are said to be *jointly normal* if their joint density function takes the form

$$
f_{XY}(x,y) = \frac{1}{2\pi\sigma_X\sigma_Y(1-\rho^2)^{1/2}} \exp\left\{-\frac{1}{2(1-\rho^2)}\left[\left(\frac{x-m_X}{\sigma_X}\right)^2\right.\right.
$$
$$
\left.\left. - 2\rho\frac{(x-m_X)(y-m_Y)}{\sigma_X\sigma_Y} + \left(\frac{y-m_Y}{\sigma_Y}\right)^2\right]\right\},  \tag{7.27}
$$
$$
(-\infty, -\infty) < (x,y) < (\infty, \infty).
$$

Equation (7.27) describes the *bivariate normal distribution*. There are five parameters associated with it: $m_X, m_Y, \sigma_X$ (greater than 0), $\sigma_Y$ (greater than 0), and $\rho$ ($|\rho| \leq 1$). A typical plot of this joint density function is given in Figure 7.8.

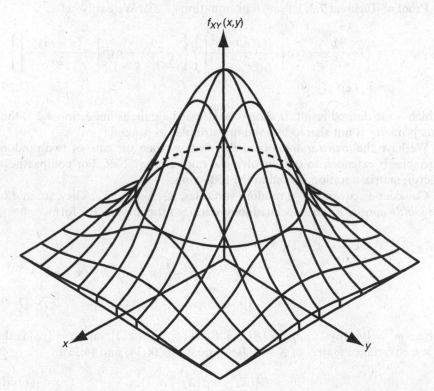

**Figure 7.8**   Bivariate normal distribution with $m_X = m_Y = 0$ and $\sigma_X = \sigma_Y$

Let us determine the marginal density function of random variable $X$. It is given by, following straightforward calculations,

$$f_X(x) = \int_{-\infty}^{\infty} f_{XY}(x, y)\mathrm{d}y = \frac{1}{(2\pi)^{1/2}\sigma_X} \exp\left[-\frac{(x - m_X)^2}{2\sigma_X^2}\right], \quad -\infty < x < \infty. \quad (7.28)$$

Thus, random variable $X$ by itself has a normal distribution $N(m_X, \sigma_X^2)$. Similar calculations show that $Y$ is also normal with distribution $N(m_Y, \sigma_Y^2)$, and $\rho = \mu_{XY}/\sigma_X\sigma_Y$ is the correlation coefficient of $X$ and $Y$. We thus see that the five parameters contained in the bivariate density function $f_{XY}(x, y)$ represent five important moments associated with the random variables. This also leads us to observe that the bivariate normal distribution is completely characterized by the first-order and second-order joint moments of $X$ and $Y$.

Another interesting and important property associated with jointly normally distributed random variables is noted in Theorem 7.3.

**Theorem 7.3:** Zero correlation implies independence when the random variables are jointly normal.

**Proof of Theorem 7.3:** let $\rho = 0$ in Equation (7.27). We easily get

$$f_{XY}(x, y) = \left\{\frac{1}{(2\pi)^{1/2}\sigma_X} \exp\left[-\frac{(x - m_X)^2}{2\sigma_X^2}\right]\right\}\left\{\frac{1}{(2\pi)^{1/2}\sigma_Y} \exp\left[-\frac{(y - m_Y)^2}{2\sigma_Y^2}\right]\right\}$$
$$= f_X(x)f_Y(y), \quad (7.29)$$

which is the desired result. It should be stressed again, as in Section 4.3.1, that this property is not shared by random variables in general.

We have the *multivariate normal distribution* when the case of two random variables is extended to that involving $n$ random variables. For compactness, vector–matrix notation is used in the following.

Consider a sequence of $n$ random variables, $X_1, X_2, \ldots, X_n$. They are said to be *jointly normal* if the associated joint density function has the form

$$f_{X_1X_2\ldots X_n}(x_1, x_2, \ldots, x_n) = f_{\mathbf{X}}(\mathbf{x})$$
$$= (2\pi)^{-n/2}|\Lambda|^{-1/2} \exp\left[-\frac{1}{2}(\mathbf{x} - \mathbf{m})^{\mathrm{T}}\Lambda^{-1}(\mathbf{x} - \mathbf{m})\right],$$
$$-\infty < \mathbf{x} < \infty, \quad (7.30)$$

where $\mathbf{m}^{\mathrm{T}} = [m_1 \ m_2 \ \ldots \ m_n] = [E\{X_1\} \ E\{X_2\} \ \ldots \ E\{X_n\}]$, and $\Lambda = [\mu_{ij}]$ is the $n \times n$ covariance matrix of $\mathbf{X}$ with [see Equations (4.34) and (4.35)]:

$$\mu_{ij} = E\{(X_i - m_i)(X_j - m_j)\}. \quad (7.31)$$

The superscripts T and −1 denote, respectively, matrix transpose and matrix inverse. Again, we see that a joint normal distribution is completely specified by the first-order and second-order joint moments.

It is instructive to derive the joint characteristic function associated with **X**. As seen from Section 4.5.3, it is defined by

$$
\begin{aligned}
\phi_{X_1 X_2 \ldots X_n}(t_1, t_2, \ldots, t_n) &= \phi_{\mathbf{X}}(\mathbf{t}) \\
&= E\{\exp[j(t_1 X_1 + \cdots + t_n X_n)]\} \\
&= \int_{-\infty}^{\infty} \cdots \int_{-\infty}^{\infty} \exp(j\mathbf{t}^T \mathbf{x}) f_{\mathbf{X}}(\mathbf{x}) d\mathbf{x},
\end{aligned}
\tag{7.32}
$$

which gives, on substituting Equation (7.30) into Equation (7.32),

$$
\phi_{\mathbf{X}}(\mathbf{t}) = \exp\left( j\mathbf{m}^T \mathbf{t} - \frac{1}{2} \mathbf{t}^T \Lambda \mathbf{t} \right),
\tag{7.33}
$$

where $\mathbf{t}^T = [t_1\ t_2 \cdots t_n]$.

Joint moments of **X** can be obtained by differentiating joint characteristic function $\phi_{\mathbf{X}}(\mathbf{t})$ with respect to **t** and setting $\mathbf{t} = \mathbf{0}$. The expectation $E\{X_1^{m_1} X_2^{m_2} \cdots X_n^{m_n}\}$, for example, is given by

$$
E\{X_1^{m_1} X_2^{m_2} \cdots X_n^{m_n}\} = j^{-(m_1 + m_2 + \cdots + m_n)} \left[ \frac{\partial^{m_1 + m_2 + \cdots + m_n}}{\partial t_1^{m_1} \partial t_2^{m_2} \cdots \partial t_n^{m_n}} \phi_{\mathbf{X}}(\mathbf{t}) \right]_{\mathbf{t}=0}
\tag{7.34}
$$

It is clear that, since joint moments of the first-order and second-order completely specify the joint normal distribution, these moments also determine joint moments of orders higher than 2. We can show that, in the case when random variables $X_1, X_2, \cdots, X_n$ have zero means, all odd-order moments of these random variables vanish, and, for $n$ even,

$$
E\{X_1 X_2 \cdots X_n\} = \sum_{m_1, \ldots, m_n} E\{X_{m_1} X_{m_2}\} E\{X_{m_2} X_{m_3}\} \cdots E\{X_{m_{n-1}} X_{m_n}\}
\tag{7.35}
$$

The sum above is taken over all possible combinations of $n/2$ pairs of the $n$ random variables. The number of terms in the summation is $(1)(3)(5) \cdots (n-3)(n-1)$.

## 7.2.4 SUMS OF NORMAL RANDOM VARIABLES

We have seen through discussions and examples that sums of random variables arise in a number of problem formulations. In the case of normal random variables, we have the following important result (Theorem 7.4).

**Theorem 7.4:** let $X_1, X_2, \ldots, X_n$ be $n$ jointly normally distributed random variables (not necessarily independent). Then random variable $Y$, where

$$Y = c_1 X_1 + c_2 X_2 + \cdots + c_n X_n, \qquad (7.36)$$

is normally distributed, where $c_1, c_2, \cdots$, and $c_n$ are constants.

**Proof of Theorem 7.4:** for convenience, the proof will be given by assuming that all $X_j, j = 1, 2, \ldots, n$, have zero means. For this case, the mean of $Y$ is clearly zero and its variance is, as seen from Equation (4.43),

$$\sigma_Y^2 = E\{Y^2\} = \sum_{i=1}^{n} \sum_{j=1}^{n} c_i c_j \mu_{ij}, \qquad (7.37)$$

where $\mu_{ij} = \text{cov}(X_i, X_j)$.

Since $X_j$ are normally distributed, their joint characteristic function is given by Equation (7.33), which is

$$\phi_{\mathbf{X}}(\mathbf{t}) = \exp\left(-\frac{1}{2} \sum_{i=1}^{n} \sum_{j=1}^{n} \mu_{ij} t_i t_j\right). \qquad (7.38)$$

The characteristic function of $Y$ is

$$\phi_Y(t) = E\{\exp(jtY)\} = E\left\{\exp\left(jt \sum_{k=1}^{n} c_k X_k\right)\right\}$$

$$= \exp\left(-\frac{1}{2} t^2 \sum_{i=1}^{n} \sum_{j=1}^{n} \mu_{ij} c_i c_j\right)$$

$$= \exp\left(-\frac{1}{2} \sigma_Y^2 t^2\right), \qquad (7.39)$$

which is the characteristic function associated with a normal random variable. Hence $Y$ is also a normal random variable.

A further generalization of the above result is given in Theorem 7.5, which we shall state without proof.

**Theorem 7.5:** let $X_1, X_2, \ldots$, and $X_n$ be $n$ normally distributed random variables (not necessarily independent). Then random variables $Y_1, Y_2, \ldots$, and $Y_m$, where

$$Y_j = \sum_{k=1}^{n} c_{jk} X_k, \quad j = 1, 2, \ldots, m, \qquad (7.40)$$

are themselves jointly normally distributed.

## 7.3  LOGNORMAL DISTRIBUTION

We have seen that normal distributions arise from sums of many random actions. Consider now another common phenomenon which is the resultant of many *multiplicative* random effects. An example of multiplicative phenomena is in fatigue studies of materials where internal material damage at a given stage of loading is a random proportion of damage at the previous stage. In biology, the distribution of the size of an organism is another example for which growth is subject to many small impulses, each of which is proportional to the momentary size. Other examples include the size distribution of particles under impact or impulsive forces, the life distribution of mechanical components, the distribution of personal incomes due to annual adjustments, and other similar phenomena.

Let us consider

$$Y = X_1 X_2 \ldots X_n. \tag{7.41}$$

We are interested in the distribution of $Y$ as $n$ becomes large, when random variables $X_j, j = 1, 2, \ldots, n$, can take only positive values.

If we take logarithms of both sides, Equation (7.41) becomes

$$\ln Y = \ln X_1 + \ln X_2 + \cdots + \ln X_n. \tag{7.42}$$

The random variable $\ln Y$ is seen as a sum of random variables $\ln X_1, \ln X_2, \ldots,$ and $\ln X_n$. It thus follows from the central limit theorem that $\ln Y$ tends to a normal distribution as $n \to \infty$. The probability distribution of $Y$ is thus determined from

$$Y = e^X, \tag{7.43}$$

where $X$ is a normal random variable.

**Definition 7.1.** Let $X$ be $N(m_X, \sigma_X^2)$. The random variable $Y$ as determined from Equation (7.43) is said to have a *lognormal* distribution.

The pdf of $Y$ is easy to determine. Since Equation (7.43) gives $Y$ as a monotonic function of $X$, Equation (5.12) immediately gives

$$f_Y(y) = \begin{cases} \dfrac{1}{y\sigma_X(2\pi)^{1/2}} \exp\left[-\dfrac{1}{2\sigma_X^2}(\ln y - m_X)^2\right], & \text{for } y \geq 0; \\ 0, & \text{elsewhere.} \end{cases} \tag{7.44}$$

Equation (7.44) shows that $Y$ has a one-sided distribution (i.e. it takes values only in the positive range of $y$). This property makes it attractive for physical

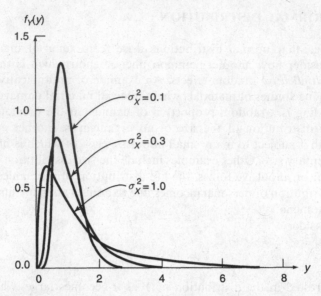

**Figure 7.9**   Lognormal distribution, $f_Y(y)$, with $m_X = 0$, for several values of $\sigma_X^2$

quantities that are restricted to having only positive values. In addition, $f_Y(y)$ takes many different shapes for different values of $m_X$ and $\sigma_X(\sigma_x > 0)$. As seen from Figure 7.9, the pdf of $Y$ is skewed to the right, this characteristic becoming more pronounced as $\sigma_X$ increases.

It is noted that parameters $m_X$ and $\sigma_X$ appearing in the pdf of $Y$ are the mean and standard deviation of $X$, or $\ln Y$, but not of $Y$. To obtain a more natural pair of parameters for $f_Y(y)$, we observe that, if medians of $X$ and $Y$ are denoted by $\theta_X$ and $\theta_Y$, respectively, the definition of the median of a random variable gives

$$0.5 = P(Y \leq \theta_Y) = P(X \leq \ln \theta_Y) = P(X \leq \theta_X),$$

or

$$\ln \theta_Y = \theta_X. \tag{7.45}$$

Since, owing to the symmetry of the normal distribution,

$$\theta_X = m_X,$$

we can write

$$m_X = \ln \theta_Y. \tag{7.46}$$

Now, writing $\sigma_X = \sigma_{\ln Y}$, the pdf of $Y$ can be written in the form

$$f_Y(y) = \begin{cases} \dfrac{1}{y\sigma_{\ln Y}(2\pi)^{1/2}} \exp\left[-\dfrac{1}{2\sigma_{\ln Y}^2}\ln^2\left(\dfrac{y}{\theta_Y}\right)\right], & \text{for } y \geq 0; \\ 0, & \text{elsewhere.} \end{cases} \qquad (7.47)$$

The mean and standard deviation of $Y$ can be found either through direct integration by using $f_Y(y)$ or by using the relationship given by Equation (7.43) together with $f_X(x)$. In terms of $\theta_Y$ and $\sigma_{\ln Y}$, they take the forms

$$\left.\begin{aligned} m_Y &= \theta_Y \exp\left(\frac{\sigma_{\ln Y}^2}{2}\right), \\ \sigma_Y^2 &= m_Y^2[\exp(\sigma_{\ln Y}^2) - 1]. \end{aligned}\right\} \qquad (7.48)$$

## 7.3.1 PROBABILITY TABULATIONS

Because of the close ties that exist between the normal distribution and the lognormal distribution through Equation (7.43), probability calculations involving a lognormal distributed random variable can be carried out with the aid of probability tables provided for normal random variables as shown below.

Consider the probability distribution function of $Y$. We have

$$F_Y(y) = P(Y \leq y) = P(X \leq \ln y) = F_X(\ln y), \quad y \geq 0. \qquad (7.49)$$

Now, since the mean of $X$ is $\ln \theta_Y$ and its variance is $\sigma_{\ln Y}^2$, we have:

$$F_Y(y) = F_U\left(\frac{\ln y - \ln \theta_Y}{\sigma_{\ln Y}}\right) = F_U\left[\frac{1}{\sigma_{\ln Y}}\ln\left(\frac{y}{\theta_Y}\right)\right], \quad y \geq 0. \qquad (7.50)$$

Since $F_U(u)$ is tabulated, Equation (7.50) can be used for probability calculations associated with $Y$ with the aid of the normal probability table.

**Example 7.5.** Problem: the annual maximum runoff $Y$ of a certain river can be modeled by a lognormal distribution. Suppose that the observed mean and standard deviation of $Y$ are $m_Y = 300\,\text{cfs}$ and $\sigma_Y = 200\,\text{cfs}$. Determine the probability $P(Y > 400\,\text{cfs})$.

Answer: using Equations (7.48), parameters $\theta_Y$ and $\sigma_{\ln Y}$ are solutions of the equations

$$\theta_Y \exp\left(\frac{\sigma_{\ln Y}^2}{2}\right) = 300,$$

$$\exp(\sigma_{\ln Y}^2) = \frac{4 \times 10^4}{9 \times 10^4} + 1,$$

resulting in

$$\left.\begin{array}{c} \theta_Y = 250, \\ \sigma_{\ln Y} = 0.61. \end{array}\right\} \tag{7.51}$$

The desired answer is, using Equation (7.50) and Table A.3,

$$P(Y > 400) = 1 - P(Y \le 400) = 1 - F_Y(400)$$

where

$$F_Y(400) = F_U\left[\frac{1}{0.61}\ln\left(\frac{400}{250}\right)\right]$$

$$= F_U(0.77) = 0.7794.$$

Hence,

$$P(Y > 400) = 1 - 0.7794 = 0.2206.$$

## 7.4  GAMMA AND RELATED DISTRIBUTIONS

The gamma distribution describes another class of useful one-sided distributions. The pdf associated with the gamma distribution is

$$f_X(x) = \begin{cases} \dfrac{\lambda^\eta}{\Gamma(\eta)} x^{\eta-1} e^{-\lambda x}, & \text{for } x \ge 0; \\ 0, & \text{elsewhere;} \end{cases} \tag{7.52}$$

where $\Gamma(\eta)$ is the well-known gamma function:

$$\Gamma(\eta) = \int_0^\infty u^{\eta-1} e^{-u} du, \tag{7.53}$$

which is widely tabulated, and

$$\Gamma(\eta) = (\eta - 1)!, \tag{7.54}$$

when $\eta$ is a positive integer.

The parameters associated with the gamma distribution are $\eta$ and $\lambda$; both are taken to be positive. Since the gamma distribution is one-sided, physical quantities that can take values only in, say, the positive range are frequently modeled by it. Furthermore, it serves as a useful model because of its versatility in the sense that a wide variety of shapes to the gamma density function can be obtained by varying the values of $\eta$ and $\lambda$. This is illustrated in Figures 7.10(a) and 7.10(b) which show plots of Equation (7.52) for several values of $\eta$ and $\lambda$. We notice from these figures that $\eta$ determines the shape of the distribution and is thus a shape parameter whereas $\lambda$ is a scale parameter for the distribution. In general, the gamma density function is unimodal, with its peak at $x = 0$ for $\eta \leq 1$, and at $x = (\eta - 1)/\lambda$ for $\eta > 1$.

As we will verify in Section 7.4.1.1, it can also be shown that the gamma distribution is an appropriate model for time required for a total of exactly $\eta$ Poisson arrivals. Because of the wide applicability of Poisson arrivals, the gamma distribution also finds numerous applications.

The distribution function of random variable $X$ having a gamma distribution is

$$F_X(x) = \int_0^x f_X(u)\mathrm{d}u = \frac{\lambda^\eta}{\Gamma(\eta)} \int_0^x u^{\eta-1} e^{-\lambda u} \mathrm{d}u;$$

$$= \frac{\Gamma(\eta, \lambda x)}{\Gamma(\eta)}, \qquad \text{for } x \geq 0; \tag{7.55}$$

$$= 0, \qquad \text{elsewhere.}$$

In the above, $\Gamma(\eta, u)$ is the incomplete gamma function,

$$\Gamma(\eta, u) = \int_0^u x^{\eta-1} e^{-x} \mathrm{d}x, \tag{7.56}$$

which is also widely tabulated.

The mean and variance of a gamma-distributed random variable $X$ take quite simple forms. After carrying out the necessary integration, we obtain

$$m_X = \frac{\eta}{\lambda}, \qquad \sigma_X^2 = \frac{\eta}{\lambda^2} \tag{7.57}$$

A number of important distributions are special cases of the gamma distribution. Two of these are discussed below in more detail.

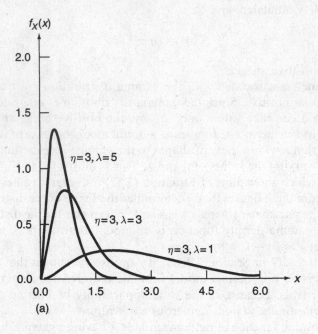

(a)

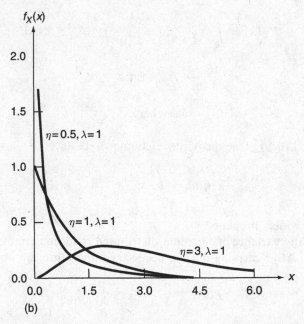

(b)

**Figure 7.10**    Gamma distribution with: (a) $\eta = 3$ and $\lambda = 5$, $\lambda = 3$, and $\lambda = 1$, and (b) $\lambda = 1$ and $\eta = 0.5$, $\eta = 1$, and $\eta = 3$

## 7.4.1  EXPONENTIAL DISTRIBUTION

When $\eta = 1$, the gamma density function given by Equation (7.52) reduces to the exponential form

$$f_X(x) = \begin{cases} \lambda e^{-\lambda x}, & \text{for } x \geq 0; \\ 0, & \text{elsewhere;} \end{cases} \qquad (7.58)$$

where $\lambda$ ($\lambda > 0$) is the parameter of the distribution. Its associated PDF, mean, and variance are obtained from Equations (7.55) and (7.57) by setting $\eta = 1$. They are

$$F_X(x) = \begin{cases} = 1 - e^{-\lambda x}, & \text{for } x \geq 0; \\ = 0, & \text{elsewhere;} \end{cases} \qquad (7.59)$$

and

$$m_X = \frac{1}{\lambda}, \quad \sigma_X^2 = \frac{1}{\lambda^2}. \qquad (7.60)$$

Among many of its applications, two broad classes stand out. First, we will show that the exponential distribution describes interarrival time when arrivals obey the Poisson distribution. It also plays a central role in reliability, where the exponential distribution is one of the most important failure laws.

### 7.4.1.1  Interarrival Time

There is a very close tie between the Poisson and exponential distributions. Let random variable $X(0, t)$ be the number of arrivals in the time interval $[0, t)$ and assume that it is Poisson distributed. Our interest now is in the time between two successive arrivals, which is, of course, also a random variable. Let this interarrival time be denoted by $T$. Its probability distribution function, $F_T(t)$, is, by definition,

$$F_T(t) = \begin{cases} P(T \leq t) = 1 - P(T > t), & \text{for } t \geq 0; \\ 0, & \text{elsewhere.} \end{cases} \qquad (7.61)$$

In terms of $X(0, t)$, the event $T > t$ is equivalent to the event that there are no arrivals during time interval $[0, t)$, or $X(0, t) = 0$. Hence, since

$P[X(0, t) = 0] = e^{-\lambda t}$ as given by Equation (6.40), we have

$$F_T(t) = \begin{cases} 1 - e^{-\lambda t}, & \text{for } t \geq 0; \\ 0, & \text{elsewhere.} \end{cases} \tag{7.62}$$

Comparing this expression with Equation (7.59), we can establish the result that the interarrival time between Poisson arrivals has an exponential distribution; the parameter $\lambda$ in the distribution of $T$ is the mean arrival rate associated with Poisson arrivals.

**Example 7.6.** Problem: referring to Example 6.11 (page 177), determine the probability that the headway (spacing measured in time) between arriving vehicles is at least 2 minutes. Also, compute the mean headway.

Answer: in Example 6.11, the parameter $\lambda$ was estimated to be 4.16 vehicles per minute. Hence, if $T$ is the headway in minutes, we have

$$P(T \geq 2) = \int_2^\infty f_T(t)\mathrm{d}t = 1 - F_T(2) = e^{-2(4.16)} = 0.00024.$$

The mean headway is

$$m_T = \frac{1}{\lambda} = \frac{1}{4.16} \text{ minutes} = 0.24 \text{ minutes.}$$

Since interarrival times for Poisson arrivals are independent, the time required for a total of $n$ Poisson arrivals is a sum of $n$ independent and exponentially distributed random variables. Let $T_j, j = 1, 2, \ldots, n$, be the interarrival time between the $(j - 1)$th and $j$th arrivals. The time required for a total of $n$ arrivals, denoted by $X_n$, is

$$X_n = T_1 + T_2 + \cdots + T_n, \tag{7.63}$$

where $T_j, j = 1, 2, \ldots, n$, are independent and exponentially distributed with the same parameter $\lambda$. In Example 4.16 (page 105), we showed that $X_n$ has a gamma distribution with $\eta = 2$ when $n = 2$. The same procedure immediately shows that, for general $n$, $X_n$ is gamma-distributed with $\eta = n$. Thus, as stated, the gamma distribution is appropriate for describing the time required for a total of $\eta$ Poisson arrivals.

**Example 7.7.** Problem: ferries depart for trips across a river as soon as nine vehicles are driven aboard. It is observed that vehicles arrive independently at an average rate of 6 per hour. Determine the probability that the time between trips will be less than 1 hour.

Answer: from our earlier discussion, the time between trips follows a gamma distribution with $\eta = 9$ and $\lambda = 6$. Hence, let $X$ be the time between trips in hours; its density function and distribution function are given by Equations (7.52) and (7.55). The desired result is, using Equation (7.55),

$$P(X \leq 1) = F_X(1) = \frac{\Gamma(\eta, \lambda)}{\Gamma(\eta)} = \frac{\Gamma(9, 6)}{\Gamma(9)}.$$

Now, $\Gamma(9) = 8!$, and the incomplete gamma function $\Gamma(9, 6)$ can be obtained by table lookup. We obtain:

$$P(X \leq 1) = 0.153.$$

An alternative computational procedure for determining $P(X \leq 1)$ inExample 7.7 can be found by noting from Equation (7.63) that random variable $X$ can be represented by a sum of $\eta$ independent random variables. Hence, according to the central limit theorem, its distribution approaches that of a normal random variable when $\eta$ is large. Thus, provided that $\eta$ is large, computations such as that required in Example 7.7 can be carried out by using Table A.3 for normal random variables. Let us again consider Example 7.7. Approximating $X$ by a normal random variable, the desired probability is [see Equation (7.25)]

$$P(X \leq 1) \simeq P\left(U \leq \frac{1 - m_X}{\sigma_X}\right),$$

where $U$ is the standardized normal random variable. The mean and standard deviation of $X$ are, using Equations (7.57),

$$m_X = \frac{\eta}{\lambda} = \frac{9}{6}$$
$$= \frac{3}{2},$$

and

$$\sigma_X = \frac{\eta^{1/2}}{\lambda} = \frac{3}{6}$$
$$= \frac{1}{2}.$$

Hence, with the aid of Table A.3,

$$P(X \leq 1) \simeq P(U \leq -1) = F_U(-1) = 1 - F_U(1)$$
$$= 1 - 0.8413 = 0.159,$$

which is quite close to the answer obtained in Example 7.7.

## 7.4.1.2   Reliability and Exponential Failure Law

One can infer from our discussion on interarrival time that many analogous situations can be treated by applying the exponential distribution. In reliability studies, the time to failure for a physical component or a system is expected to be exponentially distributed if the unit fails as soon as some single event, such as malfunction of a component, occurs, assuming such events happen independently. In order to gain more insight into failure processes, let us introduce some basic notions in reliability.

Let random variable $T$ be the time to failure of a component or system. It is useful to consider a function that gives the probability of failure during a small time increment, assuming that no failure occurred before that time. This function, denoted by $h(t)$, is called the *hazard function* or *failure rate* and is defined by

$$h(t)dt = P(t < T \le t + dt | T \ge t) \tag{7.64}$$

which gives

$$h(t) = \frac{f_T(t)}{1 - F_T(t)}. \tag{7.65}$$

In reliability studies, a hazard function appropriate for many phenomena takes the so-called 'bathtub curve', shown in Figure 7.11. The initial portion of the curve represents 'infant mortality', attributable to component defects and manufacturing imperfections. The relatively constant portion of the $h(t)$ curve represents the in-usage period in which failure is largely a result of chance failure. Wear-out failure near the end of component life is shown as the

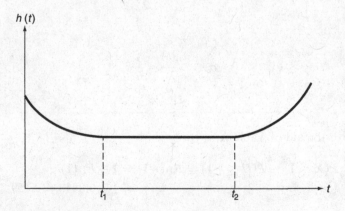

**Figure 7.11**   Typical shape of a hazard function

increasing portion of the $h(t)$ curve. System reliability can be optimized by initial 'burn-in' until time $t_1$ to avoid premature failure and by part replacement at time $t_2$ to avoid wear out.

We can now show that the exponential failure law is appropriate during the 'in-usage' period of a system's normal life. Substituting

$$f_T(t) = \lambda e^{-\lambda t}, \ t \geq 0,$$

and

$$F_T(t) = 1 - e^{-\lambda t}, \ t \geq 0,$$

into Equation (7.65), we immediately have

$$h(t) = \lambda. \tag{7.66}$$

We see from the above that parameter $\lambda$ in the exponential distribution plays the role of a (constant) failure rate.

We have seen in Example 7.7 that the gamma distribution is appropriate to describe the time required for a total of $\eta$ arrivals. In the context of failure laws, the gamma distribution can be thought of as a generalization of the exponential failure law for systems that fail as soon as exactly $\eta$ events fail, assuming events take place according to the Poisson law. Thus, the gamma distribution is appropriate as a time-to-failure model for systems having one operating unit and $\eta - 1$ standby units; these standby units go into operation sequentially, and each one has an exponential time-to-failure distribution.

## 7.4.2  CHI-SQUARED DISTRIBUTION

Another important special case of the gamma distribution is the chi-squared ($\chi^2$) distribution, obtained by setting $\lambda = 1/2$ and $\eta = n/2$ in Equation (7.52), where $n$ is a positive integer. The $\chi^2$ distribution thus contains one parameter, $n$, with pdf of the form

$$f_X(x) = \begin{cases} \dfrac{1}{2^{n/2}\Gamma(n/2)} x^{(n/2)-1} e^{-x/2}, & \text{for } x \geq 0; \\ \\ 0, & \text{elsewhere.} \end{cases} \tag{7.67}$$

The parameter $n$ is generally referred to as the *degrees of freedom*. The utility of this distribution arises from the fact that a sum of the squares of $n$ independent standardized normal random variables has a $\chi^2$ distribution with $n$ degrees of freedom; that is, if $U_1, U_2, \ldots,$ and $U_n$ are independent and distributed as $N(0, 1)$, the sum

$$X = U_1^2 + U_2^2 + \cdots + U_n^2 \tag{7.68}$$

has a $\chi^2$ distribution with $n$ degrees of freedom. One can verify this statement by determining the characteristic function of each $U_j^2$ (see Example 5.7, page 132) and using the method of characteristic functions as discussed in Section 4.5 for sums of independent random variables.

Because of this relationship, the $\chi^2$ distribution is one of our main tools in the area of statistical inference and hypothesis testing. These applications are detailed in Chapter 10.

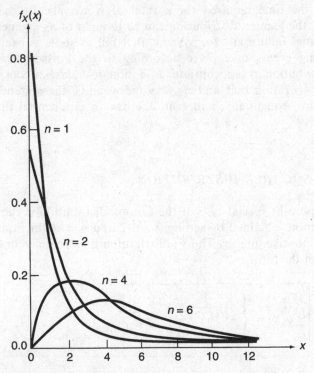

**Figure 7.12**   The $\chi^2$ distribution for $n = 1$, $n = 2$, $n = 4$, and $n = 6$

The pdf $f_X(x)$ in Equation (7.67) is plotted in Figure 7.12 for several values of $n$. It is shown that, as $n$ increases, the shape of $f_X(x)$ becomes more symmetric. In view of Equation (7.68), since $X$ can be expressed as a sum of identically distributed random variables, we expect that the $\chi^2$ distribution approaches a normal distribution as $n \to \infty$ on the basis of the central limit theorem.

The mean and variance of random variable $X$ having a $\chi^2$ distribution are easily obtained from Equation (7.57) as

$$m_X = n, \quad \sigma_X^2 = 2n. \tag{7.69}$$

## 7.5  BETA AND RELATED DISTRIBUTIONS

Whereas the lognormal and gamma distributions provide a diversity of one-sided probability distributions, the beta distribution is rich in providing varied probability distributions over a finite interval. The beta distribution is characterized by the density function

$$f_X(x) = \begin{cases} \dfrac{\Gamma(\alpha + \beta)}{\Gamma(\alpha)\Gamma(\beta)} x^{\alpha-1}(1 - x)^{\beta-1}, & \text{for } 0 \leq x \leq 1; \\[2mm] 0, & \text{elsewhere;} \end{cases} \tag{7.70}$$

where parameters $\alpha$ and $\beta$ take only positive values. The coefficient of $f_X(x)$, $\Gamma(\alpha + \beta)/[\Gamma(\alpha)\Gamma(\beta)]$, can be represented by $1/[B(\alpha, \beta)]$, where

$$B(\alpha, \beta) = \frac{\Gamma(\alpha)\Gamma(\beta)}{\Gamma(\alpha + \beta)}, \tag{7.71}$$

is known as the beta function, hence the name for the distribution given by Equation (7.70).

The parameters $\alpha$ and $\beta$ are both shape parameters; different combinations of their values permit the density function to take on a wide variety of shapes. When $\alpha, \beta > 1$, the distribution is unimodal, with its peak at $x = (\alpha - 1)/(\alpha + \beta - 2)$. It becomes U-shaped when $\alpha, \beta < 1$; it is J-shaped when $\alpha \geq 1$ and $\beta < 1$; and it takes the shape of an inverted J when $\alpha < 1$ and $\beta \geq 1$. Finally, as a special case, the uniform distribution over interval $(0,1)$ results when $\alpha = \beta = 1$. Some of these possible shapes are displayed in Figures 7.13(a) and 7.13(b).

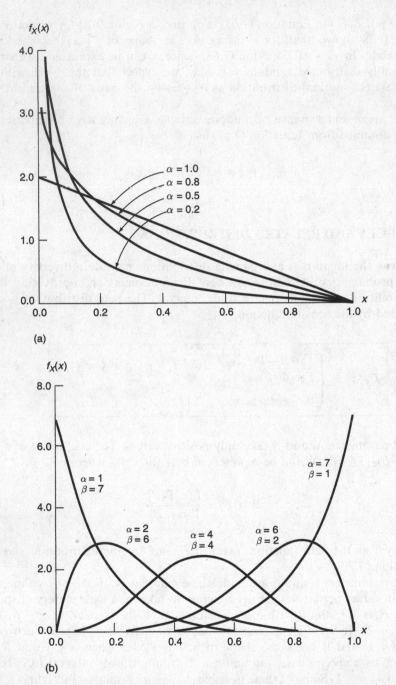

**Figure 7.13**   Beta distribution with: (a) $\beta = 2$ and $\alpha = 1.0$, $\alpha = 0.8$, $\alpha = 0.5$, and $\alpha = 0.2$; and (b) combinations of values of $\alpha$ and $\beta$ ($\alpha, \beta = 1, 2, \ldots, 7$) such that $\alpha + \beta = 8$

The mean and variance of a beta-distributed random variable $X$ are, following straightforward integrations,

$$\left. \begin{aligned} m_X &= \frac{\alpha}{\alpha + \beta}, \\ \sigma_X^2 &= \frac{\alpha\beta}{(\alpha + \beta)^2(\alpha + \beta + 1)}. \end{aligned} \right\} \tag{7.72}$$

Because of its versatility as a distribution over a finite interval, the beta distribution is used to represent a large number of physical quantities for which values are restricted to an identifiable interval. Some of the areas of application are tolerance limits, quality control, and reliability.

An interesting situation in which the beta distribution arises is as follows. Suppose a random phenomenon $Y$ can be observed independently $n$ times and, after these $n$ independent observations are ranked in order of increasing magnitude, let $y_r$ and $y_{n-s+1}$ be the values of the $r$th smallest and $s$th largest observations, respectively. If random variable $X$ is used to denote the proportion of the original $Y$ taking values between $y_r$ and $y_{n-s+1}$, it can be shown that $X$ follows a beta distribution with $\alpha = n - r - s + 1$, and $\beta = r + s$; that is,

$$f_X(x) = \begin{cases} \dfrac{\Gamma(n + 1)}{\Gamma(n - r - s + 1)\Gamma(r + s)} x^{n-r-s}(1 - x)^{r+s-1}, & \text{for } 0 \leq x \leq 1; \\ 0, & \text{elsewhere.} \end{cases} \tag{7.73}$$

This result can be found in Wilks (1942). We will not prove this result but we will use it in the next section, in Example 7.8.

### 7.5.1 PROBABILITY TABULATIONS

The probability distribution function associated with the beta distribution is

$$F_X(x) = \begin{cases} 0, & \text{for } x < 0; \\ \dfrac{\Gamma(\alpha + \beta)}{\Gamma(\alpha)\Gamma(\beta)} \displaystyle\int_0^x u^{\alpha-1}(1 - u)^{\beta-1}du, & \text{for } 0 \leq x \leq 1; \\ 1, & \text{for } x > 1; \end{cases} \tag{7.74}$$

which can be integrated directly. It also has the form of an incomplete beta function for which values for given values of $\alpha$ and $\beta$ can be found from mathematical tables. The incomplete beta function is usually denoted by

$I_x(\alpha, \beta)$. If we write $F_X(x)$ with parameters $\alpha$ and $\beta$ in the form $F(x; \alpha, \beta)$, the correspondence between $I_x(\alpha, \beta)$ and $F(x; \alpha, \beta)$ is determined as follows. If $\alpha \geq \beta$, then

$$F(x; \alpha, \beta) = I_x(\alpha, \beta). \tag{7.75}$$

If $\alpha < \beta$, then

$$F(x; \alpha, \beta) = 1 - I_{(1-x)}(\beta, \alpha). \tag{7.76}$$

Another method of evaluating $F_X(x)$ in Equation (7.74) is to note the similarity in form between $f_X(x)$ and $p_Y(k)$ of a binomial random variable $Y$ for the case where $\alpha$ and $\beta$ are positive integers. We see from Equation (6.2) that

$$p_Y(k) = \frac{n!}{k!(n-k)!} p^k (1-p)^{n-k}, \quad k = 0, 1, \ldots, n. \tag{7.77}$$

Also, $f_X(x)$ in Equation (7.70) with $\alpha$ and $\beta$ being positive integers takes the form

$$f_X(x) = \frac{(\alpha + \beta - 1)!}{(\alpha - 1)!(\beta - 1)!} x^{\alpha-1}(1 - x)^{\beta-1}, \quad \alpha, \beta = 1, 2, \ldots, \ 0 \leq x \leq 1, \tag{7.78}$$

and we easily establish the relationship

$$f_X(x) = (\alpha + \beta - 1)p_Y(k), \quad \alpha, \beta = 1, 2, \ldots, \ 0 \leq x \leq 1, \tag{7.79}$$

where $p_Y(k)$ is evaluated at $k = \alpha - 1$, with $n = \alpha + \beta - 2$, and $p = x$. For example, the value of $f_X(0.5)$ with $\alpha = 2$, and $\beta = 1$, is numerically equal to $2p_Y(1)$ with $n = 1$, and $p = 0.5$; here $p_Y(1)$ can be found from Equation (7.77) or from Table A.1 for binomial random variables.

Similarly, the relationship between $F_X(x)$ and $F_Y(k)$ can be established. It takes the form

$$F_X(x) = 1 - F_Y(k), \quad \alpha, \beta = 1, 2, \ldots, \ 0 \leq x \leq 1, \tag{7.80}$$

with $k = \alpha - 1$, $n = \alpha + \beta - 2$, and $p = x$. The PDF $F_Y(y)$ for a binomial random variable $Y$ is also widely tabulated and it can be used to advantage here for evaluating $F_X(x)$ associated with the beta distribution.

**Example 7.8.** Problem: in order to establish quality limits for a manufactured item, 10 independent samples are taken at random and the quality limits are established by using the lowest and highest sample values. What is the probability that at least 50% of the manufactured items will fail within these limits?

Answer: let $X$ be the proportion of items taking values within the established limits. Its pdf thus takes the form of Equation (7.73), with $n = 10, r = 1$, and $s = 1$.

Hence, $\alpha = 10 - 1 - 1 + 1 = 9, \beta = 1 + 1 = 2$, and

$$f_X(x) = \frac{\Gamma(11)}{\Gamma(9)\Gamma(2)} x^8(1 - x);$$

$$= \frac{10!}{8!} x^8(1 - x), \quad \text{for } 0 \le x \le 1;$$

$$= 0, \quad \text{elsewhere.}$$

The desired probability is

$$P(X > 0.50) = 1 - P(X \le 0.50) = 1 - F_X(0.50). \tag{7.81}$$

According to Equation (7.80), the value of $F_X(0.50)$ can be found from

$$F_X(0.50) = 1 - F_Y(k), \tag{7.82}$$

where $Y$ is binomial and $k = \alpha - 1 = 8$, $n = \alpha + \beta - 2 = 9$, and $p = 0.50$. Using Table A.1, we find that

$$F_Y(8) = 1 - p_Y(9) = 1 - 0.002 = 0.998. \tag{7.83}$$

Equations (7.81) and (7.82) yield

$$P(X > 0.50) = 1 - F_X(0.50) = 1 - 1 + F_Y(8) = 0.998. \tag{7.84}$$

## 7.5.2  GENERALIZED BETA DISTRIBUTION

The beta distribution can be easily generalized from one restricted to unit interval $(0, 1)$ to one covering an arbitrary interval $(a, b)$. Let $Y$ be such a generalized beta random variable. It is clear that the desired transformation is

$$Y = (b - a)X + a, \tag{7.85}$$

where $X$ is beta-distributed according to Equation (7.70). Equation (7.85) represents a monotonic transformation from $X$ and $Y$ and the procedure developed in Chapter 5 can be applied to determine the pdf of $Y$ in a straight-forward manner. Following Equation (5.12), we have

$$
f_Y(y) = \begin{cases} \dfrac{1}{(b-a)^{\alpha+\beta-1}} \dfrac{\Gamma(\alpha+\beta)}{\Gamma(\alpha)\Gamma(\beta)} (y-a)^{\alpha-1}(b-y)^{\beta-1}, & \text{for } a \le x \le b; \\ 0, & \text{elsewhere.} \end{cases} \tag{7.86}
$$

## 7.6  EXTREME-VALUE DISTRIBUTIONS

A structural engineer, concerned with the safety of a structure, is often interested in the *maximum* load and *maximum* stress in structural members. In reliability studies, the distribution of the life of a system having $n$ components in series (where the system fails if *any* component fails) is a function of the minimum time to failure of these components, whereas for a system with a parallel arrangement (where the system fails when *all* components fail) it is determined by the distribution of maximum time to failure. These examples point to our frequent concern with distributions of maximum or minimum values of a number of random variables.

To fix ideas, let $X_j, j = 1, 2, \ldots, n$, denote the $j$th gust velocity of $n$ gusts occurring in a year, and let $Y_n$ denote the annual maximum gust velocity. We are interested in the probability distribution of $Y_n$ in terms of those of $X_j$. In the following development, attention is given to the case where random variables $X_j, j = 1, 2, \ldots, n$, are independent and identically distributed with PDF $F_X(x)$ and pdf $f_X(x)$ or pmf $p_X(x)$. Furthermore, asymptotic results for $n \to \infty$ are our primary concern. For the wind-gust example given above, these conditions are not unreasonable in determining the distribution of annual maximum gust velocity. We will also determine, under the same conditions, the minimum $Z_n$ of random variables $X_1, X_2, \ldots,$ and $X_n$, which is also of interest in practical applications.

The random variables $Y_n$ and $Z_n$ are defined by

$$
\begin{aligned}
Y_n &= \max(X_1, X_2, \ldots, X_n), \\
Z_n &= \min(X_1, X_2, \ldots, X_n).
\end{aligned} \tag{7.87}
$$

The PDF of $Y_n$ is

$$
\begin{aligned}
F_{Y_n}(y) &= P(Y_n \le y) = P(\text{all } X_j \le y) \\
&= P(X_1 \le y \cap X_2 \le y \cap \cdots \cap X_n \le y).
\end{aligned}
$$

Assuming independence, we have

$$F_{Y_n}(y) = F_{X_1}(y)F_{X_2}(y)\cdots F_{X_n}(y), \tag{7.88}$$

and, if each $F_{X_j}(y) = F_X(y)$, the result is

$$F_{Y_n}(y) = [F_X(y)]^n. \tag{7.89}$$

The pdf of $Y_n$ can be easily derived from the above. When the $X_j$ are continuous, it has the form

$$f_{Y_n}(y) = \frac{dF_{Y_n}(y)}{dy} = n[F_X(y)]^{n-1}f_X(y). \tag{7.90}$$

The PDF of $Z_n$ is determined in a similar fashion. In this case,

$$\begin{aligned}
F_{Z_n}(z) &= P(Z_n \leq z) = P(\text{at least one } X_j \leq z)\\
&= P(X_1 \leq z \cup X_2 \leq z \cup \cdots \cup X_n \leq z)\\
&= 1 - P(X_1 > z \cap X_2 > z \cap \cdots \cap X_n > z).
\end{aligned}$$

When the $X_j$ are independent and identically distributed, the foregoing gives

$$\begin{aligned}
F_{Z_n}(z) &= 1 - [1 - F_{X_1}(z)][1 - F_{X_2}(z)]\cdots[1 - F_{X_n}(z)]\\
&= 1 - [1 - F_X(z)]^n.
\end{aligned} \tag{7.91}$$

If random variables $X_j$ are continuous, the pdf of $Z_n$ is

$$f_{Z_n}(z) = n[1 - F_X(z)]^{n-1}f_X(z). \tag{7.92}$$

The next step in our development is to determine the forms of $F_{Y_n}(y)$ and $F_{Z_n}(z)$ as expressed by Equations (7.89) and (7.91) as $n \to \infty$. Since the initial distribution $F_X(x)$ of each $X_j$ is sometimes unavailable, we wish to examine whether Equations (7.89) and (7.91) lead to unique distributions for $F_{Y_n}(y)$ and $F_{Z_n}(z)$, respectively, independent of the form of $F_X(x)$. This is not unlike looking for results similar to the powerful ones we obtained for the normal and lognormal distributions via the central limit theorem.

Although the distribution functions $F_{Y_n}(y)$ and $F_{Z_n}(z)$ become increasingly insensitive to exact distributional features of $X_j$ as $n \to \infty$, no unique results can be obtained that are completely independent of the form of $F_X(x)$. Some features of the distribution function $F_X(x)$ are important and, in what follows, the asymptotic forms of $F_{Y_n}(y)$ and $F_{Z_n}(z)$ are classified into three types based on general features in the distribution tails of $X_j$. Type I is sometimes referred to as Gumbel's extreme value distribution, and included in Type III is the important Weibull distribution.

## 7.6.1 TYPE-I ASYMPTOTIC DISTRIBUTIONS OF EXTREME VALUES

Consider first the Type-I asymptotic distribution of maximum values. It is the limiting distribution of $Y_n$ (as $n \to \infty$) from an initial distribution $F_X(x)$ of which the right tail is unbounded and is of an exponential type; that is, $F_X(x)$ approaches 1 at least as fast as an exponential distribution. For this case, we can express $F_X(x)$ in the form

$$F_X(x) = 1 - \exp[-g(x)], \qquad (7.93)$$

where $g(x)$ is an increasing function of $x$. A number of important distributions fall into this category, such as the normal, lognormal, and gamma distributions. Let

$$\lim_{n \to \infty} Y_n = Y. \qquad (7.94)$$

We have the following important result (Theorem 7.6).

**Theorem 7.6:** let random variables $X_1, X_2, \ldots$, and $X_n$ be independent and identically distributed with the same PDF $F_X(x)$. If $F_X(x)$ is of the form given by Equation (7.93), we have

$$\boxed{F_Y(y) = \exp\{-\exp[-\alpha(y - u)], \quad -\infty < y < \infty,} \qquad (7.95)$$

where $\alpha(\alpha > 0)$ and $u$ are two parameters of the distribution.

**Proof of Theorem 7.6:** we shall only sketch the proof here; see Gumbel (1958) for a more comprehensive and rigorous treatment.

Let us first define a quantity $u_n$, known as the characteristic value of $Y_n$, by

$$F_X(u_n) = 1 - \frac{1}{n}. \qquad (7.96)$$

It is thus the value of $X_j, j = 1, 2, \ldots, n$, at which $P(X_j \le u_n) = 1 - 1/n$. As $n$ becomes large, $F_X(u_n)$ approaches unity, or, $u_n$ is in the extreme right-hand tail of the distribution. It can also be shown that $u_n$ is the mode of $Y_n$, which can be verified, in the case of $X_j$ being continuous, by taking the derivative of $f_{Y_n}(y)$ in Equation (7.90) with respect to $y$ and setting it to zero.

If $F_X(x)$ takes the form given by Equation (7.93), we have

$$1 - \exp[-g(u_n)] = 1 - \frac{1}{n},$$

or

$$\frac{\exp[g(u_n)]}{n} = 1. \tag{7.97}$$

Now, consider $F_{Y_n}(y)$ defined by Equation (7.89). In view of Equation (7.93), it takes the form

$$
\begin{aligned}
F_{Y_n}(y) &= \{1 - \exp[-g(y)]\}^n \\
&= \left\{1 - \frac{\exp[g(u_n)]\exp[-g(y)]}{n}\right\}^n \\
&= \left\{1 - \frac{\exp\{-[g(y) - g(u_n)]\}}{n}\right\}^n.
\end{aligned}
\tag{7.98}
$$

In the above, we have introduced into the equation the factor $\exp[g(u_n)]/n$, which is unity, as shown by Equation (7.97).

Since $u_n$ is the mode or the 'most likely' value of $Y_n$, function $g(y)$ in Equation (7.98) can be expanded in powers of $(y - u_n)$ in the form

$$g(y) = g(u_n) + \alpha_n(y - u_n) + \cdots, \tag{7.99}$$

where $\alpha_n = dg(y)/dy$ is evaluated at $y = u_n$. It is positive, as $g(y)$ is an increasing function of $y$. Retaining only up to the linear term in Equation (7.99) and substituting it into Equation (7.98), we obtain

$$F_{Y_n}(y) = \left\{1 - \frac{\exp[-\alpha_n(y - u_n)]}{n}\right\}^n, \tag{7.100}$$

in which $\alpha_n$ and $u_n$ are functions only of $n$ and not of $y$. Using the identity

$$\lim_{n \to \infty}\left(1 - \frac{c}{n}\right)^n = \exp(-c),$$

for any real $c$, Equation (7.100) tends, as $n \to \infty$, to

$$F_Y(y) = \exp\{-\exp[-\alpha(y - u)]\}, \tag{7.101}$$

which was to be proved. In arriving at Equation (7.101), we have assumed that as $n \to \infty$, $F_{Y_n}(y)$ converges to $F_Y(y)$ as $Y_n$ converges to $Y$ in some probabilistic sense.

The mean and variance associated with the Type-I maximum-value distribution can be obtained through integration using Equation (7.90). We have noted that $u$ is the mode of the distribution, that is, the value of $y$ at which $f_Y(y)$ is maximum. The mean of $Y$ is

$$m_Y = u + \frac{\gamma}{\alpha}, \tag{7.102}$$

where $\gamma \simeq 0.577$ is Euler's constant; and the variance is given by

$$\sigma_Y^2 = \frac{\pi^2}{6\alpha^2}. \tag{7.103}$$

It is seen from the above that $u$ and $\alpha$ are, respectively, the location and scale parameters of the distribution. It is interesting to note that the skewness coefficient, defined by Equation (4.11), in this case is

$$\gamma_1 \simeq 1.1396,$$

which is independent of $\alpha$ and $u$. This result indicates that the Type-I maximum-value distribution has a fixed shape with a dominant tail to the right. A typical shape for $f_Y(y)$ is shown in Figure 7.14.

The Type-I asymptotic distribution for minimum values is the limiting distribution of $Z_n$ in Equation (7.91) as $n \to \infty$ from an initial distribution $F_X(x)$ of which the left tail is unbounded and is of exponential type as it decreases to zero on the left. An example of $F_X(x)$ that belongs to this class is the normal distribution.

The distribution of $Z_n$ as $n \to \infty$ can be derived by means of procedures given above for $Y_n$ through use of a symmetrical argument. Without giving details, if we let

$$\lim_{n \to \infty} Z_n = Z, \tag{7.104}$$

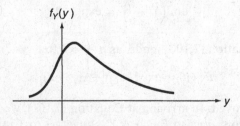

**Figure 7.14**   Typical plot of a Type-I maximum-value distribution

the PDF of $Z$ can be shown to have the form

$$F_Z(z) = 1 - \exp\{-\exp[\alpha(z - u)]\}, \quad -\infty < z < \infty \qquad (7.105)$$

where $\alpha$ and $u$ are again the two parameters of the distribution.

It is seen that Type-I asymptotic distributions for maximum and minimum values are mirror images of each other. The mode of $Z$ is $u$, and its mean, variance, and skewness coefficients are, respectively,

$$\left. \begin{array}{c} m_Z = u - \dfrac{\gamma}{\alpha} \\[2mm] \sigma_Z^2 = \dfrac{\pi^2}{6\alpha^2} \\[2mm] \gamma_1 \simeq -1.1396 \end{array} \right\} \qquad (7.106)$$

For probability calculations, values for probability distribution functions $F_Y(y)$ and $F_Z(z)$ over various ranges of $y$ and $z$ are available in, for example, Microsoft Excel 2000 (see Appendix B).

**Example 7.9.** Problem: the maximum daily gasoline demand $Y$ during the month of May at a given locality follows the Type-I asymptotic maximum-value distribution, with $m_Y = 2$ and $\sigma_Y = 1$, measured in thousands of gallons. Determine (a) the probability that the demand will exceed 4000 gallons in any day during the month of May, and (b) the daily supply level that for 95% of the time will not be exceeded by demand in any given day.

Answer: it follows from Equations (7.102) and (7.103) that parameters $\alpha$ and $u$ are determined from

$$\alpha = \frac{\pi}{\sqrt{6}\sigma_Y} = \frac{\pi}{\sqrt{6}} = 1.282,$$

$$u = m_Y - \frac{0.577}{\alpha} = 2 - \frac{0.577}{1.282} = 1.55.$$

For part (a), the solution is

$$\begin{aligned} P(Y > 4) &= 1 - F_Y(4) \\ &= 1 - \exp\{-\exp[-1.282(4 - 1.55)]\} \\ &= 1 - 0.958 = 0.042. \end{aligned}$$

For part (b), we need to determine $y$ such that

$$F_Y(y) = P(Y \le y) = 0.95,$$

or

$$\exp\{-\exp[-1.282(y - 1.55)]\} = 0.95. \qquad (7.107)$$

Taking logarithms of Equation (7.107) twice, we obtain

$$y = 3.867;$$

that is, the required supply level is 3867 gallons.

**Example 7.10.** Problem: consider the problem of estimating floods in the design of dams. Let $y_T$ denote the maximum flood associated with return period $T$. Determine the relationship between $y_T$ and $T$ if the maximum river flow follows the Type-I maximum-value distribution. Recall from Example 6.7 (page 169) that the return period $T$ is defined as the average number of years between floods for which the magnitude is greater than $y_T$.

Answer: assuming that floods occur independently, the number of years between floods with magnitudes greater than $y_T$ assumes a geometric distribution. Thus

$$T = \frac{1}{P(Y > y_T)} = \frac{1}{1 - F_Y(y_T)}. \qquad (7.108)$$

Now, from Equation (7.101),

$$F_Y(y_T) = \exp[-\exp(-b)], \qquad (7.109)$$

where $b = \alpha(y_T - u)$. The substitution of Equation (7.109) into Equation (7.108) gives the required relationship.

For values of $y_T$ where $F_Y(y_T) \to 1$, an approximation can be made by noting from Equation (7.109) that

$$\exp(-b) = -\ln F_Y(y_T) = -\{[F_Y(y_T) - 1] - \frac{1}{2}[F_Y(y_T) - 1]^2 + \cdots\}.$$

Since $F_Y(y_T)$ is close to 1, we retain only the first term in the foregoing expansion and obtain

$$1 - F_Y(y_T) \simeq \exp(-b).$$

Equation (7.108) thus gives the approximate relationship

$$y_T = u\left(1 + \frac{1}{\alpha u}\ln T\right), \qquad (7.110)$$

where $u$ is the scale factor and the value of $\alpha u$ describes the characteristics of a river; it varies from 1.5 for violent rivers to 10 for stable or mild rivers.

In closing, let us remark again that the Type-I maximum-value distribution is valid for initial distributions of such practical importance as normal, lognormal, and gamma distributions. It thus has wide applicability and is sometimes simply called the *extreme value distribution*.

## 7.6.2  TYPE-II ASYMPTOTIC DISTRIBUTIONS OF EXTREME VALUES

The Type-II asymptotic distribution of maximum values arises as the limiting distribution of $Y_n$ as $n \to \infty$ from an initial distribution of the Pareto type, that is, the PDF $F_X(x)$ of each $X_j$ is limited on the left at zero and its right tail is unbounded and approaches one according to

$$F_X(x) = 1 - ax^{-k}, \quad a, k > 0, \ x \geq 0. \tag{7.111}$$

For this class, the asymptotic distribution of $Y_n$, $F_Y(y)$, as $n \to \infty$ takes the form

$$\boxed{F_Y(y) = \exp\left[-\left(\frac{y}{v}\right)^{-k}\right], \quad v, k > 0, \ y \geq 0.} \tag{7.112}$$

Let us note that, with $F_X(x)$ given by Equation (7.111), each $X_j$ has moments only up to order $r$, where $r$ is the largest integer less than $k$. If $k > 1$, the mean of $Y$ is

$$m_Y = v\Gamma\left(1 - \frac{1}{k}\right), \tag{7.113}$$

and, if $k > 2$, the variance has the form

$$\sigma_Y^2 = v^2\left[\Gamma\left(1 - \frac{2}{k}\right) - \Gamma^2\left(1 - \frac{1}{k}\right)\right]. \tag{7.114}$$

The derivation of $F_Y(y)$ given by Equation (7.112) follows in broad outline that given for the Type-I maximum-value asymptotic distribution and will not be presented here. It has been used as a model in meteorology and hydrology (Gumbel, 1958).

A close relationship exists between the Type-I and Type-II asymptotic maximum-value distributions. Let $Y_I$ and $Y_{II}$ denote, respectively, these random

variables. It can be verified, using the techniques of transformations of random
variables, that they are related by

$$F_{Y_{II}}(y) = F_{Y_I}(\ln y), \quad y \geq 0, \tag{7.115}$$

where parameters $\alpha$ and $u$ in $F_{Y_I}(y)$ are related to parameters $k$ and $v$ in $F_{Y_{II}}(y)$ by

$$u = \ln v \quad \text{and} \quad \alpha = k. \tag{7.116}$$

When they are continuous, their pdfs obey the relationship

$$f_{Y_{II}}(y) = \frac{1}{y} f_{Y_I}(\ln y), \quad y \geq 0. \tag{7.117}$$

The Type-II asymptotic distribution of minimum values arises under analogous
conditions. With PDF $F_X(x)$ limited on the right at zero and approaching zero
on the left in a manner analogous to Equation (7.111), we have

$$F_Z(z) = 1 - \exp\left[-\left|\frac{z}{v}\right|^{-k}\right], \quad v, k > 0, \ z \leq 0. \tag{7.118}$$

However, it has not been found as useful as its counterparts in Type I and Type III
as in practice the required initial distributions are not frequently encountered.

### 7.6.3  TYPE-III ASYMPTOTIC DISTRIBUTIONS OF EXTREME VALUES

Since the Type-III maximum-value asymptotic distribution is of limited practical interest, only the minimum-value distribution will be discussed here.

The Type-III minimum-value asymptotic distribution is the limiting distribution of $Z_n$ as $n \to \infty$ for an initial distribution $F_X(x)$ in, which the left tail
increases from zero near $x = \varepsilon$ in the manner

$$F_X(x) = c(x - \varepsilon)^k, \quad c, k > 0, \ x \geq \varepsilon. \tag{7.119}$$

This class of distributions is bounded on the left at $x = \varepsilon$. The gamma distribution is such a distribution with $\varepsilon = 0$.

Again bypassing derivations, we can show the asymptotic distribution for the
minimum value to be

$$F_Z(z) = 1 - \exp\left[-\left(\frac{z - \varepsilon}{w - \varepsilon}\right)^k\right], \quad k > 0, \ w > \varepsilon, \ z \geq \varepsilon, \tag{7.120}$$

and, if it is continuous,

$$f_Z(z) = \frac{k}{w - \varepsilon}\left(\frac{z - \varepsilon}{w - \varepsilon}\right)^{k-1} \exp\left[-\left(\frac{z - \varepsilon}{w - \varepsilon}\right)^k\right], \quad k > 0, \ w > \varepsilon, \ z \geq \varepsilon. \quad (7.121)$$

The mean and variance of $Z$ are

$$\left.\begin{array}{l} m_Z = \varepsilon + (w - \varepsilon)\Gamma\left(1 + \dfrac{1}{k}\right), \\[2mm] \sigma_Z^2 = (w - \varepsilon)^2\left[\Gamma\left(1 + \dfrac{2}{k}\right) - \Gamma^2\left(1 + \dfrac{1}{k}\right)\right]. \end{array}\right\} \quad (7.122)$$

We have seen in Section 7.4.1 that the exponential distribution is used as a failure law in reliability studies, which corresponds to a constant hazard function [see Equations (7.64) and (7.66)]. The distribution given by Equations (7.120) and (7.121) is frequently used as a generalized time-to-failure model for cases in which the hazard function varies with time. One can show that the hazard function

$$h(t) = \frac{k}{w}\left(\frac{t}{w}\right)^{k-1}, \quad t \geq 0, \quad (7.123)$$

is capable of assuming a wide variety of shapes, and its associated probability density function for $T$, the time to failure, is given by

$$\boxed{f_T(t) = \frac{k}{w}\left(\frac{t}{w}\right)^{k-1} \exp[-\left(\frac{t}{w}\right)^k], \quad w, k > 0, \ t \geq 0.} \quad (7.124)$$

It is the so-called *Weibull distribution*, after Weibull, who first obtained it, heuristically (Weibull, 1939). Clearly, Equation (7.124) is a special case of Equation (7.121), with $\varepsilon = 0$.

The relationship between Type-III and Type-I minimum-value asymptotic distributions can also be established. Let $Z_I$ and $Z_{III}$ be the random variables having, respectively, Type-I and Type-III asymptotic distributions of minimum values. Then

$$F_{Z_{III}}(z) = F_{Z_I}[\ln(z - \varepsilon)], \quad z \geq \varepsilon, \quad (7.125)$$

with $u = \ln(w - \varepsilon)$, and $\alpha = k$. If they are continuous, the relationship between their pdfs is

$$f_{Z_{III}}(z) = \frac{1}{z - \varepsilon}f_{Z_I}[\ln(z - \varepsilon)], \quad z \geq \varepsilon. \quad (7.126)$$

**Table 7.1** Summary of continuous distributions

| Distribution | Probability density function | Parameters | Mean and variance |
|---|---|---|---|
| Uniform | $f_X(x) = \dfrac{1}{b-a}, a \le x \le b$ | $a, b > a$ | $\dfrac{a+b}{2}, \dfrac{(b-a)^2}{12}$ |
| Normal (Gaussian) | $f_X(x) = \dfrac{1}{(2\pi)^{1/2}\sigma}\exp\left[-\dfrac{(x-m)^2}{2\sigma^2}\right],$ $-\infty < x < \infty$ | $m, \sigma > 0$ | $m, \sigma^2$ |
| Lognormal | $f_Y(y) = \dfrac{1}{y\sigma_{\ln Y}(2\pi)^{1/2}}\exp\left[-\dfrac{1}{2\sigma_{\ln Y}^2} \times \ln^2\left(\dfrac{y}{\theta_Y}\right)\right], y \ge 0$ | $\theta_Y > 0, \sigma_{\ln Y} > 0$ | $\theta_Y \exp\left(\dfrac{\sigma_{\ln Y}^2}{2}\right), m_Y^2[\exp(\sigma_{\ln Y}^2) - 1]$ |
| Gamma | $f_X(x) = \dfrac{\lambda^\eta}{\Gamma(\eta)}x^{\eta-1}\exp(-\lambda x), x \ge 0$ | $\eta > 0, \lambda > 0$ | $\dfrac{\eta}{\lambda}, \dfrac{\eta}{\lambda^2}$ |
| Exponential | $f_X(x) = \lambda\exp(-\lambda x), x \ge 0$ | $\lambda > 0$ | $\dfrac{1}{\lambda}, \dfrac{1}{\lambda^2}$ |
| Chi-squared | $f_X(x) = \dfrac{1}{2^{\nu/2}\Gamma(\nu/2)}x^{(\nu/2)-1}\exp(-x/2),$ $x \ge 0$ | $\nu = $ positive integer | $\nu, 2\nu$ |

Beta

$$f_X(x) = \frac{\Gamma(\alpha+\beta)}{\Gamma(\alpha)\Gamma(\beta)} x^{\alpha-1}(1-x)^{\beta-1},$$
$$0 \le x \le 1$$

$\alpha > 0, \beta > 0$

$$\frac{\alpha}{\alpha+\beta}, \quad \frac{\alpha\beta}{(\alpha+\beta)^2(\alpha+\beta+1)}$$

Extreme values:

(a) Type-I max

$$f_Y(y) = \alpha \exp\{-\alpha(y-u) - \exp[-\alpha(y-u)]\}, \; -\infty < y < \infty$$

$\alpha > 0, u$

$$u + \frac{\gamma}{\alpha}, \quad \frac{\pi^2}{6\alpha^2}$$

(b) Type-I min

$$f_Z(z) = \alpha \exp\{\alpha(z-u) - \exp[\alpha(z-u)]\},$$
$$-\infty < z < \infty$$

$\alpha > 0, u$

$$u - \frac{\gamma}{\alpha}, \quad \frac{\pi^2}{6\alpha^2}$$

(c) Type-II max

$$f_Y(y) = \frac{k}{v}\left(\frac{v}{y}\right)^{k+1} \exp\left[-\left(\frac{y}{v}\right)^{-k}\right], \; y \ge 0$$

$v > 0, k > 0$

$$v\Gamma\left(1-\frac{1}{k}\right), \; v^2\left[\Gamma\left(1-\frac{2}{k}\right) - \Gamma^2\left(1-\frac{1}{k}\right)\right]$$

(d) Type-III min

$$f_Z(z) = \frac{k}{w-\varepsilon}\left(\frac{z-\varepsilon}{w-\varepsilon}\right)^{k-1} \times$$
$$\exp\left[-\left(\frac{z-\varepsilon}{w-\varepsilon}\right)^k\right], \; z \ge \varepsilon$$

$\varepsilon, w > \varepsilon, k > 0$

$$\varepsilon + (w-\varepsilon)\Gamma\left(1+\frac{1}{k}\right), \; (w-\varepsilon)^2\left[\Gamma\left(1+\frac{2}{k}\right)\right.$$
$$\left. -\Gamma^2\left(1+\frac{1}{k}\right)\right]$$

One final remark to be made is that asymptotic distributions of maximum and minimum values from the same initial distribution may not be of the same type. For example, for a gamma initial distribution, its asymptotic maximum-value distribution is of Type I whereas the minimum-value distribution falls into Type III. With reference to system time-to-failure models, a system having $n$ components in series with independent gamma life distributions for its components will have a time-to-failure distribution belonging to the Type-III asymptotic minimum-value distribution as $n$ becomes large. The corresponding model for a system having $n$ components in parallel is the Type-I asymptotic maximum-value distribution.

## 7.7 SUMMARY

As in Chapter 6, it is useful to summarize the important properties associated with some of the important continuous distributions discussed in this chapter. These are given in Table 7.1.

## REFERENCES

Gumbel, E.J., 1958, *Statistics of Extremes*, Columbia University Press, New York.
Kramer, M., 1940, "Frequency Surfaces in Two Variables Each of Which is Uniformly Distributed", *Amer. J. of Hygiene* **32** 45–64.
Lindberg, J.W., 1922, "Eine neue Herleitung des Exponentialgesetzes in der Wahrscheinlichkeifsrechnung", *Mathematische Zeitschrift* **15** 211–225.
Weibull, W., 1939, "A Statistical Theory of the Strength of Materials", *Proc. Royal Swedish Inst. for Engr. Res., Stockholm* No. 151.
Wilks, S., 1942, "Statistical Prediction with Special Reference to the Problem of Tolerance Limits", *Ann. Math. Stat.* **13** 400.

## FURTHER READING AND COMMENTS

As we mentioned in Section 7.2.1, the central limit theorem as stated may be generalized in several directions. Extensions since the 1920s include cases in which random variable $Y$ in Equation (7.14) is a sum of dependent and not necessarily identically distributed random variables. See, for example, the following two references:

Loéve, M., 1955, *Probability Theory*, Van Nostrand, New York.
Parzen, E., 1960, *Modern Probability Theory and its Applications*, John Wiley & Sons Inc., New York.

Extensive probability tables exist in addition to those given in Appendix A. Probability tables for lognormal, gamma, beta, chi-squared, and extreme-value distributions can be found in some of the references cited in Chapter 6. In particular, the following references are helpful:

Arkin, H., and Colton, R.1963, *Tables for Statisticians*, 2nd edn., Barnes and Noble, New York.

Beyer, W.H., 1996, *Handbook of Tables for Probability and Statistics*, Chemical Rubber Co., Cleveland, OH.

Hald, A., 1952, *Statistical Tables and Formulas*, John Wiley & Sons Inc. New York.

Owen, D., 1962, *Handbook of Statistical Tables*, Addision-Wesley, Reading,

Pearson, E.S., and Harley, H.O. (eds) 1954, *Biometrika Tables for Statisticians, Volume 1*, Cambridge University Press, Cambridge, England.

Additional useful references include:

Aitchison, J., and Brown, J.A.C., 1957, *The Log-normal Distribution*, Cambridge University Press, Cambridge, England.

Harter, H.L., 1964, *New Tables of the Incomplete Gamma Function Ratio and of Percentage Points of the Chi-square and Beta Distributions*, Aerospace Laboratory; US Government Printing office, Washington, DC.

National Bureau of Standards, 1954, *Tables of the Bivariate Normal Distribution and Related Functions: Applied Mathematics Series 50*, US Government Printing office, Washington, DC.

## PROBLEMS

7.1 The random variables $X$ and $Y$ are independent and uniformly distributed in interval (0.1). Determine the probability that their product $XY$ is less than 1/2.

7.2 The characteristic function (CF) of a random variable $X$ uniformly distributed in the interval $(-1, 1)$ is

$$\phi_X(t) = \frac{\sin t}{t}.$$

(a) Find the CF of $Y$, that is uniformly distributed in interval $(-a, a)$.

(b) Find the CF of $Y$ if it is uniformly distributed in interval $(a, a + b)$.

7.3 A machine component consisting of a rod-and-sleeve assembly is shown in Figure 7.15. Owing to machining inaccuracies, the inside diameter of the sleeve is uniformly distributed in the interval (1.98 cm, 2.02 cm), and the rod diameter is also uniformly distributed in the interval (1.95 cm, 2.00 cm). Assuming independence of these two distributions, find the probability that:

(a) The rod diameter is smaller than the sleeve diameter.

(b) There is at least a 0.01 cm clearance between the rod and the sleeve.

**Figure 7.15**   Rod and sleeve arrangement, for Problem 7.3

7.4  Repeat Problem 7.3 if the distribution of the rod diameter remains uniform but that of the sleeve inside diameter is $N(2\,cm, 0.0004\,cm^2)$.

7.5  The first mention of the normal distribution was made in the work of de Moivre in 1733 as one method of approximating probabilities of a binomial distribution when $n$ is large. Show that this approximation is valid and give an example showing results of this approximation.

7.6  If the distribution of temperature $T$ of a given volume of gas is $N(400, 1600)$, measured in degrees Fahrenheit, find:
    (a) $f_T(450)$;
    (b) $P(T \leq 450)$;
    (c) $P(|T - m_T| \leq 20)$;
    (d) $P(|T - m_T| \leq 20 | T \geq 300)$.

7.7  If $X$ is a random variable and distributed as $N(m, \sigma^2)$, show that

$$E\{|X - m|\} = \left(\frac{2}{\pi}\right)^{1/2} \sigma.$$

7.8  Let random variable $X$ and $Y$ be identically and normally distributed. Show that random variables $X + Y$ and $X - Y$ are independent.

7.9  Suppose that the useful lives measured in hours of two electronic devices, say $T_1$ and $T_2$, have distributions $N(40, 36)$ and $N(45, 9)$, respectively. If the electronic device is to be used for a 45-hour period, which is to be preferred? Which is preferred if it is to be used for a 48-hour period?

7.10  Verify Equation (7.13) for normal random variables.

7.11  Let random variables $X_1, X_2, \ldots, X_n$ be jointly normal with zero means. Show that

$E\{X_1 X_2 X_3\} = 0,$
$E\{X_1 X_2 X_3 X_4\} = E\{X_1 X_2\} E\{X_3 X_4\} + E\{X_1 X_3\} E\{X_2 X_4\} + E\{X_1 X_4\} E\{X_2 X_3\}.$

Generalize the results above and verify Equation (7.35).

7.12  Two rods, for which the lengths are independently, identically, and normally distributed random variables with means 4 inches and variances 0.02 square inches, are placed end to end.
    (a) What is the distribution of the total length?
    (b) What is the probability that the total length will be between 7.9 inches and 8.1 inches?

7.13  Let random variables $X_1$, $X_2$, and $X_3$ be independent and distributed according to $N(0, 1)$, $N(1, 1)$, and $N(2, 1)$, respectively. Determine probability $P(X_1 + X_2 + X_3 > 1)$.

7.14  A rope with 100 strands supports a weight of 2100 pounds. If the breaking strength of each strand is random, with mean equal to 20 pounds and standard deviation 4 pounds, and if the breaking strength of the rope is the sum of the independent breaking strengths of its strands, determine the probability that the rope will not fail under the load. (Assume there is no individual strand breakage before rope failure.)

7.15 If $X_1, X_2, \ldots, X_n$ are independent random variables, all having distribution $N(m, \sigma^2)$, determine the conditions that must be imposed on $c_1, c_2, \ldots, c_n$ such that the sum

$$Y = c_1 X_1 + c_2 X_2 + \cdots + c_n X_n$$

is also $N(m, \sigma^2)$. Can all $c$s be positive?

7.16 Let $U$ be the standardized normal random variable, and define $X = |U|$. Then, $X$ is called the folded standardized normal random variable. Determine $f_X(x)$.

7.17 The Cauchy distribution has the form

$$f_X(x) = \frac{1}{\pi(1 + x^2)}, \quad -\infty < x < \infty.$$

(a) Show that it arises from the ratio $X_1/X_2$, where $X_1$ and $X_2$ are independent and distributed as $N(0, \sigma^2)$.
(b) Show that the moments of $X$ do not exist.

7.18 Let $X_1$ and $X_2$ be independent normal random variables, both with mean 0 and standard deviation 1. Prove that:

$$Y = \arctan \frac{X_2}{X_1}$$

is uniformly distributed from $-\pi$ to $\pi$.

7.19 Verify Equations (7.48) for the lognormal distribution.

7.20 The lognormal distribution is found to be a good model for strains in structural members caused by wind loads. Let the strain be represented by $X$, with $m_X = 1$ and $\sigma_X^2 = 0.09$.
(a) Determine the probability $P(X > 1.2)$.
(b) If stress $Y$ in a structural member is related to the strain by $Y = a + bX$, with $b > 0$, determine $f_Y(y)$ and $m_Y$.

7.21 Arrivals at a rural entrance booth to the New York State Thruway are considered to be Poisson distributed with a mean arrival rate of 20 vehicles per hour. The time to process an arrival is approximately exponentially distributed with a mean time of one min.
(a) What percentage of the time is the tollbooth operator free to work on operational reports?
(b) How many cars are expected to be waiting to be processed, on average, per hour?
(c) What is the average time a driver waits in line before paying the toll?
(d) Whenever the average number of waiting vehicles reaches 5, a second tollbooth will be opened. How much will the average hourly rate of arrivals have to increase to require the addition of a second operator?

7.22 The life of a power transmission tower is exponentially distributed, with mean life 25 years. If three towers, operated independently, are being erected at the same time, what is the probability that at least 2 will still stand after 35 years?

7.23 For a gamma-distributed random variable, show that:
(a) Its mean and variance are those given by Equation (7.57).
(b) It has a positive skewness.

7.24 Show that, if $\eta$ is a positive integer, the probability distribution function (PDF) of a gamma-distributed random variable $X$ can be written as

$$F_X(x) = \begin{cases} 0, & \text{for } x \leq 0; \\ 1 - \sum_{k=0}^{\eta-1} \dfrac{(\lambda x)^k e^{-\lambda x}}{k!}, & \text{for } x > 0. \end{cases}$$

Recognize that the terms in the sum take the form of the Poisson mass function and therefore can be calculated with the aid of probability tables for Poisson distributions.

7.25 The system shown in Figure 7.16 has three redundant components, A–C. Let their operating lives (in hours) be denoted by $T_1$, $T_2$, and $T_3$, respectively. If the redundant parts come into operation only when the online component fails (cold redundancy), then the operating life of the system, $T$, is $T = T_1 + T_2 + T_3$.
Let $T_1$, $T_2$, and $T_3$ be independent random variables, each distributed as

$$f_{T_j}(t_j) = \begin{cases} \frac{1}{100} e^{-t_j/100}, & \text{for } t_j \geq 0, \; j = 1, 2, 3; \\ 0, & \text{otherwise.} \end{cases}$$

Determine the probability that the system will operate at least 300 hours.

7.26 We showed in Section 7.4.1 that an exponential failure law leads to a constant failure rate. Show that the converse is also true; that is, if $h(t)$ as defined by Equation (7.65) is a constant then the time to failure $T$ is exponentially distributed.

7.27 A shifted exponential distribution is defined as an exponential distribution shifted to the right by an amount $a$; that is, if random variable $X$ has an exponential distribution with

$$f_X(x) = \begin{cases} \lambda e^{-\lambda x}, & \text{for } x \geq 0; \\ 0, & \text{elsewhere;} \end{cases}$$

random variable $Y$ has a shifted exponential distribution if $f_Y(y)$ has the same shape as $f_X(x)$ but its nonzero portion starts at point $a$ rather than zero. Determine the relationship between $X$ and $Y$ and probability density function (pdf) $f_Y(y)$. What are the mean and variance of $Y$?

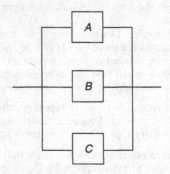

**Figure 7.16**  System of components, for Problem 7.25

7.28 Let random variable $X$ be $\chi^2$-distributed with parameter $n$. Show that the limiting distribution of

$$\frac{X - n}{(2n)^{1/2}}$$

as $n \to \infty$ is $N(0, 1)$.

7.29 Let $X_1, X_2, \ldots, X_n$ be independent random variables with common PDF $F_X(x)$ and pdf $f_X(x)$. Equations (7.89) and (7.91) give, respectively, the PDFs of their maximum and minimum values. Let $X_{(j)}$ be the random variable denoting the $j$th-smallest value of $X_1, X_2, \ldots, X_n$. Show that the PDF of $X_{(j)}$ has the form

$$F_{X_{(j)}}(x) = \sum_{k=j}^{n} \binom{n}{k} [F_X(x)]^k [1 - F_X(x)]^{n-k}, \quad j = 1, 2, \ldots, n.$$

7.30 Ten points are distributed uniformly and independently in interval $(0, 1)$. Find:
(a) The probability that the point lying farthest to the right is to the left of $3/4$.
(b) The probability that the point lying next farthest to the right is to the right of $1/2$.

7.31 Let the number of arrivals in a time interval obey the distribution given in Problem 6.32, which corresponds to a Poisson-type distribution with a time-dependent mean rate of arrival. Show that the pdf of time between arrivals is given by

$$f_T(t) = \begin{cases} \left(\dfrac{v}{w}\right) t^{v-1} \exp\left(-\dfrac{t^v}{w}\right), & \text{for } t \geq 0; \\ 0, & \text{elsewhere.} \end{cases}$$

As we see from Equation (7.124), it is the Weibull distribution.

7.32 A multiple-member structure in a parallel arrangement, as shown in Figure 7.17, supports a load $s$. It is assumed that all members share the load equally, that their resistances are random and identically distributed with common PDF $F_R(r)$, and that they act independently. If a member fails when the load it supports exceeds its resistance, show that the probability that failure will occur to $n - k$ members among $n$ initially existing members is

$$\left[1 - F_R\left(\frac{s}{n}\right)\right]^n, \quad k = n,$$

and

$$\sum_{j=1}^{n-k} \binom{n}{j} \left[F_R\left(\frac{s}{n}\right)\right]^j p^n_{(n-j)k}(s), \quad k = 0, 1, \ldots, n-1,$$

where

$$p^n_{kk}(s) = \left[1 - F_R\left(\frac{s}{k}\right)\right]^k,$$

$$p^i_{jk}(s) = \sum_{r=1}^{j-k} \binom{j}{r} \left[F_R\left(\frac{s}{j}\right) - F_R\left(\frac{s}{i}\right)\right]^r p^j_{(j-r)k}(s), \quad n \geq i > j > k.$$

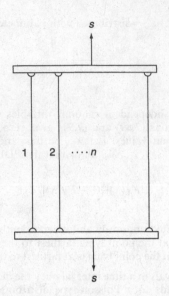

**Figure 7.17**  Structure under load $s$, for Problem 7.32

7.33  What is the probability sought in Problem 7.32 if the load is also a random variable $S$ with pdf $f_S(s)$?

7.34  Let $n = 3$ in Problem 7.32. Determine the probabilities of failure in zero, one, two, and three members in Problem 7.32 if $R$ follows a uniform distribution over interval $(80, 100)$, and $s = 270$. Is partial failure (one-member or two-member failure) possible in this case?

7.35  To show that, as a time-to-failure model, the Weibull distribution corresponds to a wide variety of shapes of the hazard function, graph the hazard function in Equation (7.123) and the corresponding Weibull distribution in Equation (7.124) for the following combinations of parameter values: $k = 0.5, 1, 2$, and 3; and $w = 1$ and 2.

7.36  The ranges of $n$ independent test flights of a supersonic aircraft are assumed to be identically distributed with PDF $F_X(x)$ and pdf $f_X(x)$. If *range span* is defined as the distance between the maximum and minimum ranges of these $n$ values, determine the pdf of the range span in terms of $F_X(x)$ or $f_X(x)$. Expressing it mathematically, the pdf of interest is that of $S$, where

$$S = Y - Z,$$

with

$$Y = \max(X_1, X_2, \ldots, X_n),$$

and

$$Z = \min(X_1, X_2, \ldots, X_n).$$

Note that random variables $Y$ and $Z$ are not independent.

# Part B

# Statistical Inference, Parameter Estimation, and Model Verification

# 8

# Observed Data and Graphical Representation

Referring to Figure 1.1 in Chapter 1, we are concerned in this and subsequent chapters with step D → E of the basic cycle in probabilistic modeling, that is, parameter estimation and model verification on the basis of observed data. In Chapters 6 and 7, our major concern has been the selection of an appropriate model (probability distribution) to represent a physical or natural phenomenon based on our understanding of its underlying properties. In order to specify the model completely, however, it is required that the parameters in the distribution be assigned. We now consider this problem of parameter estimation using available data. Included in this discussion are techniques for assessing the reasonableness of a selected model and the problem of selecting a model from among a number of contending distributions when no single one is preferred on the basis of the underlying physical characteristics of a given phenomenon.

Let us emphasize at the outset that, owing to the probabilistic nature of the situation, the problem of parameter estimation is precisely that – an estimation problem. A sequence of observations, say $n$ in number, is a *sample* of observed values of the underlying random variable. If we were to repeat the sequence of $n$ observations, the random nature of the experiment should produce a different sample of observed values. Any reasonable rule for extracting parameter estimates from a set of $n$ observations will thus give different estimates for different sets of observations. In other words, no single sequence of observations, finite in number, can be expected to yield true parameter values. What we are basically interested in, therefore, is to obtain relevant information about the distribution parameters by actually observing the underlying random phenomenon and using these observed numerical values in a systematic way.

---

*Fundamentals of Probability and Statistics for Engineers* T.T. Soong © 2004 John Wiley & Sons, Ltd
ISBNs: 0-470-86813-9 (HB)   0-470-86814-7 (PB)

## 8.1   HISTOGRAM AND FREQUENCY DIAGRAMS

Given a set of independent observations $x_1, x_2, \ldots,$ and $x_n$ of a random variable $X$, a useful first step is to organize and present them properly so that they can be easily interpreted and evaluated. When there are a large number of observed data, a *histogram* is an excellent graphical representation of the data, facilitating (a) an evaluation of adequacy of the assumed model, (b) estimation of percentiles of the distribution, and (c) estimation of the distribution parameters.

Let us consider, for example, a chemical process that is producing batches of a desired material; 200 observed values of the percentage yield, $X$, representing a relatively large sample size, are given in Table 8.1 (Hill, 1975). The sample values vary from 64 to 76. Dividing this range into 12 equal intervals and plotting the total number of observed yields in each interval as the height of a rectangle over the interval results in the histogram as shown in Figure 8.1. A *frequency diagram* is obtained if the ordinate of the histogram is divided by the total number of observations, 200 in this case, and by the interval width $\Delta$ (which happens to be one in this example). We see that the histogram or the frequency diagram gives an immediate impression of the range, relative frequency, and scatter associated with the observed data.

In the case of a discrete random variable, the histogram and frequency diagram as obtained from observed data take the shape of a bar chart as opposed to connected rectangles in the continuous case. Consider, for example, the distribution of the number of accidents per driver during a six-year time span in California. The data

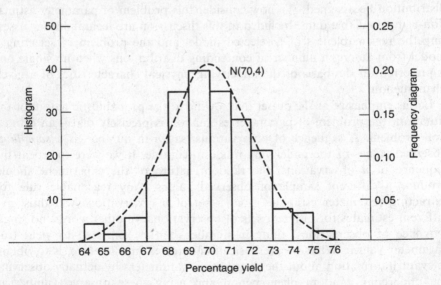

**Figure 8.1**   Histogram and frequency diagram for percentage yield
(data source: Hill, 1975)

**Table 8.1** Chemical yield data (data source: Hill, 1975)

| Batch no. | Yield (%) | Batch no. | Yield (%) | Batch no. | Yield (%) | Batch no. | Yield (%) | Batch no. | Yield (%) |
|---|---|---|---|---|---|---|---|---|---|
| 1 | 68.4 | 41 | 68.7 | 81 | 68.5 | 121 | 73.3 | 161 | 70.5 |
| 2 | 69.1 | 42 | 69.1 | 82 | 71.4 | 122 | 75.8 | 162 | 68.8 |
| 3 | 71.0 | 43 | 69.3 | 83 | 68.9 | 123 | 70.4 | 163 | 72.9 |
| 4 | 69.3 | 44 | 69.4 | 84 | 67.6 | 124 | 69.0 | 164 | 69.0 |
| 5 | 72.9 | 45 | 71.1 | 85 | 72.2 | 125 | 72.2 | 165 | 68.1 |
| 6 | 72.5 | 46 | 69.4 | 86 | 69.0 | 126 | 69.8 | 166 | 67.7 |
| 7 | 71.1 | 47 | 75.6 | 87 | 69.4 | 127 | 68.3 | 167 | 67.1 |
| 8 | 68.6 | 48 | 70.1 | 88 | 73.0 | 128 | 68.4 | 168 | 68.1 |
| 9 | 70.6 | 49 | 69.0 | 89 | 71.9 | 129 | 70.0 | 169 | 71.7 |
| 10 | 70.9 | 50 | 71.8 | 90 | 70.7 | 130 | 70.9 | 170 | 69.0 |
| 11 | 68.7 | 51 | 70.1 | 91 | 67.0 | 131 | 72.6 | 171 | 72.0 |
| 12 | 69.5 | 52 | 64.7 | 92 | 71.1 | 132 | 70.1 | 172 | 71.5 |
| 13 | 72.6 | 53 | 68.2 | 93 | 71.8 | 133 | 68.9 | 173 | 74.9 |
| 14 | 70.5 | 54 | 71.3 | 94 | 67.3 | 134 | 64.6 | 174 | 78.7 |
| 15 | 68.5 | 55 | 71.6 | 95 | 71.9 | 135 | 72.5 | 175 | 69.0 |
| 16 | 71.0 | 56 | 70.1 | 96 | 70.3 | 136 | 73.5 | 176 | 70.8 |
| 17 | 74.4 | 57 | 71.8 | 97 | 70.0 | 137 | 68.6 | 177 | 70.0 |
| 18 | 68.8 | 58 | 72.5 | 98 | 70.3 | 138 | 68.6 | 178 | 70.3 |
| 19 | 72.4 | 59 | 71.1 | 99 | 72.9 | 139 | 64.7 | 179 | 67.5 |
| 20 | 69.2 | 60 | 67.1 | 100 | 68.5 | 140 | 65.9 | 180 | 71.7 |
| 21 | 69.5 | 61 | 70.6 | 101 | 69.8 | 141 | 69.3 | 181 | 74.0 |
| 22 | 69.8 | 62 | 68.0 | 102 | 67.9 | 142 | 70.3 | 182 | 67.6 |
| 23 | 70.3 | 63 | 69.1 | 103 | 69.8 | 143 | 70.7 | 183 | 71.1 |
| 24 | 69.0 | 64 | 71.7 | 104 | 66.5 | 144 | 65.7 | 184 | 64.6 |
| 25 | 66.4 | 65 | 72.2 | 105 | 67.5 | 145 | 71.1 | 185 | 74.0 |
| 26 | 72.3 | 66 | 69.7 | 106 | 71.0 | 146 | 70.4 | 186 | 67.9 |
| 27 | 74.4 | 67 | 68.3 | 107 | 72.8 | 147 | 69.2 | 187 | 68.5 |
| 28 | 69.2 | 68 | 68.7 | 108 | 68.1 | 148 | 73.7 | 188 | 73.4 |
| 29 | 71.0 | 69 | 73.1 | 109 | 73.6 | 149 | 68.5 | 189 | 70.4 |
| 30 | 66.5 | 70 | 69.0 | 110 | 68.0 | 150 | 68.5 | 190 | 70.7 |
| 31 | 69.2 | 71 | 69.8 | 111 | 69.6 | 151 | 70.7 | 191 | 71.6 |
| 32 | 69.0 | 72 | 69.6 | 112 | 70.6 | 152 | 72.3 | 192 | 66.9 |
| 33 | 69.4 | 73 | 70.2 | 113 | 70.0 | 153 | 71.4 | 193 | 72.6 |
| 34 | 71.5 | 74 | 68.4 | 114 | 68.5 | 154 | 69.2 | 194 | 72.2 |
| 35 | 68.0 | 75 | 68.7 | 115 | 68.0 | 155 | 73.9 | 195 | 69.1 |
| 36 | 68.2 | 76 | 72.0 | 116 | 70.0 | 156 | 70.2 | 196 | 71.3 |
| 37 | 71.1 | 77 | 71.9 | 117 | 69.2 | 157 | 69.6 | 197 | 67.9 |
| 38 | 72.0 | 78 | 74.1 | 118 | 70.3 | 158 | 71.6 | 198 | 66.1 |
| 39 | 68.3 | 79 | 69.3 | 119 | 67.2 | 159 | 69.7 | 199 | 70.8 |
| 40 | 70.6 | 80 | 69.0 | 120 | 70.7 | 160 | 71.2 | 200 | 69.5 |

given in Table 8.2 are six-year accident records of 7842 California drivers (Burg, 1967, 1968). Based upon this set of observations, the histogram has the form given in Figure 8.2. The frequency diagram is obtained in this case simply by dividing the ordinate of the histogram by the total number of observations, which is 7842.

Returning now to the chemical yield example, the frequency diagram as shown in Figure 8.1 has the familiar properties of a probability density function (pdf). Hence, probabilities associated with various events can be estimated. For example, the probability of a batch having less than 68% yield can be read off from the frequency diagram by summing over the areas to the left of 68%, giving 0.13 (0.02 + 0.01 + 0.025 + 0.075). Similarly, the probability of a batch having yields greater than 72% is 0.18 (0.105 + 0.035 + 0.03 + 0.01). Let us remember, however, these are probabilities calculated based on the observed data. A different set of data obtained from the same chemical process would in general lead to a different frequency diagram and hence different values for these probabilities. Consequently, they are, at best, estimates of probabilities $P(X < 68)$ and $P(X > 72)$ associated with the underlying random variable $X$.

A remark on the choice of the number of intervals for plotting the histograms and frequency diagrams is in order. For this example, the choice of 12 intervals is convenient on account of the range of values spanned by the observations and of the fact that the resulting resolution is adequate for calculations of probabilities carried out earlier. In Figure 8.3, a histogram is constructed using 4 intervals instead of 12 for the same example. It is easy to see that it projects quite a different, and less accurate, visual impression of data behavior. It is thus important to choose the number of intervals *consistent* with the information one wishes to extract from the mathematical model. As a practical guide, Sturges (1926) suggests that an approximate value for the number of intervals, $k$, be determined from

$$k = 1 + 3.3 \log_{10} n, \tag{8.1}$$

where $n$ is the sample size.

From the modeling point of view, it is reasonable to select a normal distribution as the probabilistic model for percentage yield $X$ by observing that its random variations are the resultant of numerous independent random sources in the chemical manufacturing process. Whether or not this is a reasonable selection can be

**Table 8.2**  Six-year accident record for 7842 California drivers (data source: Burg, 1967, 1968)

| Number of accidents | Number of drivers |
|---|---|
| 0 | 5147 |
| 1 | 1859 |
| 2 | 595 |
| 3 | 167 |
| 4 | 54 |
| 5 | 14 |
| >5 | 6 |
| | Total = 7842 |

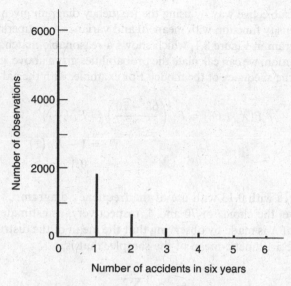

**Figure 8.2**  Histogram from six-year accident data (data source: Burg, 1967, 1968)

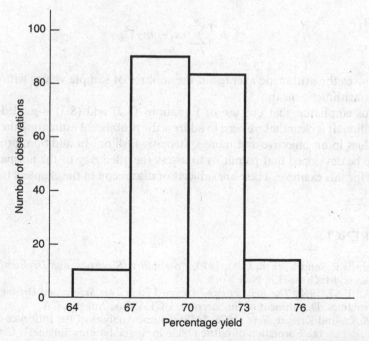

**Figure 8.3**  Histogram for percentage yield with four intervals (data source: Hill, 1975)

evaluated in a subjective way by using the frequency diagram given in Figure 8.1. The normal density function with mean 70 and variance 4 is superimposed on the frequency diagram in Figure 8.1, which shows a reasonable match. Based on this normal distribution, we can calculate the probabilities given above, giving a further assessment of the adequacy of the model. For example, with the aid of Table A.3,

$$P(X < 68) = F_U\left(\frac{68 - 70}{2}\right) = F_U(-1)$$
$$= 1 - F_U(1)$$
$$= 0.159,$$

which compares with 0.13 with use of the frequency diagram.

In the above, the choice of 70 and 4, respectively, as estimates of the mean and variance of $X$ is made by observing that the mean of the distribution should be close to the arithmetic mean of the sample, that is,

$$m_X \cong \frac{1}{n}\sum_{j=1}^{n} x_j, \tag{8.2}$$

and the variance can be approximated by

$$\sigma_X^2 \cong \frac{1}{n}\sum_{j=1}^{n}(x_j - m_X)^2, \tag{8.3}$$

which gives the arithmetic average of the squares of sample values with respect to their arithmetic mean.

Let us emphasize that our use of Equations (8.2) and (8.3) is guided largely by intuition. It is clear that we need to address the problem of estimating the parameter values in an objective and more systematic fashion. In addition, procedures need to be developed that permit us to assess the adequacy of the normal model chosen for this example. These are subjects of discussion in the chapters to follow.

## REFERENCES

Benjamin, J.R., and Cornell, C.A., 1970, *Probability, Statistics, and Decision for Civil Engineers*, McGraw-Hill, New York.

Burg, A., 1967, 1968, *The Relationship between Vision Test Scores and Driving Record*, two volumes. Department of Engineering, UCLA, Los Angeles, CA.

Chen, K.K., and Krieger, R.R., 1976, "A Statistical Analysis of the Influence of Cyclic Variation on the Formation of Nitric Oxide in Spark Ignition Engines", *Combustion Sci. Tech.* **12** 125–134.

Dunham, J.W., Brekke, G.N., and Thompson, G.N., 1952, *Live Loads on Floors in Buildings: Building Materials and Structures Report 133*, National Bureau of Standards, Washington, DC.

Ferreira Jr, J., 1974, "The Long-term Effects of Merit-rating Plans for Individual Motorist", *Oper. Research* **22** 954–978.

Hill, W.J., 1975, *Statistical Analysis for Physical Scientists: Class Notes*, State University of New York, Buffalo, NY.

Jelliffe, R.W., Buell, J., Kalaba, R., Sridhar, R., and Rockwell, R., 1970, "A Mathematical Study of the Metabolic Conversion of Digitoxin to Digoxin in Man", *Math. Biosci.* **6** 387–403.

Link, V.F., 1972, *Statistical Analysis of Blemishes in a SEC Image Tube*, masters thesis, State University of New York, Buffalo, NY.

Sturges, H.A., 1926, "The Choice of a Class Interval", *J. Am. Stat. Assoc.* **21** 65–66.

## PROBLEMS

8.1 It has been shown that the frequency diagram gives a graphical representation of the probability density function. Use the data given in Table 8.1 and construct a diagram that approximates the probability distribution function of percentage yield $X$.

8.2 In parts (a)–(l) below, observations or sample values of size $n$ are given for a random phenomenon.

(i) If not already given, plot the histogram and frequency diagram associated with the designated random variable $X$.

(ii) Based on the shape of these diagrams and on your understanding of the underlying physical situation, suggest one probability distribution (normal, Poisson, gamma, etc.) that may be appropriate for $X$. Estimate parameter value(s) by means of Equations (8.2) and (8.3) and, for the purposes of comparison, plot the proposed probability density function (pdf) or probability mass function (pmf) and superimpose it on the frequency diagram.

(a) $X$ is the maximum annual flood flow of the Feather River at Oroville, CA. Data given in Table 8.3 are records of maximum flood flows in 1000 cfs for the years 1902 to 1960 (source: Benjamin and Cornell, 1970).

(b) $X$ is the number of accidents per driver during a six-year time span in California. Data are given in Table 8.2 for 7842 drivers.

(c) $X$ is the time gap in seconds between cars on a stretch of highway. Table 8.4 gives measurements of time gaps in seconds between successive vehicles at a given location ($n = 100$).

(d) $X$ is the sum of two successive gaps in Part (c) above.

(e) $X$ is the number of vehicles arriving per minute at a toll booth on New York State Thruway. Measurements of 105 one-minute arrivals are given in Table 8.5.

(f) $X$ is the number of five-minute arrivals in Part (e) above.

(g) $X$ is the amount of yearly snowfall in inches in Buffalo, NY. Given in Table 8.6 are recorded snowfalls in inches from 1909 to 2002.

(h) $X$ is the peak combustion pressure in kPa per cycle. In spark ignition engines, cylinder pressure during combustion varies from cycle to cycle. The histogram of peak combustion pressure in kPa is shown in Figure 8.4 for 280 samples (source: Chen and Krieger, 1976).

**Table 8.3**    Maximum flood flows (in 1000 cfs), 1902–60 (source: Benjamin and Cornell, 1970).

| Year | Flood | Year | Flood | Year | Flood |
|------|-------|------|-------|------|-------|
| 1902 | 42  | 1922 | 36  | 1942 | 110 |
| 1903 | 102 | 1923 | 22  | 1943 | 108 |
| 1904 | 118 | 1924 | 42  | 1944 | 25  |
| 1905 | 81  | 1925 | 64  | 1945 | 60  |
| 1906 | 128 | 1926 | 56  | 1946 | 54  |
| 1907 | 230 | 1927 | 94  | 1947 | 46  |
| 1908 | 16  | 1928 | 185 | 1948 | 37  |
| 1909 | 140 | 1929 | 14  | 1949 | 17  |
| 1910 | 31  | 1930 | 80  | 1950 | 46  |
| 1911 | 75  | 1931 | 12  | 1951 | 92  |
| 1912 | 16  | 1932 | 23  | 1952 | 13  |
| 1913 | 17  | 1933 | 9   | 1953 | 59  |
| 1914 | 122 | 1934 | 20  | 1954 | 113 |
| 1915 | 81  | 1935 | 59  | 1955 | 55  |
| 1916 | 42  | 1936 | 85  | 1956 | 203 |
| 1917 | 80  | 1937 | 19  | 1957 | 83  |
| 1918 | 28  | 1938 | 185 | 1958 | 102 |
| 1919 | 66  | 1939 | 8   | 1959 | 35  |
| 1920 | 23  | 1940 | 152 | 1960 | 135 |
| 1921 | 62  | 1941 | 84  |      |     |

**Table 8.4**    Time gaps between vehicles (in seconds)

| 4.1 | 3.5 | 2.2 | 2.7 | 2.7 | 4.1 | 3.4 | 1.8 | 3.1 | 2.1 |
|-----|-----|-----|-----|-----|-----|-----|-----|-----|-----|
| 2.1 | 1.7 | 2.3 | 3.0 | 4.1 | 3.2 | 2.2 | 2.3 | 1.5 | 1.1 |
| 2.5 | 4.7 | 1.8 | 4.8 | 1.8 | 4.0 | 4.9 | 3.1 | 5.7 | 5.7 |
| 3.1 | 2.0 | 2.9 | 5.9 | 2.1 | 3.0 | 4.4 | 2.1 | 2.6 | 2.7 |
| 3.2 | 2.5 | 1.7 | 2.0 | 2.7 | 1.2 | 9.0 | 1.8 | 2.1 | 5.4 |
| 2.1 | 3.8 | 4.5 | 3.3 | 2.1 | 2.1 | 7.1 | 4.7 | 3.1 | 1.7 |
| 2.2 | 3.1 | 1.7 | 3.1 | 2.3 | 8.1 | 5.7 | 2.2 | 4.0 | 2.7 |
| 1.5 | 1.7 | 4.0 | 6.4 | 1.5 | 2.2 | 1.2 | 5.1 | 2.7 | 2.4 |
| 1.7 | 1.2 | 2.7 | 7.0 | 3.9 | 5.2 | 2.7 | 3.5 | 2.9 | 1.2 |
| 1.5 | 2.7 | 2.9 | 4.1 | 3.1 | 1.9 | 4.8 | 4.0 | 3.0 | 2.7 |

(i)    $X_1, X_2$, and $X_3$ are annual premiums paid by low-risk, medium-risk, and high-risk drivers. The frequency diagram for each group is given in Figure 8.5. (simulated results, over 50 years, are from Ferreira, 1974).

(j)    $X$ is the number of blemishes in a certain type of image tube for television, 58 data points are used for construction of the histogram shown in Figure 8.6. (source: Link, 1972).

(k)    $X$ is the difference between observed and computed urinary digitoxin excretion, in micrograms per day. In a study of metabolism of digitoxin to digoxin in patients, long-term studies of urinary digitoxin excretion were carried out on four patients. A histogram of the difference between

**Table 8.5**   Arrivals per minute at a New York State Thruway toll booth

| 9  | 9  | 11 | 15 | 6  | 11 | 9  | 6  | 11 | 8  | 10 |
|----|----|----|----|----|----|----|----|----|----|----|
| 3  | 9  | 8  | 5  | 7  | 15 | 7  | 14 | 6  | 6  | 16 |
| 6  | 8  | 10 | 6  | 10 | 11 | 9  | 7  | 7  | 11 | 10 |
| 3  | 8  | 4  | 7  | 15 | 6  | 7  | 7  | 8  | 7  | 5  |
| 13 | 12 | 11 | 10 | 8  | 14 | 3  | 15 | 13 | 5  | 7  |
| 12 | 7  | 10 | 4  | 16 |    | 7  | 11 | 11 | 13 | 10 |
| 9  | 10 | 11 | 6  | 6  |    | 8  | 9  | 5  | 5  | 5  |
| 11 | 6  | 7  | 9  | 5  |    | 12 | 12 | 4  | 13 | 4  |
| 12 | 16 | 10 | 14 | 15 |    | 16 | 10 | 8  | 10 | 6  |
| 18 | 13 | 6  | 9  | 4  |    | 13 | 14 | 6  | 10 | 10 |

**Table 8.6**   Annual snowfall, in inches, in Buffalo, NY, 1909–2002

| Year | Snowfall | Year | Snowfall | Year | Snowfall |
|------|----------|------|----------|------|----------|
| 1909–1910 | 126.4 | 1939–1940 | 77.8 | 1969–1970 | 120.5 |
| 1910–1911 | 82.4 | 1940–1941 | 79.3 | 1970–1971 | 97.0 |
| 1911–1912 | 78.1 | 1941–1942 | 89.6 | 1971–1972 | 109.9 |
| 1912–1913 | 51.1 | 1942–1943 | 85.5 | 1972–1973 | 78.8 |
| 1913–1914 | 90.9 | 1943–1944 | 58.0 | 1973–1974 | 88.7 |
| 1914–1915 | 76.2 | 1944–1945 | 120.7 | 1974–1975 | 95.6 |
| 1915–1916 | 104.5 | 1945–1946 | 110.5 | 1975–1976 | 82.5 |
| 1916–1917 | 87.4 | 1946–1947 | 65.4 | 1976–1977 | 199.4 |
| 1917–1918 | 110.5 | 1947–1948 | 39.9 | 1977–1978 | 154.3 |
| 1918–1919 | 25.0 | 1948–1949 | 40.1 | 1978–1979 | 97.3 |
| 1919–1920 | 69.3 | 1949–1950 | 88.7 | 1979–1980 | 68.4 |
| 1920–1921 | 53.5 | 1950–1951 | 71.4 | 1980–1981 | 60.9 |
| 1921–1922 | 39.8 | 1951–1952 | 83.0 | 1981–1982 | 112.4 |
| 1922–1923 | 63.6 | 1952–1953 | 55.9 | 1982–1983 | 52.4 |
| 1923–1924 | 46.7 | 1953–1954 | 89.9 | 1983–1984 | 132.5 |
| 1924–1925 | 72.9 | 1954–1955 | 84.6 | 1984–1985 | 107.2 |
| 1925–1926 | 74.6 | 1955–1956 | 105.2 | 1985–1986 | 114.7 |
| 1926–1927 | 83.6 | 1956–1957 | 113.7 | 1986–1987 | 67.5 |
| 1927–1928 | 80.7 | 1957–1958 | 124.7 | 1987–1988 | 56.4 |
| 1928–1929 | 60.3 | 1958–1959 | 114.5 | 1988–1989 | 67.4 |
| 1929–1930 | 79.0 | 1959–1960 | 115.6 | 1989–1990 | 93.7 |
| 1930–1931 | 74.4 | 1960–1961 | 102.4 | 1990–1991 | 57.5 |
| 1931–1932 | 49.6 | 1961–1962 | 101.4 | 1991–1992 | 92.8 |
| 1932–1933 | 54.7 | 1962–1963 | 89.8 | 1992–1993 | 93.2 |
| 1933–1934 | 71.8 | 1963–1964 | 71.5 | 1993–1994 | 112.7 |
| 1934–1935 | 49.1 | 1964–1965 | 70.9 | 1994–1995 | 74.6 |
| 1935–1936 | 103.9 | 1965–1966 | 98.3 | 1995–1996 | 141.4 |
| 1936–1937 | 51.6 | 1966–1967 | 55.5 | 1996–1997 | 97.6 |
| 1937–1938 | 82.4 | 1967–1968 | 66.1 | 1997–1998 | 75.6 |
| 1938–1939 | 83.6 | 1968–1969 | 78.4 | 1998–1999 | 100.5 |
|  |  |  |  | 1999–2000 | 63.6 |
|  |  |  |  | 2000–2001 | 158.7 |
|  |  |  |  | 2001–2002 | 132.4 |

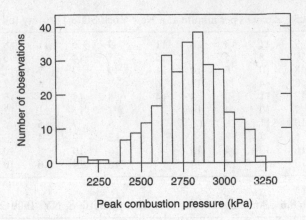

**Figure 8.4**   Histogram for Problem 8.2(h) (source: Chen and Krieger, 1976)

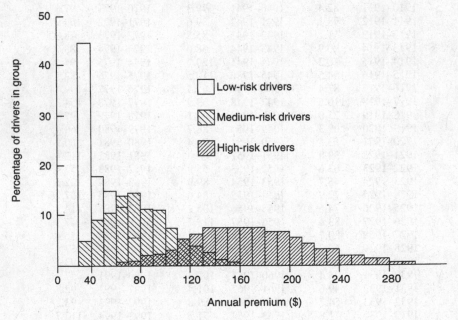

**Figure 8.5**   Frequency diagrams for Problem 8.2(i) (source: Ferreira, 1974)

observed and computed urinary digitoxin excretion in micrograms per day
is given in Figure 8.7 ($n = 100$) (source: Jelliffe *et al.*, 1970).

(l)  $X$ is the live load in pounds per square feet (psf) in warehouses. The
histogram in Figure 8.8 represents 220 measurements of live loads on
different floors of a warehouse over bays of areas of approximately 400
square feet (source: Dunham, 1952).

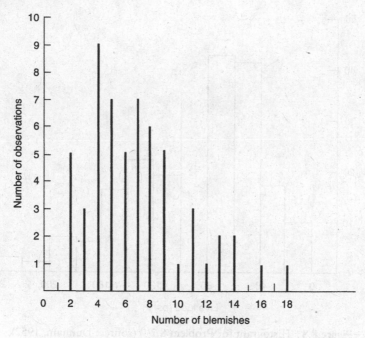

**Figure 8.6** Histogram for Problem 8.2(j) (source: Link, 1972)

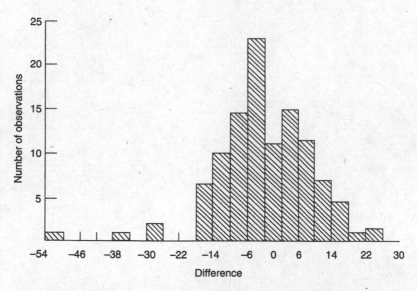

**Figure 8.7** Histogram for Problem 8.2(k) (source: Jelliffe *et al.*, 1970). Note: the horizontal axis shows the difference between the observed and computed urinary digitoxin excretion, in micrograms per day

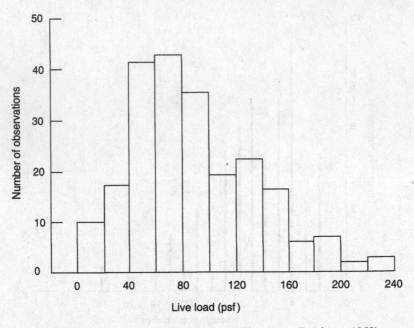

**Figure 8.8**   Histogram for Problem 8.2(l) (source: Dunham, 1952)

# 9

# Parameter Estimation

Suppose that a probabilistic model, represented by probability density function (pdf) $f(x)$, has been chosen for a physical or natural phenomenon for which parameters $\theta_1, \theta_2, \ldots$ are to be estimated from independently observed data $x_1, x_2, \ldots, x_n$. Let us consider for a moment a single parameter $\theta$ for simplicity and write $f(x; \theta)$ to mean a specified probability distribution where $\theta$ is the unknown parameter to be estimated. The parameter estimation problem is then one of determining an appropriate function of $x_1, x_2, \ldots, x_n$, say $h(x_1, x_2, \ldots, x_n)$, which gives the 'best' estimate of $\theta$. In order to develop systematic estimation procedures, we need to make more precise the terms that were defined rather loosely in the preceding chapter and introduce some new concepts needed for this development.

## 9.1 SAMPLES AND STATISTICS

Given an independent data set $x_1, x_2, \ldots, x_n$, let

$$\hat{\theta} = h(x_1, x_2, \ldots, x_n) \tag{9.1}$$

be an estimate of parameter $\theta$. In order to ascertain its general properties, it is recognized that, if the experiment that yielded the data set were to be repeated, we would obtain different values for $x_1, x_2, \ldots, x_n$. The function $h(x_1, x_2, \ldots, x_n)$ when applied to the new data set would yield a different value for $\hat{\theta}$. We thus see that estimate $\hat{\theta}$ is itself a random variable possessing a probability distribution, which depends both on the functional form defined by $h$ and on the distribution of the underlying random variable $X$. The appropriate representation of $\hat{\theta}$ is thus

$$\hat{\Theta} = h(X_1, X_2, \ldots, X_n), \tag{9.2}$$

where $X_1, X_2, \ldots, X_n$ are random variables, representing a *sample* from random variable $X$, which is referred to in this context as the *population*. In practically

---

*Fundamentals of Probability and Statistics for Engineers* T.T. Soong © 2004 John Wiley & Sons, Ltd
ISBNs: 0-470-86813-9 (HB)   0-470-86814-7 (PB)

all applications, we shall assume that sample $X_1, X_2, \ldots, X_n$ possesses the following properties:

- Property 1: $X_1, X_2, \ldots, X_n$ are independent.
- Property 2: $f_{X_j}(x) = f_X(x)$ for all $x, j = 1, 2, \ldots, n$.

The random variables $X_1, \ldots, X_n$ satisfying these conditions are called a *random sample* of size $n$. The word 'random' in this definition is usually omitted for the sake of brevity. If $X$ is a random variable of the discrete type with probability mass function (pmf) $p_X(x)$, then $p_{X_j}(x) = p_X(x)$ for each $j$.

A specific set of observed values $(x_1, x_2, \ldots, x_n)$ is a set of *sample values* assumed by the sample. The problem of parameter estimation is one class in the broader topic of statistical inference in which our object is to make inferences about various aspects of the underlying population distribution on the basis of observed sample values. For the purpose of clarification, the interrelationships among $X, (X_1, X_2, \ldots, X_n)$, and $(x_1, x_2, \ldots, x_n)$ are schematically shown in Figure 9.1.

Let us note that the properties of a sample as given above imply that certain conditions are imposed on the manner in which observed data are obtained. Each datum point must be observed from the population independently and under identical conditions. In sampling a population of percentage yield, as discussed in Chapter 8, for example, one would avoid taking adjacent batches if correlation between them is to be expected.

A *statistic* is any function of a given sample $X_1, X_2, \ldots, X_n$ that does not depend on the unknown parameter. The function $h(X_1, X_2, \ldots, X_n)$ in Equation (9.2) is thus a statistic for which the value can be determined once the sample values have been observed. It is important to note that a statistic, being a function of random variables, is a random variable. When used to estimate a distribution parameter, its statistical properties, such as mean, variance, and distribution, give information concerning the quality of this particular estimation procedure. Certain statistics play an important role in statistical estimation theory; these include sample mean, sample variance, order statistics, and other sample moments. Some properties of these important statistics are discussed below.

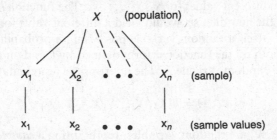

**Figure 9.1**  Population, sample, and sample values

### 9.1.1   SAMPLE MEAN

The statistic

$$\overline{X} = \frac{1}{n}\sum_{i=1}^{n} X_i \tag{9.3}$$

is called the *sample mean* of population $X$. Let the population mean and variance be, respectively,

$$\left.\begin{array}{l} E\{X\} = m, \\ \text{var}\{X\} = \sigma^2. \end{array}\right\} \tag{9.4}$$

The mean and variance of $\overline{X}$, the sample mean, are easily found to be

$$E\{\overline{X}\} = \frac{1}{n}\sum_{i=1}^{n} E\{X_i\} = \frac{1}{n}(nm) = m, \tag{9.5}$$

and, owing to independence,

$$\text{var}\{\overline{X}\} = E\{(\overline{X} - m)^2\} = E\left\{\left[\frac{1}{n}\sum_{i=1}^{n}(X_i - m)\right]^2\right\}$$
$$= \frac{1}{n^2}(n\sigma^2) = \frac{\sigma^2}{n}, \tag{9.6}$$

which is inversely proportional to sample size $n$. As $n$ increases, the variance of $\overline{X}$ decreases and the distribution of $\overline{X}$ becomes sharply peaked at $E\{\overline{X}\} = m$. Hence, it is intuitively clear that statistic $\overline{X}$ provides a good procedure for estimating population mean $m$. This is another statement of the law of large numbers that was discussed in Example 4.12 (page 96) and Example 4.13 (page 97).

Since $\overline{X}$ is a sum of independent random variables, its distribution can also be determined either by the use of techniques developed in Chapter 5 or by means of the method of characteristic functions given in Section 4.5. We further observe that, on the basis of the central limit theorem (Section 7.2.1), sample mean $\overline{X}$ approaches a normal distribution as $n \to \infty$. More precisely, random variable

$$(\overline{X} - m)\left(\frac{\sigma}{n^{1/2}}\right)^{-1}$$

approaches $N(0, 1)$ as $n \to \infty$.

## 9.1.2   SAMPLE VARIANCE

The statistic

$$S^2 = \frac{1}{n-1} \sum_{i=1}^{n} (X_i - \overline{X})^2 \qquad (9.7)$$

is called the *sample variance* of population $X$. The mean of $S^2$ can be found by expanding the squares in the sum and taking termwise expectations. We first write Equation (9.7) as

$$S^2 = \frac{1}{n-1} \sum_{i=1}^{n} [(X_i - m) - (\overline{X} - m)]^2$$

$$= \frac{1}{n-1} \sum_{i=1}^{n} \left[ (X_i - m) - \frac{1}{n} \sum_{j=1}^{n} (X_j - m) \right]^2$$

$$= \frac{1}{n} \sum_{i=1}^{n} (X_i - m)^2 - \frac{1}{n(n-1)} \sum_{\substack{i,j=1 \\ i \neq j}}^{n} (X_i - m)(X_j - m).$$

Taking termwise expectations and noting mutual independence, we have

$$E\{S^2\} = \sigma^2, \qquad (9.8)$$

where $m$ and $\sigma^2$ are defined in Equations (9.4). We remark at this point that the reason for using $1/(n-1)$ rather than $1/n$ in Equation (9.7) is to make the mean of $S^2$ equal to $\sigma^2$. As we shall see in the next section, this is a desirable property for $S^2$ if it is to be used to estimate $\sigma^2$, the true variance of $X$.

The variance of $S^2$ is found from

$$\mathrm{var}\{S^2\} = E\{(S^2 - \sigma^2)^2\}. \qquad (9.9)$$

Upon expanding the right-hand side and carrying out expectations term by term, we find that

$$\mathrm{var}\{S^2\} = \frac{1}{n} \left[ \mu_4 - \frac{n-3}{n-1} \sigma^4 \right], \qquad (9.10)$$

where $\mu_4$ is the fourth central moment of $X$; that is,

$$\mu_4 = E\{(X - m)^4\}. \qquad (9.11)$$

Equation (9.10) shows again that the variance of $S^2$ is an inverse function of $n$.

In principle, the distribution of $S^2$ can be derived with use of techniques advanced in Chapter 5. It is, however, a tedious process because of the complex nature of the expression for $S^2$ as defined by Equation (9.7). For the case in which population $X$ is distributed according to $N(m, \sigma^2)$, we have the following result (Theorem 9.1).

**Theorem 9.1:** Let $S^2$ be the sample variance of size $n$ from normal population $N(m, \sigma^2)$, then $(n-1)S^2/\sigma^2$ has a chi-squared $(\chi^2)$ distribution with $(n-1)$ degrees of freedom.

**Proof of Theorem 9.1:** the chi-squared distribution is given in Section 7.4.2. In order to sketch a proof for this theorem, let us note from Section 7.4.2 that random variable $Y$,

$$Y = \frac{1}{\sigma^2} \sum_{i=1}^{n} (X_i - m)^2, \tag{9.12}$$

has a chi-squared distribution of $n$ degrees of freedom since each term in the sum is a squared normal random variable and is independent of other random variables in the sum. Now, we can show that the difference between $Y$ and $(n-1)S^2/\sigma^2$ is

$$Y - \frac{(n-1)S^2}{\sigma^2} = \left[ (\overline{X} - m)\left(\frac{\sigma}{n^{1/2}}\right)^{-1} \right]^2. \tag{9.13}$$

Since the right-hand side of Equation (9.13) is a random variable having a chi-squared distribution with one degree of freedom, Equation (9.13) leads to the result that $(n-1)S^2/\sigma^2$ is chi-squared distributed with $(n-1)$ degrees of freedom provided that independence exists between $(n-1)S^2/\sigma^2$ and

$$\left[ (\overline{X} - m)\left(\frac{\sigma}{n^{1/2}}\right)^{-1} \right]^2$$

The proof of this independence is not given here but can be found in more advanced texts (e.g. Anderson and Bancroft, 1952).

### 9.1.3  SAMPLE MOMENTS

The $k$th sample moment is

$$M_k = \frac{1}{n} \sum_{i=1}^{n} X_i^k. \tag{9.14}$$

Following similar procedures as given above, we can show that

$$
\left.\begin{aligned}
E\{M_k\} &= \alpha_k, \\
\mathrm{var}\{M_k\} &= \frac{1}{n}(\alpha_{2k} - \alpha_k^2),
\end{aligned}\right\}
\tag{9.15}
$$

where $\alpha_k$ is the $k$th moment of population $X$.

### 9.1.4   ORDER STATISTICS

A sample $X_1, X_2, \ldots, X_n$ can be ranked in order of increasing numerical magnitude. Let $X_{(1)}, X_{(2)}, \ldots, X_{(n)}$ be such a rearranged sample, where $X_{(1)}$ is the smallest and $X_{(n)}$ the largest. Then $X_{(k)}$ is called the $k$th-order statistic. Extreme values $X_{(1)}$ and $X_{(n)}$ are of particular importance in applications, and their properties have been discussed in Section 7.6.

In terms of the probability distribution function (PDF) of population $X$, $F_X(x)$, it follows from Equations (7.89) and (7.91) that the PDFs of $X_{(1)}$ and $X_{(n)}$ are

$$
F_{X_{(1)}}(x) = 1 - [1 - F_X(x)]^n,
\tag{9.16}
$$

$$
F_{X_{(n)}}(x) = F_X^n(x).
\tag{9.17}
$$

If $X$ is continuous, the pdfs of $X_{(1)}$ and $X_{(n)}$ are of the form [see Equations (7.90) and (7.92)]

$$
f_{X_{(1)}}(x) = n[1 - F_X(x)]^{n-1}f_X(x),
\tag{9.18}
$$

$$
f_{X_{(n)}}(x) = nF_X^{n-1}(x)f_X(x).
\tag{9.19}
$$

The means and variances of order statistics can be obtained through integration, but they are not expressible as simple functions of the moments of population $X$.

## 9.2   QUALITY CRITERIA FOR ESTIMATES

We are now in a position to propose a number of criteria under which the quality of an estimate can be evaluated. These criteria define generally desirable properties for an estimate to have as well as provide a guide by which the quality of one estimate can be compared with that of another.

Before proceeding, a remark is in order regarding the notation to be used. As seen in Equation (9.2), our objective in parameter estimation is to determine a statistic

$$\hat{\Theta} = h(X_1, X_2, \ldots, X_n), \tag{9.20}$$

which gives a good estimate of parameter $\theta$. This statistic will be called an *estimator* for $\theta$, for which properties, such as mean, variance, or distribution, provide a measure of quality of this estimator. Once we have observed sample values $x_1, x_2, \ldots, x_n$, the observed estimator,

$$\hat{\theta} = h(x_1, x_2, \ldots, x_n), \tag{9.21}$$

has a numerical value and will be called an *estimate* of parameter $\theta$.

## 9.2.1  UNBIASEDNESS

An estimator $\hat{\Theta}$ is said to be an *unbiased* estimator for $\theta$ if

$$\boxed{E\{\hat{\Theta}\} = \theta,} \tag{9.22}$$

for all $\theta$. This is clearly a desirable property for $\hat{\Theta}$, which states that, on average, we expect $\hat{\Theta}$ to be close to true parameter value $\theta$. Let us note here that the requirement of unbiasedness may lead to other undesirable consequences. Hence, the overall quality of an estimator does not rest on any single criterion but on a set of criteria.

We have studied two statistics, $\overline{X}$ and $S^2$, in Sections 9.1.1 and 9.1.2. It is seen from Equations (9.5) and (9.8) that, if $\overline{X}$ and $S^2$ are used as estimators for the population mean $m$ and population variance $\sigma^2$, respectively, they are unbiased estimators. This nice property for $S^2$ suggests that the sample variance defined by Equation (9.7) is preferred over the more natural choice obtained by replacing $1/(n-1)$ by $1/n$ in Equation (9.7). Indeed, if we let

$$S^{2*} = \frac{1}{n} \sum_{i=1}^{n} (X_i - \overline{X})^2, \tag{9.23}$$

its mean is

$$E\{S^{2*}\} = \frac{n-1}{n} \sigma^2,$$

and estimator $S^{2*}$ has a bias indicated by the coefficient $(n-1)/n$.

## 9.2.2   MINIMUM VARIANCE

It seems natural that, if $\hat{\Theta} = h(X_1, X_2, \ldots, X_n)$ is to qualify as a good estimator for $\theta$, not only its mean should be close to true value $\theta$ but also there should be a good probability that any of its observed values $\hat{\theta}$ will be close to $\theta$. This can be achieved by selecting a statistic in such a way that not only is $\hat{\Theta}$ unbiased but also its variance is as small as possible. Hence, the second desirable property is one of minimum variance.

**Definition 9.1.** let $\hat{\Theta}$ be an unbiased estimator for $\theta$. It is an *unbiased minimum-variance* estimator for $\theta$ if, for all other unbiased estimators $\Theta^*$ of $\theta$ from the same sample,

$$\boxed{\text{var}\{\hat{\Theta}\} \leq \text{var}\{\Theta^*\},} \tag{9.24}$$

for all $\theta$.

Given two unbiased estimators for a given parameter, the one with smaller variance is preferred because smaller variance implies that observed values of the estimator tend to be closer to its mean, the true parameter value.

**Example 9.1.** Problem: we have seen that $\overline{X}$ obtained from a sample of size $n$ is an unbiased estimator for population mean $m$. Does the quality of $\overline{X}$ improve as $n$ increases?

Answer: we easily see from Equation (9.5) that the mean of $\overline{X}$ is independent of the sample size; it thus remains unbiased as $n$ increases. Its variance, on the other hand, as given by Equation (9.6) is

$$\text{var}\{\overline{X}\} = \frac{\sigma^2}{n}, \tag{9.25}$$

which decreases as $n$ increases. Thus, based on the minimum variance criterion, the quality of $\overline{X}$ as an estimator for $m$ improves as $n$ increases.

**Example 9.2.** Part 1. Problem: based on a fixed sample size $n$, is $\overline{X}$ the best estimator for $m$ in terms of unbiasedness and minimum variance?

Approach: in order to answer this question, it is necessary to show that the variance of $\overline{X}$ as given by Equation (9.25) is the smallest among *all* unbiased estimators that can be constructed from the sample. This is certainly difficult to do. However, a powerful theorem (Theorem 9.2) shows that it is possible to determine the minimum achievable variance of any unbiased estimator obtained from a given sample. This lower bound on the variance thus permits us to answer questions such as the one just posed.

**Theorem 9.2:** *the Cramér–Rao inequality*. Let $X_1, X_2, \ldots, X_n$ denote a sample of size $n$ from a population $X$ with pdf $f(x; \theta)$, where $\theta$ is the unknown parameter, and let $\hat{\Theta} = h(X_1, X_2, \ldots, X_n)$ be an unbiased estimator for $\theta$. Then, the variance of $\hat{\Theta}$ satisfies the inequality

$$\text{var}\{\hat{\Theta}\} \geq \left\{ n E\left\{ \left[ \frac{\partial \ln f(X; \theta)}{\partial \theta} \right]^2 \right\} \right\}^{-1}, \qquad (9.26)$$

if the indicated expectation and differentiation exist. An analogous result with $p(X; \theta)$ replacing $f(X; \theta)$ is obtained when $X$ is discrete.

**Proof of Theorem 9.2:** the joint probability density function (jpdf) of $X_1, X_2, \ldots,$ and $X_n$ is, because of their mutual independence, $f(x_1; \theta) f(x_2; \theta) \cdots f(x_n; \theta)$. The mean of statistic $\hat{\Theta}$, $\hat{\Theta} = h(X_1, X_2, \ldots, X_n)$, is

$$E\{\hat{\Theta}\} = E\{h(X_1, X_2, \ldots, X_n)\},$$

and, since $\hat{\Theta}$ is unbiased, it gives

$$\theta = \int_{-\infty}^{\infty} \cdots \int_{-\infty}^{\infty} h(x_1, \ldots, x_n) f(x_1; \theta) \cdots f(x_n; \theta) \, dx_1 \cdots dx_n. \qquad (9.27)$$

Another relation we need is the identity:

$$1 = \int_{-\infty}^{\infty} f(x_i; \theta) \, dx_i, \quad i = 1, 2, \ldots, n. \qquad (9.28)$$

Upon differentiating both sides of each of Equations (9.27) and (9.28) with respect to $\theta$, we have

$$1 = \int_{-\infty}^{\infty} \cdots \int_{-\infty}^{\infty} h(x_1, \ldots, x_n) \left[ \sum_{j=1}^{n} \frac{1}{f(x_j; \theta)} \frac{\partial f(x_j; \theta)}{\partial \theta} \right] f(x_1; \theta) \cdots f(x_n; \theta) \, dx_1 \cdots dx_n$$

$$\qquad (9.30)$$

$$= \int_{-\infty}^{\infty} \cdots \int_{-\infty}^{\infty} h(x_1, \ldots, x_n) \left[ \sum_{j=1}^{n} \frac{\partial \ln f(x_j; \theta)}{\partial \theta} \right] f(x_1; \theta) \cdots f(x_n; \theta) \, dx_1 \cdots dx_n,$$

$$0 = \int_{-\infty}^{\infty} \frac{\partial f(x_i; \theta)}{\partial \theta} \, dx_i$$

$$\qquad (9.30)$$

$$= \int_{-\infty}^{\infty} \frac{\partial \ln f(x_i; \theta)}{\partial \theta} f(x_i; \theta) \, dx_i, \quad i = 1, 2, \ldots, n.$$

Let us define a new random variable $Y$ by

$$Y = \sum_{j=1}^{n} \frac{\partial \ln f(X_j; \theta)}{\partial \theta}.$$ (9.31)

Equation (9.30) shows that

$$E\{Y\} = 0.$$

Moreover, since $Y$ is a sum of $n$ independent random variables, each with mean zero and variance $E\{[\partial \ln f(X; \theta)/\partial \theta]^2\}$, the variance of $Y$ is the sum of the $n$ variances and has the form

$$\sigma_Y^2 = nE\left\{ \left[ \frac{\partial \ln f(X; \theta)}{\partial \theta} \right]^2 \right\}.$$ (9.32)

Now, it follows from Equation (9.29) that

$$1 = E\{\hat{\Theta} Y\}.$$ (9.33)

Recall that

$$E\{\hat{\Theta} Y\} = E\{\hat{\Theta}\} E\{Y\} + \rho_{\hat{\Theta} Y} \sigma_{\hat{\Theta}} \sigma_Y,$$

or

$$1 = \theta(0) + \rho_{\hat{\Theta} Y} \sigma_{\hat{\Theta}} \sigma_Y.$$ (9.34)

As a consequence of property $\rho^2 \leq 1$, we finally have

$$\frac{1}{\sigma_{\hat{\Theta}}^2 \sigma_Y^2} \leq 1,$$

or, using Equation (9.32),

$$\sigma_{\hat{\Theta}}^2 \geq \frac{1}{\sigma_Y^2} = \left\{ nE\left\{ \left[ \frac{\partial \ln f(X; \theta)}{\partial \theta} \right]^2 \right\} \right\}^{-1}.$$ (9.35)

The proof is now complete.

In the above, we have assumed that differentiation with respect to $\theta$ under an integral or sum sign are permissible. Equation (9.26) gives a lower bound on the variance of any unbiased estimator and it expresses a fundamental limitation on the accuracy with which a parameter can be estimated. We also note that this lower bound is, in general, a function of $\theta$, the true parameter value.

Several remarks in connection with the Cramér–Rao lower bound (CRLB) are now in order.

• Remark 1: the expectation in Equation (9.26) is equivalent to $-E\{\partial^2 \ln f(X;\theta)/\partial\theta^2\}$, or

$$\sigma_{\hat{\Theta}}^2 \geq -\left\{ nE\left[\frac{\partial^2 \ln f(X;\theta)}{\partial\theta^2}\right]\right\}^{-1}. \tag{9.36}$$

This alternate expression offers computational advantages in some cases.

• Remark 2: the result given by Equation (9.26) can be extended easily to multiple parameter cases. Let $\theta_1, \theta_2, \ldots,$ and $\theta_m$ $(m \leq n)$ be the unknown parameters in $f(x; \theta_1, \ldots, \theta_m)$, which are to be estimated on the basis of a sample of size $n$. In vector notation, we can write

$$\boldsymbol{\theta}^{\mathrm{T}} = [\theta_1 \quad \theta_2 \quad \cdots \quad \theta_m], \tag{9.37}$$

with corresponding vector unbiased estimator

$$\hat{\boldsymbol{\Theta}}^{\mathrm{T}} = [\hat{\Theta}_1 \quad \hat{\Theta}_2 \quad \cdots \quad \hat{\Theta}_m]. \tag{9.38}$$

Following similar steps in the derivation of Equation (9.26), we can show that the Cramér–Rao inequality for multiple parameters is of the form

$$\mathrm{cov}\{\hat{\boldsymbol{\Theta}}\} \geq \frac{\Lambda^{-1}}{n}, \tag{9.39}$$

where $\Lambda^{-1}$ is the inverse of matrix $\Lambda$ for which the elements are

$$\Lambda_{ij} = E\left\{\frac{\partial \ln f(X;\boldsymbol{\theta})}{\partial\theta_i}\frac{\partial \ln f(X;\boldsymbol{\theta})}{\partial\theta_j}\right\}, \quad i,j = 1,2,\ldots,m. \tag{9.40}$$

Equation (9.39) implies that

$$\mathrm{var}\{\hat{\Theta}_j\} \geq \frac{(\Lambda^{-1})_{jj}}{n} \geq \frac{1}{n\Lambda_{jj}}, \quad j = 1,2,\ldots,m, \tag{9.41}$$

where $(\Lambda^{-1})_{jj}$ is the $jj$th element of $\Lambda^{-1}$.

• Remark 3: the CRLB can be transformed easily under a transformation of the parameter. Suppose that, instead of $\theta$, parameter $\phi = g(\theta)$ is of interest, which is a one-to-one transformation and differentiable with respect to $\theta$; then,

$$\text{CRLB for var}\{\hat{\Phi}\} = \left[\frac{dg(\theta)}{d\theta}\right]^2 [\text{CRLB for var}(\hat{\Theta})], \qquad (9.42)$$

where $\hat{\Phi}$ is an unbiased estimator for $\phi$.

- Remark 4: given an unbiased estimator $\hat{\Theta}$ for parameter $\theta$, the ratio of its CRLB to its variance is called the *efficiency* of $\hat{\Theta}$. The efficiency of any unbiased estimator is thus always less than or equal to 1. An unbiased estimator with efficiency equal to 1 is said to be *efficient*. We must point out, however, that efficient estimators exist only under certain conditions.

We are finally in the position to answer the question posed in Example 9.2.

**Example 9.2.** part 2. Answer: first, we note that, in order to apply the CRLB, pdf $f(x;\theta)$ of population $X$ must be known. Suppose that $f(x;m)$ for this example is $N(m, \sigma^2)$. We have

$$\ln f(X;m) = \ln\left\{\frac{1}{(2\pi)^{1/2}\sigma}\exp\left[\frac{-(X-m)^2}{2\sigma^2}\right]\right\}$$

$$= \ln\left[\frac{1}{(2\pi)^{1/2}\sigma}\right] - \frac{(X-m)^2}{2\sigma^2},$$

and

$$\frac{\partial \ln f(X;m)}{\partial m} = \frac{X-m}{\sigma^2}.$$

Thus,

$$E\left\{\left[\frac{\partial \ln f(X;m)}{\partial m}\right]^2\right\} = \frac{1}{\sigma^4}E\{(X-m)^2\} = \frac{1}{\sigma^2}.$$

Equation (9.26) then shows that the CRLB for the variance of any unbiased estimator for $m$ is $\sigma^2/n$. Since the variance of $\overline{X}$ is $\sigma^2/n$, it has the minimum variance among all unbiased estimators for $m$ when population $X$ is distributed normally.

**Example 9.3.** Problem: consider a population $X$ having a normal distribution $N(0, \sigma^2)$ where $\sigma^2$ is an unknown parameter to be estimated from a sample of size $n > 1$. (a) Determine the CRLB for the variance of any unbiased estimator for $\sigma^2$. (b) Is sample variance $S^2$ an efficient estimator for $\sigma^2$?

Answer: let us denote $\sigma^2$ by $\theta$. Then,

$$f(X; \theta) = \frac{1}{(2\pi\theta)^{1/2}} \exp\left(\frac{-X^2}{2\theta}\right),$$

and

$$\ln f(X; \theta) = -\frac{X^2}{2\theta} - \frac{1}{2}\ln 2\pi\theta,$$

$$\frac{\partial \ln f(X; \theta)}{\partial \theta} = \frac{X^2}{2\theta^2} - \frac{1}{2\theta},$$

$$\frac{\partial^2 \ln f(X; \theta)}{\partial \theta^2} = -\frac{X^2}{\theta^3} + \frac{1}{2\theta^2},$$

$$E\left\{\frac{\partial^2 \ln f(X; \theta)}{\partial \theta^2}\right\} = -\frac{\theta}{\theta^3} + \frac{1}{2\theta^2} = -\frac{1}{2\theta^2}.$$

Hence, according to Equation (9.36), the CRLB for the variance of any unbiased estimator for $\theta$ is $2\theta^2/n$.

For $S^2$, it has been shown in Section 9.1.2 that it is an unbiased estimator for $\theta$ and that its variance is [see Equation (9.10)]

$$\text{var}\{S^2\} = \frac{1}{n}\left(\mu_4 - \frac{n-3}{n-1}\sigma^4\right)$$

$$= \frac{1}{n}\left(3\sigma^4 - \frac{n-3}{n-1}\sigma^4\right)$$

$$= \frac{2\sigma^4}{n-1} = \frac{2\theta^2}{n-1},$$

since $\mu_4 = 3\sigma^4$ when $X$ is normally distributed. The efficiency of $S^2$, denoted by $e(S^2)$, is thus

$$e(S^2) = \frac{\text{CRLB}}{\text{var}(S^2)} = \frac{n-1}{n}.$$

We see that the sample variance is not an efficient estimator for $\theta$ in this case. It is, however, *asymptotically efficient* in the sense that $e(S^2) \to 1$ as $n \to \infty$.

**Example 9.4.** Problem: determine the CRLB for the variance of any unbiased estimator for $\theta$ in the lognormal distribution

$$
f(x;\theta) = \begin{cases} \dfrac{1}{x(2\pi\theta)^{1/2}}\exp\left(-\dfrac{1}{2\theta}\ln^2 x\right), & \text{for } x \geq 0, \text{ and } \theta > 0; \\ 0, & \text{elsewhere.} \end{cases}
$$

Answer: we have

$$
\frac{\partial \ln f(X;\theta)}{\partial \theta} = -\frac{1}{2\theta} + \frac{\ln^2 X}{2\theta^2},
$$

$$
\frac{\partial^2 \ln f(X;\theta)}{\partial \theta^2} = \frac{1}{2\theta^2} - \frac{\ln^2 X}{\theta^3},
$$

$$
E\left\{\frac{\partial^2 \ln f(X;\theta)}{\partial \theta^2}\right\} = \frac{1}{2\theta^2} - \frac{\theta}{\theta^3} = -\frac{1}{2\theta^2}.
$$

It thus follows from Equation (9.36) that the CRLB is $2\theta^2/n$.

Before going to the next criterion, it is worth mentioning again that, although unbiasedness as well as small variance is desirable it does not mean that we should discard all biased estimators as inferior. Consider two estimators for a parameter $\theta$, $\hat{\Theta}_1$ and $\hat{\Theta}_2$, the pdfs of which are depicted in Figure 9.2(a). Although $\hat{\Theta}_2$ is biased, because of its smaller variance, the probability of an observed value of $\hat{\Theta}_2$ being closer to the true value $\theta$ can well be higher than that associated with an observed value of $\hat{\Theta}_1$. Hence, one can argue convincingly that $\hat{\Theta}_2$ is the better estimator of the two. A more dramatic situation is shown in Figure 9.2(b). Clearly, based on a particular sample of size $n$, an observed value of $\hat{\Theta}_2$ will likely be closer to the true value $\theta$ than that of $\hat{\Theta}_1$ even though $\hat{\Theta}_1$ is again unbiased. It is worthwhile for us to reiterate our remark advanced in Section 9.2.1 – that the quality of an estimator does not rest on any single criterion but on a combination of criteria.

**Example 9.5.** To illustrate the point that unbiasedness can be outweighed by other considerations, consider the problem of estimating parameter $\theta$ in the binomial distribution

$$
p_X(k) = \theta^k(1-\theta)^{1-k}, \quad k = 0, 1. \tag{9.43}
$$

Let us propose two estimators, $\hat{\Theta}_1$ and $\hat{\Theta}_2$, for $\theta$ given by

$$
\left.\begin{aligned} \hat{\Theta}_1 &= \overline{X}, \\ \hat{\Theta}_2 &= \frac{n\overline{X}+1}{n+2}, \end{aligned}\right\} \tag{9.44}
$$

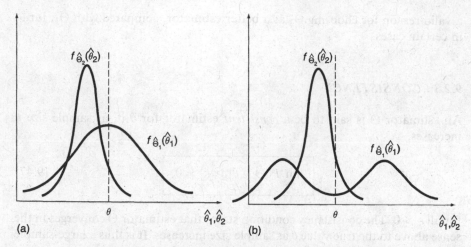

**Figure 9.2**  Probability density functions of $\hat{\Theta}_1$ and $\hat{\Theta}_2$

where $\overline{X}$ is the sample mean based on a sample of size $n$. The choice of $\hat{\Theta}_1$ is intuitively obvious since $E\{\overline{X}\} = \theta$, and the choice of $\hat{\Theta}_2$ is based on a prior probability argument that is not our concern at this point.

Since

$$E\{\overline{X}\} = \theta,$$

and

$$\sigma_{\overline{X}}^2 = \frac{\theta(1-\theta)}{n}$$

we have

$$\left.\begin{array}{l} E\{\hat{\Theta}_1\} = \theta, \\[2mm] E\{\hat{\Theta}_2\} = \dfrac{n\theta + 1}{n + 2}, \end{array}\right\} \tag{9.45}$$

and

$$\left.\begin{array}{l} \sigma_{\hat{\Theta}_1}^2 = \dfrac{\theta(1-\theta)}{n}, \\[3mm] \sigma_{\hat{\Theta}_2}^2 = \dfrac{n^2}{(n+2)^2}\sigma_{\overline{X}}^2 = \dfrac{n\theta(1-\theta)}{(n+2)^2}. \end{array}\right\} \tag{9.46}$$

We see from the above that, although $\hat{\Theta}_2$ is a biased estimator, its variance is smaller than that of $\hat{\Theta}_1$, particularly when $n$ is of a moderate value. This is a valid reason for choosing $\hat{\Theta}_2$ as a better estimator, compared with $\hat{\Theta}_1$, for $\theta$, in certain cases.

### 9.2.3  CONSISTENCY

An estimator $\hat{\Theta}$ is said to be a *consistent* estimator for $\theta$ if, as sample size $n$ increases,

$$\lim_{n \to \infty} P[|\hat{\Theta} - \theta| \geq \varepsilon] = 0, \qquad (9.47)$$

for all $\varepsilon > 0$. The consistency condition states that estimator $\hat{\Theta}$ converges in the sense above to the true value $\theta$ as sample size increases. It is thus a large-sample concept and is a good quality for an estimator to have.

**Example 9.6.** Problem: show that estimator $S^2$ in Example 9.3 is a consistent estimator for $\sigma^2$.

Answer: using the Chebyshev inequality defined in Section 4.2, we can write

$$P\{|S^2 - \sigma^2| \geq \varepsilon\} \leq \frac{1}{\varepsilon^2} E\{(S^2 - \sigma^2)^2\}.$$

We have shown that $E\{S^2\} = \sigma^2$, and $\mathrm{var}\{S^2\} = 2\theta^2/(n-1)$. Hence,

$$\lim_{n \to \infty} P\{|S^2 - \sigma^2| \geq \varepsilon\} \leq \lim_{n \to \infty} \frac{1}{\varepsilon^2} \left( \frac{2\theta^2}{n-1} \right) = 0.$$

Thus $S^2$ is a consistent estimator for $\sigma^2$.

Example 9.6 gives an expedient procedure for checking whether an estimator is consistent. We shall state this procedure as a theorem below (Theorem 9.3). It is important to note that this theorem gives a *sufficient*, but not *necessary*, condition for consistency.

**Theorem 9.3:** Let $\hat{\Theta}$ be an estimator for $\theta$ based on a sample of size $n$. Then, if

$$\lim_{n \to \infty} E\{\hat{\Theta}\} = \theta, \quad \text{and} \quad \lim_{n \to \infty} \mathrm{var}\{\hat{\Theta}\} = 0, \qquad (9.48)$$

estimator $\hat{\Theta}$ is a consistent estimator for $\theta$.

The proof of Theorem 9.3 is essentially given in Example 9.6 and will not be repeated here.

## 9.2.4  SUFFICIENCY

Let $X_1, X_2, \ldots, X_n$ be a sample of a population $X$ the distribution of which depends on unknown parameter $\theta$. If $Y = h(X_1, X_2, \ldots, X_n)$ is a statistic such that, for any other statistic

$$Z = g(X_1, X_2, \ldots, X_n),$$

the conditional distribution of $Z$, given that $Y = y$ does not depend on $\theta$, then $Y$ is called a *sufficient statistic* for $\theta$. If also $E\{Y\} = \theta$, then $Y$ is said to be a *sufficient estimator* for $\theta$.

In words, the definition for sufficiency states that, if $Y$ is a sufficient statistic for $\theta$, all sample information concerning $\theta$ is contained in $Y$. A sufficient statistic is thus of interest in that if it can be found for a parameter then an estimator based on this statistic is able to make use of all the information that the sample contains regarding the value of the unknown parameter. Moreover, an important property of a sufficient estimator is that, starting with any unbiased estimator of a parameter $\theta$ that is not a function of the sufficient estimator, it is possible to find an unbiased estimator based on the sufficient statistic that has a variance smaller than that of the initial estimator. Sufficient estimators thus have variances that are smaller than any other unbiased estimators that do not depend on sufficient statistics.

If a sufficient statistic for a parameter $\theta$ exists, Theorem 9.4, stated here without proof, provides an easy way of finding it.

**Theorem 9.4:** *Fisher–Neyman factorization criterion*. Let

$$Y = h(X_1, X_2, \ldots, X_n)$$

be a statistic based on a sample of size $n$. Then $Y$ is a sufficient statistic for $\theta$ if and only if the joint probability density function of $X_1, X_2, \ldots,$ and $X_n, f_X(x_1; \theta) \cdots f_X(x_n; \theta)$, can be factorized in the form

$$\prod_{j=1}^{n} f_X(x_j; \theta) = g_1[h(x_1, \ldots, x_n), \theta] g_2(x_1, \ldots, x_n). \tag{9.49}$$

If $X$ is discrete, we have

$$\prod_{j=1}^{n} p_X(x_j; \theta) = g_1[h(x_1, \ldots, x_n), \theta] g_2(x_1, \ldots, x_n). \tag{9.50}$$

The sufficiency of the factorization criterion was first pointed out by Fisher (1922). Neyman (1935) showed that it is also necessary.

The foregoing results can be extended to the multiple parameter case. Let $\boldsymbol{\theta}^T = [\theta_1 \ldots \theta_m]$, $m \leq n$, be the parameter vector. Then $Y_1 = h_1(X_1, \ldots, X_n), \ldots,$ $Y_r = h_r(X_1, \ldots, X_n), r \geq m$, is a set of sufficient statistics for $\boldsymbol{\theta}$ if and only if

$$\prod_{j=1}^{n} f_X(x_j; \boldsymbol{\theta}) = g_1[\boldsymbol{h}(x_1, \ldots, x_n), \boldsymbol{\theta}]g_2(x_1, \ldots, x_n), \qquad (9.51)$$

where $\boldsymbol{h}^T = [h_1 \cdots h_r]$. A similar expression holds when $X$ is discrete.

**Example 9.7.** Let us show that statistic $\overline{X}$ is a sufficient statistic for $\theta$ in Example 9.5. In this case,

$$\prod_{j=1}^{n} p_X(x_j; \theta) = \prod_{j=1}^{n} \theta^{x_j}(1-\theta)^{1-x_j}$$
$$= \theta^{\Sigma x_j}(1-\theta)^{n-\Sigma x_j}. \qquad (9.52)$$

We see that the joint probability mass function (jpmf) is a function of $\Sigma x_j$ and $\theta$. If we let

$$Y = \sum_{j=1}^{n} X_j,$$

the jpmf of $X_1, \ldots,$ and $X_n$ takes the form given by Equation (9.50), with

$$g_1 = \theta^{\Sigma x_j}(1-\theta)^{n-\Sigma x_j},$$

and

$$g_2 = 1.$$

In this example,

$$\sum_{j=1}^{n} X_j$$

is thus a sufficient statistic for $\theta$. We have seen in Example 9.5 that both $\hat{\Theta}_1$ and $\hat{\Theta}_2$, where $\hat{\Theta}_1 = \overline{X}$, and $\hat{\Theta}_2 = (n\overline{X} + 1)/(n + 2)$, are based on this sufficient statistic. Furthermore, $\hat{\Theta}_1$, being unbiased, is a sufficient estimator for $\theta$.

**Example 9.8.** Suppose $X_1, X_2, \ldots,$ and $X_n$ are a sample taken from a Poisson distribution; that is,

$$p_X(k; \theta) = \frac{\theta^k e^{-\theta}}{k!}, \quad k = 0, 1, 2, \ldots, \qquad (9.53)$$

where $\theta$ is the unknown parameter. We have

$$\prod_{j=1}^{n} p_X(x_j; \theta) = \frac{\theta^{\Sigma x_j} e^{-n\theta}}{\prod x_j!}, \tag{9.54}$$

which can be factorized in the form of Equation (9.50) by letting

$$g_1 = \theta^{\Sigma x_j} e^{-n\theta},$$

and

$$g_2 = \frac{1}{\prod x_j!}$$

It is seen that

$$Y = \sum_{j=1}^{n} X_j$$

is a sufficient statistic for $\theta$.

## 9.3 METHODS OF ESTIMATION

Based on the estimation criteria defined in Section 9.2, some estimation techniques that yield 'good', and sometimes 'best', estimates of distribution parameters are now developed.

Two approaches to the parameter estimation problem are discussed in what follows: point estimation and interval estimation. In point estimation, we use certain prescribed methods to arrive at a value for $\hat{\theta}$ as a function of the observed data that we accept as a 'good' estimate of $\theta$ – good in terms of unbiasedness, minimum variance, etc., as defined by the estimation criteria.

In many scientific studies it is more useful to obtain information about a parameter beyond a single number as its estimate. Interval estimation is a procedure by which bounds on the parameter value are obtained that not only give information on the numerical value of the parameter but also give an indication of the level of confidence one can place on the possible numerical value of the parameter on the basis of a sample. Point estimation will be discussed first, followed by the development of methods of interval estimation.

### 9.3.1 POINT ESTIMATION

We now proceed to present two general methods of finding point estimators for distribution parameters on the basis of a sample from a population.

### 9.3.1.1 Method of Moments

The oldest systematic method of point estimation was proposed by Pearson (1894) and was extensively used by him and his co-workers. It was neglected for a number of years because of its general lack of optimum properties and because of the popularity and universal appeal associated with the method of maximum likelihood, to be discussed in Section 9.3.1.2. The moment method, however, appears to be regaining its acceptance, primarily because of its expediency in terms of computational labor and the fact that it can be improved upon easily in certain cases.

The method of moments is simple in concept. Consider a selected probability density function $f(x; \theta_1, \theta_2, \ldots, \theta_m)$ for which parameters $\theta_j, j = 1, 2, \ldots, m$, are to be estimated based on sample $X_1, X_2, \ldots, X_n$ of $X$. The theoretical or population moments of $X$ are

$$\alpha_i = \int_{-\infty}^{\infty} x^i f(x; \theta_1, \ldots, \theta_m) \mathrm{d}x, \quad i = 1, 2, \ldots. \tag{9.55}$$

They are, in general, functions of the unknown parameters; that is,

$$\alpha_i = \alpha_i(\theta_1, \theta_2, \ldots, \theta_m). \tag{9.56}$$

However, sample moments of various orders can be found from the sample by [see Equation (9.14)]

$$M_i = \frac{1}{n} \sum_{j=1}^{n} X_j^i, \quad i = 1, 2, \ldots. \tag{9.57}$$

The method of moments suggests that, in order to determine estimators $\hat{\Theta}_1, \ldots,$ and $\hat{\Theta}_m$ from the sample, we equate a sufficient number of sample moments to the corresponding population moments. By establishing and solving as many resulting moment equations as there are parameters to be estimated, estimators for the parameter are obtained. Hence, the procedure for determining $\hat{\Theta}_1, \hat{\Theta}_2, \ldots,$ and $\hat{\Theta}_m$ consists of the following steps:

- Step 1: let

$$\boxed{\alpha_i(\hat{\Theta}_1, \ldots, \hat{\Theta}_m) = M_i, \quad i = 1, 2, \ldots, m.} \tag{9.58}$$

These yield $m$ moment equations in $m$ unknowns $\hat{\Theta}_j, j = 1, \ldots, m$.

- Step 2: solve for $\hat{\Theta}_j, j = 1, \ldots, m$, from this system of equations. These are called the *moment estimators* for $\theta_1, \ldots,$ and $\theta_m$.

Let us remark that it is not necessary to consider $m$ *consecutive* moment equations as indicated by Equations (9.58); any convenient set of $m$ equations that lead to the solution for $\hat{\Theta}_j, j = 1, \ldots, m$, is sufficient. Lower-order moment equations are preferred, however, since they require less manipulation of observed data.

An attractive feature of the method of moments is that the moment equations are straightforward to establish, and there is seldom any difficulty in solving them. However, a shortcoming is that such desirable properties as unbiasedness or efficiency are not generally guaranteed for estimators so obtained.

However, consistency of moment estimators can be established under general conditions. In order to show this, let us consider a single parameter $\theta$ whose moment estimator $\hat{\Theta}$ satisfies the moment equation

$$\alpha_i(\hat{\Theta}) = M_i, \tag{9.59}$$

for some $i$. The solution of Equation (9.59) for $\hat{\Theta}$ can be represented by $\hat{\Theta} = \hat{\Theta}(M_i)$, for which the Taylor's expansion about $\alpha_i(\theta)$ gives

$$\hat{\Theta} = \theta + \hat{\Theta}^{(1)}[\alpha_i(\theta)][M_i - \alpha_i(\theta)] + \frac{\hat{\Theta}^{(2)}[\alpha_i(\theta)]}{2!}[M_i - \alpha_i(\theta)]^2 + \cdots, \tag{9.60}$$

where superscript $(k)$ denotes the $k$th derivative with respect to $M_i$. Upon performing successive differentiations of Equation (9.59) with respect to $M_i$, Equation (9.60) becomes

$$\hat{\Theta} - \theta = [M_i - \alpha_i(\theta)]\left[\frac{d\alpha_i(\theta)}{d\theta}\right]^{-1} - \frac{1}{2}[M_i - \alpha_i(\theta)]^2\left[\frac{d^2\alpha_i(\theta)}{d\theta^2}\right]\left[\frac{d\alpha_i(\theta)}{d\theta}\right]^{-3} + \cdots. \tag{9.61}$$

The bias and variance of $\hat{\Theta}$ can be found by taking the expectation of Equation (9.61) and the expectation of the square of Equation (9.61), respectively. Up to the order of $1/n$, we find

$$\left.\begin{aligned}
E\{\hat{\Theta}\} - \theta &= -\frac{1}{2n}(\alpha_{2i} - \alpha_i^2)\left(\frac{d^2\alpha_i}{d\theta^2}\right)\left(\frac{d\alpha_i}{d\theta}\right)^{-3}, \\[2mm]
\text{var}\{\hat{\Theta}\} &= \frac{1}{n}(\alpha_{2i} - \alpha_i^2)\left(\frac{d\alpha_i}{d\theta}\right)^{-2}.
\end{aligned}\right\} \tag{9.62}$$

Assuming that all the indicated moments and their derivatives exist, Equations (9.62) show that

$$\lim_{n \to \infty} E\{\hat{\Theta}\} = \theta,$$

and

$$\lim_{n\to\infty} \text{var}\{\hat{\Theta}\} = 0,$$

and hence $\hat{\Theta}$ is consistent.

**Example 9.9.** Problem: let us select the normal distribution as a model for the percentage yield discussed in Chapter 8; that is,

$$f(x; m, \sigma^2) = \frac{1}{(2\pi)^{1/2}\sigma} \exp\left[-\frac{(x-m)^2}{2\sigma^2}\right], \quad -\infty < x < \infty. \tag{9.63}$$

Estimate parameters $\theta_1 = m$, and $\theta_2 = \sigma^2$, based on the 200 sample values given in Table 8.1, page 249.

Answer: following the method of moments, we need two moment equations, and the most convenient ones are obviously

$$\alpha_1 = M_1 = \overline{X},$$

and

$$\alpha_2 = M_2.$$

Now,

$$\alpha_1 = \theta_1.$$

Hence, the first of these moment equations gives

$$\hat{\Theta}_1 = \overline{X} = \frac{1}{n}\sum_{j=1}^{n} X_j. \tag{9.64}$$

The properties of this estimator have already been discussed in Example 9.2. It is unbiased and has minimum variance among all unbiased estimators for $m$. We see that the method of moments produces desirable results in this case.

The second moment equation gives

$$\hat{\Theta}_1^2 + \hat{\Theta}_2 = M_2 = \frac{1}{n}\sum_{j=1}^{n} X_j^2,$$

or

$$\hat{\Theta}_2 = M_2 - M_1^2 = \frac{1}{n}\sum_{j=1}^{n} (X_j - \overline{X})^2. \tag{9.65}$$

This, as we have shown, is a biased estimator for $\sigma^2$.

Estimates $\hat{\theta}_1$ and $\hat{\theta}_2$ of $\theta_1 = m$ and $\theta_2 = \sigma^2$ based on the sample values given by Table 8.1 are, following Equations (9.64) and (9.65),

$$\hat{\theta}_1 = \frac{1}{200} \sum_{j=1}^{200} x_j \cong 70,$$

$$\hat{\theta}_2 = \frac{1}{200} \sum_{j=1}^{200} (x_j - \hat{\theta}_1)^2 \cong 4,$$

where $x_j$, $j = 1, 2, \ldots, 200$, are sample values given in Table 8.1.

**Example 9.10.** Problem: consider the binomial distribution

$$p_X(k; p) = p^k (1 - p)^{1-k}, \quad k = 0, 1. \tag{9.66}$$

Estimate parameter $p$ based on a sample of size $n$.

Answer: the method of moments suggests that we determine the estimator for $p$, $\hat{P}$, by equating $\alpha_1$ to $M_1 = \overline{X}$. Since

$$\alpha_1 = E\{X\} = p,$$

we have

$$\hat{P} = \overline{X}. \tag{9.67}$$

The mean of $\hat{P}$ is

$$E\{\hat{P}\} = \frac{1}{n} \sum_{j=1}^{n} E\{X_j\} = p. \tag{9.68}$$

Hence it is an unbiased estimator. Its variance is given by

$$\mathrm{var}\{\hat{P}\} = \mathrm{var}\{\overline{X}\} = \frac{\sigma^2}{n} = \frac{p(1 - p)}{n}. \tag{9.69}$$

It is easy to derive the CRLB for this case and show that $\hat{P}$ defined by Equation (9.67) is also efficient.

**Example 9.11.** Problem: a set of 214 observed gaps in traffic on a section of Arroyo Seco Freeway is given in Table 9.1. If the exponential density function

$$f(t; \lambda) = \lambda e^{-\lambda t}, \quad t \geq 0, \tag{9.70}$$

is proposed for the gap, determine parameter $\lambda$ from the data.

**Table 9.1**  Observed traffic gaps on Arroyo Seco Freeway,
for Example 9.11 (Source: Gerlough, 1955)

| Gap length (s) | Gaps (No.) | Gap length (s) | Gaps (No.) |
|---|---|---|---|
| 0–1 | 18 | 16–17 | 6 |
| 1–2 | 25 | 17–18 | 4 |
| 2–3 | 21 | 18–19 | 3 |
| 3–4 | 13 | 19–20 | 3 |
| 4–5 | 11 | 20–21 | 1 |
| 5–6 | 15 | 21–22 | 1 |
| 6–7 | 16 | 22–23 | 1 |
| 7–8 | 12 | 23–24 | 0 |
| 8–9 | 11 | 24–25 | 1 |
| 9–10 | 11 | 25–26 | 0 |
| 10–11 | 8 | 26–27 | 1 |
| 11–12 | 12 | 27–28 | 1 |
| 12–13 | 6 | 28–29 | 1 |
| 13–14 | 3 | 29–30 | 2 |
| 14–15 | 3 | 30–31 | 1 |
| 15–16 | 3 | | |

Answer: in this case,

$$\alpha_1 = \frac{1}{\lambda},$$

and, following the method of moments, the simplest estimator, $\hat{\Lambda}$, for $\lambda$ is obtained from

$$\alpha_1 = \overline{X}, \quad \text{or} \quad \hat{\Lambda} = \frac{1}{\overline{X}}. \tag{9.71}$$

Hence, the desired estimate is

$$\hat{\lambda} = \left( \frac{1}{214} \sum_{j=1}^{214} x_j \right)^{-1}$$

$$= \frac{214}{18(0.5) + 25(1.5) + \cdots + 1(30.5)} \tag{9.72}$$

$$= 0.13 \, \text{s}^{-1}.$$

Let us note that, although $\overline{X}$ is an unbiased estimator for $\alpha_1$, the estimator for $\lambda$ obtained above is not unbiased since

$$E\left\{ \frac{1}{\overline{X}} \right\} \neq \frac{1}{E\{\overline{X}\}}.$$

**Example 9.12.** Suppose that population $X$ has a uniform distribution over the range $(0, \theta)$ and we wish to estimate parameter $\theta$ from a sample of size $n$.

The density function of $X$ is

$$f(x; \theta) = \begin{cases} \dfrac{1}{\theta}, & \text{for } 0 \leq x \leq \theta; \\ 0, & \text{elsewhere}; \end{cases} \qquad (9.73)$$

and the first moment is

$$\alpha_1 = \frac{\theta}{2}. \qquad (9.74)$$

It follows from the method of moments that, on letting $\alpha_1 = \overline{X}$, we obtain

$$\hat{\Theta} = 2\overline{X} = \frac{2}{n} \sum_{j=1}^{n} X_j. \qquad (9.75)$$

Upon little reflection, the validity of this estimator is somewhat questionable because, by definition, all values assumed by $X$ are supposed to lie within interval $(0, \theta)$. However, we see from Equation (9.75) that it is possible that some of the samples are greater than $\hat{\Theta}$. Intuitively, a better estimator might be

$$\hat{\Theta} = X_{(n)}, \qquad (9.76)$$

where $X_{(n)}$ is the $n$th-order statistic. As we will see, this would be the outcome following the method of maximum likelihood, to be discussed in the next section.

Since the method of moments requires only $\alpha_i$, the moments of population $X$, the knowledge of its pdf is not necessary. This advantage is demonstrated in Example 9.13.

**Example 9.13.** Problem: consider measuring the length $r$ of an object with use of a sensing instrument. Owing to inherent inaccuracies in the instrument, what is actually measured is $X$, as shown in Figure 9.3, where $X_1$ and $X_2$ are identically and normally distributed with mean zero and unknown variance $\sigma^2$. Determine a moment estimator $\hat{\Theta}$ for $\theta = r^2$ on the basis of a sample of size $n$ from $X$.

Answer: now, random variable $X$ is

$$X = [(r + X_1)^2 + X_2^2]^{1/2}. \qquad (9.77)$$

The pdf of $X$ with unknown parameters $\theta$ and $\sigma^2$ can be found by using techniques developed in Chapter 5. It is, however, unnecessary here since some

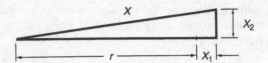

**Figure 9.3**    Measurement $X$, for Example 9.13

moments of $X$ can be directly generated from Equation (9.77). We remark that, although an estimator for $\sigma^2$ is not required, it is nevertheless an unknown parameter and must be considered together with $\theta$. In the applied literature, an unknown parameter for which the value is of no interest is sometimes referred to as a *nuisance parameter*.

Two moment equations are needed in this case. However, we see from Equation (9.77) that the odd-order moments of $X$ are quite complicated. For simplicity, the second-order and fourth-order moment equations will be used. We easily obtain from Equation (9.77)

$$\left.\begin{array}{l} \alpha_2 = \theta + 2\sigma^2, \\ \alpha_4 = \theta^2 + 8\theta\sigma^2 + 8\sigma^4. \end{array}\right\} \tag{9.78}$$

The two moment equations are

$$\left.\begin{array}{l} \hat{\Theta} + 2\widehat{\Sigma^2} = M_2, \\ \hat{\Theta}^2 + 8\hat{\Theta}\widehat{\Sigma^2} + 8\widehat{\Sigma^2}^2 = M_4. \end{array}\right\} \tag{9.79}$$

Solving for $\hat{\Theta}$, we have

$$\hat{\Theta} = (2M_2^2 - M_4)^{1/2}. \tag{9.80}$$

Incidentally, a moment estimator $\widehat{\Sigma^2}$ for $\sigma^2$, if needed, is obtained from Equations (9.79) to be

$$\widehat{\Sigma^2} = \frac{1}{2}(M_2 - \hat{\Theta}). \tag{9.81}$$

**Combined Moment Estimators.**    Let us take another look at Example 9.11 for the purpose of motivating the following development. In this example, an estimator for $\lambda$ has been obtained by using the first-order moment equation. Based on the same sample, one can obtain additional moment estimators for $\lambda$ by using higher-order moment equations. For example, since $\alpha_2 = 2/\lambda^2$, the second-order moment equation,

$$\alpha_2 = M_2,$$

produces a moment estimator $\hat{\Lambda}$ for $\lambda$ in the form

$$\hat{\Lambda} = \left(\frac{2}{M_2}\right)^{1/2}. \tag{9.82}$$

Although this estimator may be inferior to $1/\overline{X}$ in terms of the quality criteria we have established, an interesting question arises: given two or more moment estimators, can they be combined to yield an estimator superior to any of the individual moment estimators?

In what follows, we consider a combined moment estimator derived from an optimal linear combination of a set of moment estimators. Let $\hat{\Theta}^{(1)}, \hat{\Theta}^{(2)}, \ldots,$ $\hat{\Theta}^{(p)}$ be $p$ moment estimators for the same parameter $\theta$. We seek a combined estimator $\Theta_p^*$ in the form

$$\Theta_p^* = w_1 \hat{\Theta}^{(1)} + \cdots + w_p \hat{\Theta}^{(p)}, \tag{9.83}$$

where coefficients $w_1, \ldots,$ and $w_p$ are to be chosen in such a way that it is unbiased if $\hat{\Theta}^{(j)}$, $j = 1, 2, \ldots, p$, are unbiased and the variance of $\Theta_p^*$ is minimized.

The unbiasedness condition requires that

$$w_1 + \cdots + w_p = 1. \tag{9.84}$$

We thus wish to determine coefficients $w_j$ by minimizing

$$Q = \text{var}\{\Theta_p^*\} = \text{var}\left\{\sum_{j=1}^{p} w_j \hat{\Theta}^{(j)}\right\}, \tag{9.85}$$

subject to Equation (9.84).

Let $\boldsymbol{u}^T = [1 \quad \cdots \quad 1]$, $\hat{\boldsymbol{\Theta}}^T = [\hat{\Theta}^{(1)} \quad \cdots \quad \hat{\Theta}^{(p)}]$, and $\boldsymbol{w}^T = [w_1 \quad \cdots \quad w_p]$. Equations (9.84) and (9.85) can be written in the vector–matrix form

$$\boldsymbol{w}^T \boldsymbol{u} = 1, \tag{9.86}$$

and

$$Q(\boldsymbol{w}) = \text{var}\left\{\sum_{j=1}^{p} w_j \hat{\Theta}^{(j)}\right\} = \boldsymbol{w}^T \Lambda \boldsymbol{w}, \tag{9.87}$$

where $\Lambda = [\lambda_{ij}]$ with $\lambda_{ij} = \text{cov}\{\hat{\Theta}^{(i)}, \hat{\Theta}^{(j)}\}$.

In order to minimize Equation (9.87) subject to Equation (9.86), we consider

$$Q_1(\boldsymbol{w}) = \boldsymbol{w}^T \Lambda \boldsymbol{w} - \boldsymbol{w}^T \boldsymbol{u} \lambda - \lambda \boldsymbol{u}^T \boldsymbol{w} \tag{9.88}$$

where $\lambda$ is the Lagrange multiplier. Taking the first variation of Equation (9.88) and setting it to zero we obtain

$$\delta Q_1(w) = \delta w^\mathrm{T}(\Lambda w - u\lambda) + (w^\mathrm{T}\Lambda - \lambda u^\mathrm{T})\delta w = 0$$

as a condition of extreme. Since $\delta w$ and $\delta w^\mathrm{T}$ are arbitrary, we require that

$$\Lambda w - u\lambda = 0 \quad \text{and} \quad w^\mathrm{T}\Lambda - \lambda u^\mathrm{T} = 0, \tag{9.89}$$

and either of these two relations gives

$$w^\mathrm{T} = \lambda u^\mathrm{T}\Lambda^{-1}. \tag{9.90}$$

The constraint Equation (9.86) is now used to determine $\lambda$. It implies that

$$w^\mathrm{T}u = \lambda u^\mathrm{T}\Lambda^{-1}u = 1,$$

or

$$\lambda = \frac{1}{u^\mathrm{T}\Lambda^{-1}u}. \tag{9.91}$$

Hence, we have from Equations (9.90) and (9.91)

$$w^\mathrm{T} = \frac{u^\mathrm{T}\Lambda^{-1}}{u^\mathrm{T}\Lambda^{-1}u}. \tag{9.92}$$

The variance of $\Theta_p^*$ is

$$\mathrm{var}\{\Theta_p^*\} = w^\mathrm{T}\Lambda w = \frac{1}{u^\mathrm{T}\Lambda^{-1}u}, \tag{9.93}$$

in view of Equation (9.92).

Several attractive features are possessed by $\Theta_p^*$. For example, we can show that its variance is smaller than or equal to that of any of the simple moment estimators $\hat{\Theta}^{(j)}, j = 1, 2, \ldots, p$, and furthermore (see Soong, 1969),

$$\mathrm{var}\{\Theta_p^*\} \le \mathrm{var}\{\Theta_q^*\}, \tag{9.94}$$

if $p \ge q$.

**Example 9.14.** Consider the problem of estimating parameter $\theta$ in the log-normal distribution

$$f(x; \theta) = \frac{1}{x(2\pi\theta)^{1/2}} \exp\left[-\frac{1}{2\theta}\ln^2 x\right], \quad x \ge 0, \ \theta > 0, \tag{9.95}$$

from a sample of size $n$.

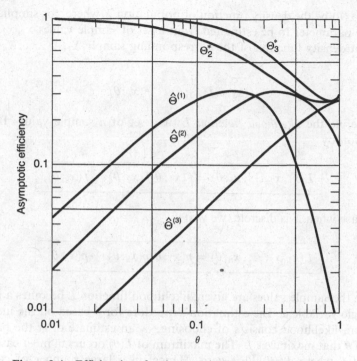

**Figure 9.4** Efficiencies of estimators in Example 9.14 as $n \to \infty$

Three moment estimators for $\theta - \hat{\Theta}^{(1)}$, $\hat{\Theta}^{(2)}$, and $\hat{\Theta}^{(3)}$ – can be found by means of establishing and solving the first three moment equations. Let $\Theta_2^*$ be the combined moment estimator of $\hat{\Theta}^{(1)}$ and $\hat{\Theta}^{(2)}$, and let $\Theta_3^*$ be the combined estimator of all three. As we have obtained the CRLB for the variance of any unbiased estimator for $\theta$ in Example 9.4, the efficiency of each of the above estimators can be calculated. Figure 9.4 shows these efficiencies as $n \to \infty$. As we can see, a significant increase in efficiency can result by means of combining even a small number of moment estimators.

### 9.3.1.2   Method of Maximum Likelihood

First introduced by Fischer in 1922, the method of maximum likelihood has become the most important general method of estimation from a theoretical point of view. Its greatest appeal stems from the fact that some very general properties associated with this procedure can be derived and, in the case of large samples, they are optimal properties in terms of the criteria set forth in Section 9.2.

Let $f(x; \theta)$ be the density function of population $X$ where, for simplicity, $\theta$ is the only parameter to be estimated from a set of sample values $x_1, x_2, \ldots, x_n$. The joint density function of the corresponding sample $X_1, X_2, \ldots, X_n$ has the form

$$f(x_1; \theta)f(x_2; \theta) \cdots f(x_n; \theta).$$

We define the *likelihood function* $L$ of a set of $n$ sample values from the population by

$$\boxed{L(x_1, x_2, \ldots, x_n; \theta) = f(x_1; \theta)f(x_2; \theta) \cdots f(x_n; \theta).} \qquad (9.96)$$

In the case when $X$ is discrete, we write

$$\boxed{L(x_1, x_2, \ldots, x_n; \theta) = p(x_1; \theta)p(x_2; \theta) \cdots p(x_n; \theta).} \qquad (9.97)$$

When the sample values are given, likelihood function $L$ becomes a function of a single variable $\theta$. The estimation procedure for $\theta$ based on the method of maximum likelihood consists of choosing, as an estimate of $\theta$, the particular value of $\theta$ that maximizes $L$. The maximum of $L(\theta)$ occurs in most cases at the value of $\theta$ where $dL(\theta)/d\theta$ is zero. Hence, in a large number of cases, the *maximum likelihood estimate* (MLE) $\hat{\theta}$ of $\theta$ based on sample values $x_1, x_2, \ldots,$ and $x_n$ can be determined from

$$\frac{dL(x_1, x_2, \ldots, x_n; \hat{\theta})}{d\hat{\theta}} = 0. \qquad (9.98)$$

As we see from Equations (9.96) and (9.97), function $L$ is in the form of a product of many functions of $\theta$. Since $L$ is always nonnegative and attains its maximum for the same value of $\hat{\theta}$ as $\ln L$, it is generally easier to obtain MLE $\hat{\theta}$ by solving

$$\boxed{\frac{d \ln L(x_1, \ldots, x_n; \hat{\theta})}{d\hat{\theta}} = 0,} \qquad (9.99)$$

because $\ln L$ is in the form of a sum rather than a product.

Equation (9.99) is referred to as the *likelihood equation*. The desired solution is one where root $\hat{\theta}$ is a function of $x_j, j = 1, 2, \ldots, n$, if such a root exists. When several roots of Equation (9.99) exist, the MLE is the root corresponding to the global maximum of $L$ or $\ln L$.

To see that this procedure is plausible, we observe that the quantity

$$L(x_1, x_2, \ldots, x_n; \theta) dx_1 dx_2 \cdots dx_n$$

is the probability that sample $X_1, X_2, \ldots, X_n$ takes values in the region defined by $(x_1 + dx_1, x_2 + dx_2, \ldots, x_n + dx_n)$. Given the sample values, this probability gives a measure of likelihood that they are from the population. By choosing a value of $\theta$ that maximizes $L$, or $\ln L$, we in fact say that we prefer the value of $\theta$ that makes as probable as possible the event that the sample values indeed come from the population.

The extension to the case of several parameters is straightforward. In the case of $m$ parameters, the likelihood function becomes

$$L(x_1, \ldots, x_n; \theta_1, \ldots, \theta_m),$$

and the MLEs of $\theta_j, j = 1, \ldots, m$, are obtained by solving simultaneously the system of likelihood equations

$$\frac{\partial \ln L}{\partial \hat{\theta}_j} = 0, \quad j = 1, 2, \ldots, m. \tag{9.100}$$

A discussion of some of the important properties associated with a maximum likelihood estimator is now in order. Let us represent the solution of the likelihood equation, Equation (9.99), by

$$\hat{\theta} = h(x_1, x_2, \ldots, x_n). \tag{9.101}$$

The *maximum likelihood estimator* $\hat{\Theta}$ for $\theta$ is then

$$\hat{\Theta} = h(X_1, X_2, \ldots, X_n). \tag{9.102}$$

The universal appeal enjoyed by maximum likelihood estimators stems from the optimal properties they possess when the sample size becomes large. Under mild regularity conditions imposed on the pdf or pmf of population $X$, two notable properties are given below, without proof.

**Property 9.1:** *consistency and asymptotic efficiency*. Let $\hat{\Theta}$ be the maximum likelihood estimator for $\theta$ in pdf $f(x; \theta)$ on the basis of a sample of size $n$. Then, as $n \to \infty$,

$$E\{\hat{\Theta}\} \to \theta, \tag{9.103}$$

and

$$\text{var}\{\hat{\Theta}\} \to \left\{ nE \left\{ \left[ \frac{\partial \ln f(X; \theta)}{\partial \theta} \right]^2 \right\} \right\}^{-1}. \tag{9.104}$$

Analogous results are obtained when population $X$ is discrete. Furthermore, the distribution of $\hat{\Theta}$ tends to a normal distribution as $n$ becomes large.

This important result shows that MLE $\hat{\Theta}$ is consistent. Since the variance given by Equation (9.104) is equal to the Cramér–Rao lower bound, it is efficient as $n$ becomes large, or *asymptotically efficient*. The fact that MLE $\hat{\Theta}$ is normally distributed as $n \to \infty$ is also of considerable practical interest as probability statements can be made regarding any observed value of a maximum likelihood estimator as $n$ becomes large.

Let us remark, however, these important properties are *large-sample* properties. Unfortunately, very little can be said in the case of a small sample size; it may be biased and nonefficient. This lack of reasonable small-sample properties can be explained in part by the fact that maximum likelihood estimation is based on finding the mode of a distribution by attempting to select the true parameter value. Estimators, in contrast, are generally designed to approach the true value rather than to produce an exact hit. Modes are therefore not as desirable as the mean or median when the sample size is small.

**Property 9.2:** *invariance property*. It can be shown that, if $\hat{\Theta}$ is the MLE of $\theta$, then the MLE of a function of $\theta$, say $g(\theta)$, is $g(\hat{\Theta})$, where $g(\theta)$ is assumed to represent a one-to-one transformation and be differentiable with respect to $\theta$.

This important invariance property implies that, for example, if $\hat{\Sigma}$ is the MLE of the standard deviation $\sigma$ in a distribution, then the MLE of the variance $\sigma^2, \widehat{\Sigma^2}$, is $\hat{\Sigma}^2$.

Let us also make an observation on the solution procedure for solving likelihood equations. Although it is fairly simple to establish Equation (9.99) or Equations (9.100), they are frequently highly nonlinear in the unknown estimates, and close-form solutions for the MLE are sometimes difficult, if not impossible, to achieve. In many cases, iterations or numerical schemes are necessary.

**Example 9.15.** Let us consider Example 9.9 again and determine the MLEs of $m$ and $\sigma^2$. The logarithm of the likelihood function is

$$\ln L = -\frac{1}{2\sigma^2} \sum_{j=1}^{n} (x_j - m)^2 - \frac{1}{2} n \ln \sigma^2 - \frac{1}{2} n \ln 2\pi. \qquad (9.105)$$

Let $\theta_1 = m$, and $\theta_2 = \sigma^2$, as before; the likelihood equations are

$$\frac{\partial \ln L}{\partial \hat{\theta}_1} = \frac{1}{\hat{\theta}_2} \sum_{j=1}^{n} (x_j - \hat{\theta}_1) = 0,$$

$$\frac{\partial \ln L}{\partial \hat{\theta}_2} = \frac{1}{2\hat{\theta}_2^2} \sum_{j=1}^{n} (x_j - \hat{\theta}_1)^2 - \frac{n}{2\hat{\theta}_2} = 0.$$

Solving the above equations simultaneously, the MLEs of $m$ and $\sigma^2$ are found to be

$$\hat{\theta}_1 = \frac{1}{n}\sum_{j=1}^{n} x_j,$$

and

$$\hat{\theta}_2 = \frac{1}{n}\sum_{j=1}^{n}(x_j - \hat{\theta}_1)^2.$$

The maximum likelihood estimators for $m$ and $\sigma^2$ are, therefore,

$$\left.\begin{aligned}
\hat{\Theta}_1 &= \frac{1}{n}\sum_{j=1}^{n} X_j = \overline{X}, \\
\hat{\Theta}_2 &= \frac{1}{n}\sum_{j=1}^{n}(X_j - \overline{X})^2 = \frac{n-1}{n}S^2,
\end{aligned}\right\} \tag{9.106}$$

which coincide with their moment estimators in this case. Although $\hat{\Theta}_2$ is biased, consistency and asymptotic efficiency for both $\hat{\Theta}_1$ and $\hat{\Theta}_2$ can be easily verified.

**Example 9.16.** Let us determine the MLE of $\theta$ considered in Example 9.12. Now,

$$f(x;\theta) = \begin{cases} \dfrac{1}{\theta}, & \text{for } 0 \le x \le \theta; \\ 0, & \text{elsewhere.} \end{cases} \tag{9.107}$$

The likelihood function becomes

$$L(x_1, x_2, \ldots, x_n; \theta) = \left(\frac{1}{\theta}\right)^n, \quad 0 \le x_i \le \theta, \text{for all } i. \tag{9.108}$$

A plot of $L$ is given in Figure 9.5. However, we note from the condition associated with Equation (9.108) that all sample values $x_i$ must be smaller than or equal to $\theta$, implying that only the portion of the curve to the right of $\max(x_1, \ldots, x_n)$ is applicable. Hence, the maximum of $L$ occurs at $\theta = \max(x_1, x_2, \ldots, x_n)$, or, the MLE for $\theta$ is

$$\hat{\theta} = \max(x_1, x_2, \ldots, x_n), \tag{9.109}$$

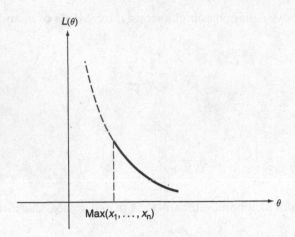

**Figure 9.5**   Likelihood function, $L(\theta)$, for Example 9.16

and the maximum likelihood estimator for $\theta$ is

$$\hat{\Theta} = \max(X_1, X_2, \ldots, X_n) = X_{(n)}. \qquad (9.110)$$

This estimator is seen to be different from that obtained by using the moment method [Equation (9.75)] and, as we already commented in Example 9.12, it is a more logical choice.

Let us also note that we did not obtain Equation (9.109) by solving the likelihood equation. The likelihood equation does not apply in this case as the maximum of $L$ occurs at the boundary and the derivative is not zero there.

It is instructive to study some of the properties of $\hat{\Theta}$ given by Equation (9.110). The pdf of $\hat{\Theta}$ is given by [see Equation (9.19)]

$$f_{\hat{\Theta}}(x) = nF_X^{n-1}(x)f_X(x). \qquad (9.111)$$

With $f_X(x)$ given by Equation (9.107) and

$$F_X(x) = \begin{cases} 0, & \text{for } x < 0; \\ \dfrac{x}{\theta}, & \text{for } 0 \le x \le \theta; \\ 1, & \text{for } x > \theta; \end{cases} \qquad (9.112)$$

we have

$$f_{\hat{\Theta}}(x) = \begin{cases} \dfrac{nx^{n-1}}{\theta^n}, & \text{for } 0 \le x \le \theta; \\ 0, & \text{elsewhere.} \end{cases} \qquad (9.113)$$

The mean and variance of $\hat{\Theta}$ are

$$E\{\hat{\Theta}\} = \int_0^\theta x f_{\hat{\Theta}}(x)\mathrm{d}x = \frac{n}{n+1}\theta, \qquad (9.114)$$

$$\mathrm{var}\{\hat{\Theta}\} = \int_0^\theta \left(x - \frac{n}{n+1}\theta\right)^2 f_{\hat{\Theta}}(x)\mathrm{d}x = \left[\frac{n}{(n+1)^2(n+2)}\right]\theta^2. \qquad (9.115)$$

We see that $\hat{\Theta}$ is biased but consistent.

**Example 9.17.** Let us now determine the MLE of $\theta = r^2$ in Example 9.13. To carry out this estimation procedure, it is now necessary to determine the pdf of $X$ given by Equation (9.77). Applying techniques developed in Chapter 5, we can show that $X$ is characterized by the *Rice distribution* with pdf given by (see Benedict and Soong, 1967)

$$f_X(x;\theta,\sigma^2) = \begin{cases} \dfrac{x}{\sigma^2}\mathrm{I}_0\left(\dfrac{\theta^{1/2}x}{\sigma^2}\right)\exp\left(-\dfrac{x^2+\theta}{2\sigma^2}\right), & \text{for } x \geq 0; \\ 0, & \text{elsewhere}; \end{cases} \qquad (9.116)$$

where $\mathrm{I}_0$ is the modified zeroth-order Bessel function of the first kind.

Given a sample of size $n$ from population $X$, the likelihood function takes the form

$$L = \prod_{j=1}^n f_X(x_j;\theta,\sigma^2). \qquad (9.117)$$

The MLEs of $\theta$ and $\sigma^2$, $\hat{\theta}$ and $\widehat{\sigma^2}$, satisfy the likelihood equations

$$\frac{\partial \ln L}{\partial \hat{\theta}} = 0, \quad \text{and} \quad \frac{\partial \ln L}{\partial \widehat{\sigma^2}} = 0, \qquad (9.118)$$

which, upon simplifying, can be written as

$$\frac{1}{n\hat{\theta}^{1/2}}\sum_{j=1}^n \frac{x_j \mathrm{I}_1(y_j)}{\mathrm{I}_0(y_j)} - 1 = 0, \qquad (9.119)$$

and

$$\widehat{\sigma^2} = \frac{1}{2}\left(\frac{1}{n}\sum_{j=1}^n x_j^2 - \hat{\theta}\right), \qquad (9.120)$$

where $I_1$ is the modified first-order Bessel function of the first kind, and

$$y_j = \frac{x_j \hat{\theta}^{1/2}}{\widehat{\sigma^2}}.$$ (9.121)

As we can see, although likelihood equations can be established, they are complicated functions of $\hat{\theta}$ and $\widehat{\sigma^2}$, and we must resort to numerical means for their solutions. As we have pointed out earlier, this difficulty is often encountered when using the method of maximum likelihood. Indeed, Example 9.13 shows that the method of moments offers considerable computational advantage in this case.

The variances of the maximum likelihood estimators for $\theta$ and $\sigma^2$ can be obtained, in principle, from Equations (9.119) and (9.120). We can also show that their variances can be larger than those associated with the moment estimators obtained in Example 9.13 for moderate sample sizes (see Benedict and Soong, 1967). This observation serves to remind us again that, although maximum likelihood estimators possess optimal asymptotic properties, they may perform poorly when the sample size is small.

## 9.3.2   INTERVAL ESTIMATION

We now examine another approach to the problem of parameter estimation. As stated in the introductory text of Section 9.3, the interval estimation provides, on the basis of a sample from a population, not only information on the parameter values to be estimated, but also an indication of the level of confidence that can be placed on possible numerical values of the parameters. Before developing the theory of interval estimation, an example will be used to demonstrate that a method that appears to be almost intuitively obvious could lead to conceptual difficulties.

Suppose that five sample values $-3, 2, 1.5, 0.5$, and $2.1$ – are observed from a normal distribution having an unknown mean $m$ and a known variance $\sigma^2 = 9$. From Example 9.15, we see that the MLE of $m$ is the sample mean $\overline{X}$ and thus

$$\hat{m} = \frac{1}{5}(3 + 2 + 1.5 + 0.5 + 2.1) = 1.82.$$ (9.122)

Our additional task is to determine the upper and lower limits of an interval such that, with a specified level of confidence, the true mean $m$ will lie in this interval.

The maximum likelihood estimator for $m$ is $\overline{X}$, which, being a sum of normal random variables, is normal with mean $m$ and variance $\sigma^2/n = 9/5$.

The standardized random variable $U$, defined by

$$U = \frac{\sqrt{5}(\overline{X} - m)}{3}, \tag{9.123}$$

is then $N(0, 1)$ and it has pdf

$$f_U(u) = \frac{1}{(2\pi)^{1/2}} e^{-u^2/2}, \quad -\infty < u < \infty. \tag{9.124}$$

Suppose we specify that the probability of $U$ being in interval $(-u_1, u_1)$ is equal to 0.95. From Table A.3 we find that $u_1 = 1.96$ and

$$P(-1.96 < U < 1.96) = \int_{-1.96}^{1.96} f_U(u) du = 0.95, \tag{9.125}$$

or, on substituting Equation (9.123) into Equation (9.125),

$$P(\overline{X} - 2.63 < m < \overline{X} + 2.63) = 0.95, \tag{9.126}$$

and, using Equation (9.122), the observed interval is

$$P(-0.81 < m < 4.45) = 0.95. \tag{9.127}$$

Equation (9.127) gives the desired result but it must be interpreted carefully. The mean $m$, although unknown, is nevertheless deterministic; and it either lies in an interval or it does not. However, we see from Equation (9.126) that the interval is a function of statistic $\overline{X}$. Hence, the proper way to interpret Equations (9.126) and (9.127) is that the probability of the *random interval* $(\overline{X} - 2.63, \overline{X} + 2.63)$ covering the distribution's true mean $m$ is 0.95, and Equation (9.127) gives the observed interval based upon the given sample values.

Let us place the concept illustrated by the example above in a more general and precise setting, through Definition 9.2.

**Definition 9.2.** Suppose that a sample $X_1, X_2, \ldots, X_n$ is drawn from a population having pdf $f(x; \theta)$, $\theta$ being the parameter to be estimated. Further suppose that $L_1(X_1, \ldots, X_n)$ and $L_2(X_1, \ldots, X_n)$ are two statistics such that $L_1 < L_2$ with probability 1. The interval $(L_1, L_2)$ is called a $[100(1 - \alpha)]\%$ *confidence interval* for $\theta$ if $L_1$ and $L_2$ can be selected such that

$$P(L_1 < \theta < L_2) = 1 - \alpha. \tag{9.128}$$

Limits $L_1$ and $L_2$ are called, respectively, the *lower* and *upper confidence limits* for $\theta$, and $1 - \alpha$ is called the *confidence coefficient*. The value of $1 - \alpha$ is generally taken as 0.90, 0.95, 0.99, and 0.999.

We now make several remarks concerning the foregoing definition.

- Remark 1: as we see from Equation (9.126), confidence limits are functions of a given sample. The confidence interval thus will generally vary in position and width from sample to sample.
- Remark 2: for a given sample, confidence limits are not unique. In other words, many pairs of statistics $L_1$ and $L_2$ exist that satisfy Equation (9.128). For example, in addition to the pair $(-1.96, 1.96)$, there are many other pairs of values (not symmetric about zero) that could give the probability 0.95 in Equation (9.125). However, it is easy to see that this particular pair gives the minimum-width interval.
- Remark 3: in view of the above, it is thus desirable to define a set of quality criteria for interval estimators so that the 'best' interval can be obtained. Intuitively, the 'best' interval is the shortest interval. Moreover, since interval width $L = L_2 - L_1$ is a random variable, we may like to choose 'minimum expected interval width' as a good criterion. Unfortunately, there may not exist statistics $L_1$ and $L_2$ that give rise to an expected interval width that is minimum for all values of $\theta$.
- Remark 4: just as in point estimation, sufficient statistics also play an important role in interval estimation, as Theorem 9.5 demonstrates.

**Theorem 9.5:** let $L_1$ and $L_2$ be two statistics based on a sample $X_1, \ldots, X_n$ from a population $X$ with pdf $f(x; \theta)$ such that $P(L_1 < \theta < L_2) = 1 - \alpha$. Let $Y = h(X_1, \ldots, X_n)$ be a sufficient statistic. Then there exist two functions $R_1$ and $R_2$ of $Y$ such that $P(R_1 < \theta < R_2) = 1 - \alpha$ and such that two interval widths $L = L_2 - L_1$ and $R = R_2' - R_1$ have the same distribution.

This theorem shows that, if a minimum interval width exists, it can be obtained by using functions of sufficient statistics as confidence limits.

The construction of confidence intervals for some important cases will be carried out in the following sections. The method consists essentially of finding an appropriate random variable for which values can be calculated on the basis of observed sample values and the parameter value but for which the distribution does not depend on the parameter. More general methods for obtaining confidence intervals are discussed in Mood (1950, chapter 11) and Wilks (1962, chapter 12).

### 9.3.2.1 Confidence Interval for $m$ in $N(m, \sigma^2)$ with Known $\sigma^2$

The confidence interval given by Equation (9.126) is designed to estimate the mean of a normal population with known variance. In general terms, the procedure shows that we first determine a (symmetric) interval in $U$ to achieve a confidence coefficient of $1 - \alpha$. Writing $u_{\alpha/2}$ for the value of $U$ above which the area under $f_U(u)$ is $\alpha/2$, that is, $P(U > u_{\alpha/2}) = \alpha/2$ (see Figure 9.6), we have

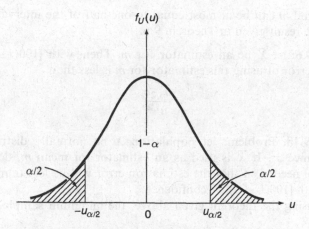

**Figure 9.6**   $[100(1 - \alpha)]\%$ confidence limits for $U$

$$P(-u_{\alpha/2} < U < u_{\alpha/2}) = 1 - \alpha. \tag{9.129}$$

Hence, using the transformation given by Equation (9.123), we have the general result

$$P\left(\overline{X} - \frac{u_{\alpha/2}\sigma}{n^{1/2}} < m < \overline{X} + \frac{u_{\alpha/2}\sigma}{n^{1/2}}\right) = 1 - \alpha. \tag{9.130}$$

This result can also be used to estimate means of nonnormal populations with known variances if the sample size is large enough to justify use of the central limit theorem.

It is noteworthy that, in this case, the position of the interval is a function of $\overline{X}$ and therefore is a function of the sample. The width of the interval, in contrast, is a function only of sample size $n$, being inversely proportional to $n^{1/2}$.

The $[100(1 - \alpha)]\%$ confidence interval for $m$ given in Equation (9.130) also provides an estimate of the accuracy of our point estimator $\overline{X}$ for $m$. As we see from Figure 9.7, the true mean $m$ lies within the indicated interval with $[100(1 - \alpha)]\%$ confidence. Since $\overline{X}$ is at the center of the interval, the distance

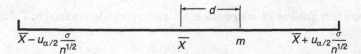

**Figure 9.7**   Error in point estimator $\overline{X}$ for $m$

between $\overline{X}$ and $m$ can be at most equal to one-half of the interval width. We thus have the result given in Theorem 9.6.

**Theorem 9.6:** let $\overline{X}$ be an estimator for $m$. Then, with $[100(1 - \alpha)]\%$ confidence, the error of using this estimator for $m$ is less than

$$\frac{u_{\alpha/2}\sigma}{n^{1/2}}$$

**Example 9.18.** Problem: let population $X$ be normally distributed with known variance $\sigma^2$. If $\overline{X}$ is used as an estimator for mean $m$, determine the sample size $n$ needed so that the estimation error will be less than a specified amount $\varepsilon$ with $[100(1 - \alpha)]\%$ confidence.

Answer: using the theorem given above, the minimum sample size $n$ must satisfy

$$\varepsilon = \frac{u_{\alpha/2}\sigma}{n^{1/2}}.$$

Hence, the solution for $n$ is

$$n = \left(\frac{u_{\alpha/2}\sigma}{\varepsilon}\right)^2. \tag{9.131}$$

### 9.3.2.2 Confidence Interval for $m$ in $N(m, \sigma^2)$ with Unknown $\sigma^2$

The difference between this problem and the preceding one is that, since $\sigma$ is not known, we can no longer use

$$U = (\overline{X} - m)\left(\frac{\sigma}{n^{1/2}}\right)^{-1}$$

as the random variable for confidence limit calculations regarding mean $m$. Let us then use sample variance $S^2$ as an unbiased estimator for $\sigma^2$ and consider the random variable

$$Y = (\overline{X} - m)\left(\frac{S}{n^{1/2}}\right)^{-1}. \tag{9.132}$$

The random variable $Y$ is now a function of random variables $\overline{X}$ and $S$. In order to determine its distribution, we first state Theorem 9.7.

**Theorem 9.7:** *Student's* t-*distribution*. Consider a random variable $T$ defined by

$$T = U\left(\frac{V}{n}\right)^{-1/2}. \tag{9.133}$$

If $U$ is $N(0, 1)$, $V$ is $\chi^2$-distributed with $n$ degrees of freedom, and $U$ and $V$ are independent, then the pdf of $T$ has the form

$$f_T(t) = \frac{\Gamma[(n+1)/2]}{\Gamma(n/2)(n\pi)^{1/2}} \left(1 + \frac{t^2}{n}\right)^{-(n+1)/2}, \quad -\infty < t < \infty. \tag{9.134}$$

This distribution is known as *Student's t-distribution* with $n$ degrees of freedom; it is named after W.S. Gosset, who used the pseudonym 'Student' in his research publications.

**Proof of Theorem 9.7:** the proof is straightforward following methods given in Chapter 5. Sine $U$ and $V$ are independent, their jpdf is

$$f_{UV}(u, v) = \begin{cases} \left(\dfrac{1}{(2\pi)^{1/2}} e^{-u^2/2}\right)\left[\dfrac{1}{2^{n/2}\Gamma(n/2)} v^{(n/2)-1} e^{-v/2}\right], & \text{for } -\infty < u < \infty, \text{ and } v > 0, \\ 0, & \text{elsewhere.} \end{cases} \tag{9.135}$$

Consider the transformation from $U$ and $V$ to $T$ and $V$. The method discussed in Section 5.3 leads to

$$f_{TV}(t, v) = f_{UV}[g_1^{-1}(t, v), g_2^{-1}(t, v)]|J|, \tag{9.136}$$

where

$$g_1^{-1}(t, v) = t\left(\frac{v}{n}\right)^{1/2}, \quad g_2^{-1}(t, v) = v, \tag{9.137}$$

and the Jacobian is

$$J = \begin{vmatrix} \dfrac{\partial g_1^{-1}}{\partial t} & \dfrac{\partial g_1^{-1}}{\partial v} \\ \dfrac{\partial g_2^{-1}}{\partial t} & \dfrac{\partial g_2^{-1}}{\partial v} \end{vmatrix} = \begin{vmatrix} \left(\dfrac{v}{n}\right)^{1/2} & t\left[2n\left(\dfrac{v}{n}\right)^{1/2}\right]^{-1} \\ 0 & 1 \end{vmatrix} = \left(\frac{v}{n}\right)^{1/2}. \tag{9.138}$$

The substitution of Equations (9.135), (9.137), and (9.138) into Equation (9.136) gives the jpdf $f_{TV}(t, v)$ of $T$ and $V$. The pdf of $T$ as given by Equation (9.134) is obtained by integrating $f_{TV}(t, v)$ with respect to $v$.

It is seen from Equation (9.134) that the $t$-distribution is symmetrical about the origin. As $n$ increases, it approaches that of a standardized normal random variable.

Returning to random variable $Y$ defined by Equation (9.132), let

$$U = (\overline{X} - m)\left(\frac{\sigma}{n^{1/2}}\right)^{-1}$$

and

$$V = \frac{(n-1)S^2}{\sigma^2}$$

Then

$$Y = U\left(\frac{V}{n-1}\right)^{-1/2}, \qquad (9.139)$$

where $U$ is clearly distributed according to $N(0, 1)$. We also see from Section 9.1.2 that $(n-1)S^2/\sigma^2$ has the chi-squared distribution with $(n-1)$ degrees of freedom. Furthermore, although we will not verify it here, it can be shown that $\overline{X}$ and $S^2$ are independent. In accordance with Theorem 9.7, random variable $Y$ thus has a $t$-distribution with $(n-1)$ degrees of freedom.

The random variable $Y$ can now be used to establish confidence intervals for mean $m$. We note that the value of $Y$ depends on the unknown mean $m$, but its distribution does not.

The $t$-distribution is tabulated in Table A.4 in Appendix A. Let $t_{n,\,\alpha/2}$ be the value such that

$$P(T > t_{n,\alpha/2}) = \frac{\alpha}{2},$$

with $n$ representing the number of degrees of freedom (see Figure 9.8). We have the result

$$P(-t_{n-1,\alpha/2} < Y < t_{n-1,\alpha/2}) = 1 - \alpha. \qquad (9.140)$$

Upon substituting Equation (9.132) into Equation (9.140), a $[100(1 - \alpha)]\%$ confidence interval for mean $m$ is thus given by

$$\boxed{P\left(\overline{X} - \frac{t_{n-1,\alpha/2}S}{n^{1/2}} < m < \overline{X} + \frac{t_{n-1,\alpha/2}S}{n^{1/2}}\right) = 1 - \alpha.} \qquad (9.141)$$

Since both $\overline{X}$ and $S$ are functions of the sample, both the position and the width of the confidence interval given above will vary from sample to sample.

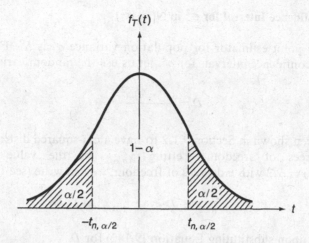

**Figure 9.8**  [$100(1 - \alpha)$]% confidence limits for $T$ with $n$ degrees of freedom

**Example 9.19.** Problem: let us assume that the annual snowfall in the Buffalo area is normally distributed. Using the snowfall record from 1970–79 as given in Problem 8.2(g) (Table 8.6, page 257), determine a 95% confidence interval for mean $m$.

Answer: for this example, $\alpha = 0.05$, $n = 10$, the observed sample mean is

$$\bar{x} = \frac{1}{10}(120.5 + 97.0 + \cdots + 97.3) = 112.4,$$

and the observed sample variance is

$$s^2 = \frac{1}{9}[(120.5 - 112.4)^2 + (97.0 - 112.4)^2 + \cdots + (97.3 - 112.4)^2]$$

$$= 1414.3.$$

Using Table A.4, we find that $t_{9,0.025} = 2.262$. Substituting all the values given above into Equation (9.141) gives

$$P(85.5 < m < 139.3) = 0.95.$$

It is clear that this interval would be different if we had incorporated more observations into our calculations or if we had chosen a different set of yearly snowfall data.

### 9.3.2.3  Confidence Interval for $\sigma^2$ in $N(m, \sigma^2)$

An unbiased point estimator for population variance $\sigma^2$ is $S^2$. For the construction of confidence intervals for $\sigma^2$, let us use the random variable

$$D = \frac{(n - 1)S^2}{\sigma^2}, \tag{9.142}$$

which has been shown in Section 9.1.2 to have a chi-squared distribution with $(n - 1)$ degrees of freedom. Letting $\chi^2_{n, \alpha/2}$ be the value such that $P(D > \chi^2_{n, \alpha/2}) = \alpha/2$ with $n$ degrees of freedom, we can write (see Figure 9.9)

$$P(\chi^2_{n-1, 1-(\alpha/2)} < D < \chi^2_{n-1, \alpha/2}) = 1 - \alpha, \tag{9.143}$$

which gives, upon substituting Equation (9.142) for $D$,

$$P\left[ \frac{(n - 1)S^2}{\chi^2_{n-1, \alpha/2}} < \sigma^2 < \frac{(n - 1)S^2}{\chi^2_{n-1, 1-(\alpha/2)}} \right] = 1 - \alpha. \tag{9.144}$$

Let us note that the $[100(1 - \alpha)]\%$ confidence interval for $\sigma^2$ as defined by Equation (9.144) is not the minimum-width interval on the basis of a given sample. As we see in Figure 9.9, a shift to the left, leaving area $\alpha/2 - \varepsilon$ to the left and area $\alpha/2 + \varepsilon$ to the right under the $f_D(d)$ curve, where $\varepsilon$ is an appropriate amount, will result in a smaller confidence interval. This is because the width needed at the left to give an increase of $\varepsilon$ in the area is less than the corresponding width eliminated at the right. The minimum interval width for a given

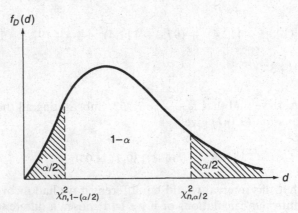

**Figure 9.9**  $[100(1 - \alpha)]\%$ confidence limits for $D$ with $n$ degrees of freedom

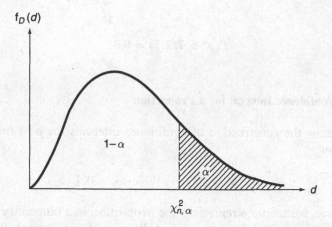

**Figure 9.10**   One-sided $[100(1 - \alpha)]\%$ confidence limit for $D$ with $n$ degrees of freedom

number of degrees of freedom can be determined by interpolation from tabulated values of the PDF of the chi-squared distribution.

Table A.5 in Appendix A gives selected values of $\chi^2_{n,\alpha}$ for various values of $n$ and $\alpha$. For convenience, Equation (9.144) is commonly used for constructing two-sided confidence intervals for $\sigma^2$ of a normal population. If a one-sided confidence interval is desired, it is then given by (see Figure 9.10)

$$P\left[\sigma^2 > \frac{(n-1)S^2}{\chi^2_{n-1,\alpha}}\right] = 1 - \alpha. \tag{9.145}$$

**Example 9.20.** Consider Example 9.19 again; let us determine both two-sided and one-sided 95% confidence intervals for $\sigma^2$.

As seen from Example 9.19, the observed sample variance $s^2$, is

$$s^2 = 1414.3.$$

The values of $\chi^2_{9,0.975}$, $\chi^2_{9,0.025}$, and $\chi^2_{9,0.05}$ are obtained from Table A.5 to be as follows:

$$\chi^2_{9,0.975} = 2.700, \quad \chi^2_{9,0.025} = 19.023, \quad \chi^2_{9,0.05} = 16.919.$$

Equations (9.144) and (9.145) thus lead to, with $n = 10$ and $\alpha = 0.05$,

$$P(669.12 < \sigma^2 < 4714.33) = 0.95,$$

and

$$P(\sigma^2 > 752.3) = 0.95.$$

### 9.3.2.4  Confidence Interval for a Proportion

Consider now the construction of confidence intervals for $p$ in the binomial distribution

$$p_X(k) = p^k(1 - p)^{1-k}, \quad k = 0, 1.$$

In the above, parameter $p$ represents the proportion in a binomial experiment. Given a sample of size $n$ from population $X$, we see from Example 9.10 that an unbiased and efficient estimator for $p$ is $\overline{X}$. For large $n$, random variable $\overline{X}$ is approximately normal with mean $p$ and variance $p(1 - p)/n$.
  Defining

$$U = (\overline{X} - p)\left[\frac{p(1-p)}{n}\right]^{-1/2}, \tag{9.146}$$

random variable $U$ tends to $N(0, 1)$ as $n$ becomes large. In terms of $U$, we have the same situation as in Section 9.3.2.1 and Equation (9.129) gives

$$P(-u_{\alpha/2} < U < u_{\alpha/2}) = 1 - \alpha. \tag{9.147}$$

The substitution of Equation (9.146) into Equation (9.147) gives

$$P\left[-u_{\alpha/2} < (\overline{X} - p)\left[\frac{p(1-p)}{n}\right]^{-1/2} < u_{\alpha/2}\right] = 1 - \alpha. \tag{9.148}$$

In order to determine confidence limits for $p$, we need to solve for $p$ satisfying the equation

$$|\overline{X} - p|\left[\frac{p(1-p)}{n}\right]^{-1/2} \le u_{\alpha/2},$$

or, equivalently

$$(\overline{X} - p)^2 \le \frac{u_{\alpha/2}^2 p(1-p)}{n}. \tag{9.149}$$

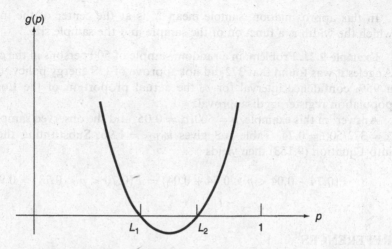

**Figure 9.11** Parabola defined by Equation (9.150)

Upon transposing the right-hand side, we have

$$p^2\left(1 + \frac{u_{\alpha/2}^2}{n}\right) - p\left(2\overline{X} + \frac{u_{\alpha/2}^2}{n}\right) + \overline{X}^2 \leq 0. \qquad (9.150)$$

In Equation (9.150), the left-hand side defines a parabola, as shown in Figure 9.11, and two roots $L_1$ and $L_2$ of Equation (9.150) with the equal sign define the interval within which the parabola is negative. Hence, solving the quadratic equation defined by Equation (9.150), we have

$$L_{1,2} = \left\{\left(\overline{X} + \frac{u_{\alpha/2}^2}{2n}\right) \mp \left(\frac{u_{\alpha/2}}{n^{1/2}}\right)\left[\overline{X}(1 - \overline{X}) + \frac{u_{\alpha/2}^2}{4n}\right]^{1/2}\right\}\left(1 + \frac{u_{\alpha/2}^2}{n}\right)^{-1}. \qquad (9.151)$$

For large $n$, they can be approximated by

$$L_{1,2} = \overline{X} \mp u_{\alpha/2}\left[\frac{\overline{X}(1 - \overline{X})}{n}\right]^{1/2} \qquad (9.152)$$

An approximate $[100(1 - \alpha)]\%$ confidence interval for $p$ is thus given by, for large $n$,

$$P\left(\overline{X} - u_{\alpha/2}\left[\frac{\overline{X}(1 - \overline{X})}{n}\right]^{1/2} < p < \overline{X} + u_{\alpha/2}\left[\frac{\overline{X}(1 - \overline{X})}{n}\right]^{1/2}\right) = 1 - \alpha \qquad (9.153)$$

In this approximation, sample mean $\overline{X}$ is at the center of the interval for which the width is a function of the sample and the sample size.

**Example 9.21.** Problem: in a random sample of 500 persons in the city of Los Angeles it was found that 372 did not approve of US energy policy. Determine a 95% confidence interval for $p$, the actual proportion of the Los Angeles population registering disapproval.

Answer: in this example, $n = 500$, $\alpha = 0.05$, and the observed sample mean is $\bar{x} = 372/500 = 0.74$. Table A.3 gives $u_{0.025} = 1.96$. Substituting these values into Equation (9.153) then yields

$$P(0.74 - 0.04 < p < 0.74 + 0.04) = P(0.70 < p < 0.78) = 0.95.$$

# REFERENCES

Anderson, R.L., and Bancroft, T.A., 1952, *Statistical Theory in Research*, McGraw-Hill, New York.

Benedict, T.R., and Soong, T.T., 1967, "The Joint Estimation of Signal and Noise from the Sum Envelope", *IEEE Trans. Information Theory*, **IT-13** 447–454.

Fisher, R.A., 1922, "On the Mathematical Foundation of Theoretical Statistics", *Phil. Trans. Roy. Soc. London, Ser. A* **222** 309–368.

Gerlough, D.L., 1955, "The Use of Poisson Distribution in Traffic", in *Poisson and Traffic*, The Eno Foundation for Highway Traffic Control, Saugatuk, CT.

Mood, A.M., 1950, *Introduction to the Theory of Statistics*, McGraw-Hill, New York.

Neyman, J., 1935, "Su un Teorema Concernente le Cosiddeti Statistiche Sufficienti", *Giorn. Inst. Ital. Atturi.* **6** 320–334.

Pearson, K., 1894, "Contributions to the Mathematical Theory of Evolution", *Phil. Trans. Roy. Soc. London, Ser. A* **185** 71–78.

Soong, T.T., 1969, "An Extension of the Moment Method in Statistical Estimation", *SIAM J. Appl. Math.* **17** 560–568.

Wilks, S.S., 1962, *Mathematical Statistics*, John Wiley & Sons Inc., New York.

# FURTHER READING AND COMMENTS

The Cramér–Rao inequality is named after two well-known statisticians, H. Cramér and C.R. Rao, who independently established this result in the following references. However, this inequality was first stated by Fisher in 1922 (see the Reference section). In fact, much of the foundation of parameter estimation and statistical inference in general, such as concepts of consistency, efficiency, and sufficiency, was laid down by Fisher in a series of publications, given below.

Cramér, H., 1946, *Mathematical Methods of Statistics*, Princeton University Press, Princeton, NJ.

Fisher, R.A., 1924, "On a Distribution Yielding the Error Functions of Several Well-known Statistics", *Proc. Int. Math. Congress, Vol. II*, Toronto, 805–813.

Fisher, R.A., 1925, *Statistical Methods for Research Workers*, 14th edn, Hafner, New York, (1st edn, 1925).

Fisher, R.A., 1925, "Theory of Statistical Estimation", *Proc. Camb. Phil. Soc.* **22** 700–725.

Rao, C.R., 1945, "Information and Accuracy Attainable in the Estimation of Statistical Parameters", *Bull. Calcutta Math. Soc.* **37** 81–91.

## PROBLEMS

The following notations and abbreviations are used in some statements of the problems:

$\overline{X}$ = sample mean
$\overline{x}$ = observed sample mean
$S^2$ = sample variance
$s^2$ = observed sample variance
CRLB = Cramér-Rao lower bound
ME = moment estimator, or moment estimate
MLE = maximum likelihood estimator, or maximum likelihood estimate
pdf = probability density function
pmf = probability mass function

9.1 In order to make enrollment projections, a survey is made of numbers of children in 100 families in a school district; the result is given in Table 9.2. Determine $\overline{x}$, the observed sample mean, and $s^2$, the observed sample variance, on the basis of these 100 sample values.

**Table 9.2**  Data for Problem 9.1

| Children (No.) | Families (No.) |
|---|---|
| 0 | 21 |
| 1 | 24 |
| 2 | 30 |
| 3 | 16 |
| 4 | 4 |
| 5 | 4 |
| 6 | 0 |
| 7 | 1 |
| | $n = 100$ |

9.2 Verify that the variance of sample variance $S^2$ as defined by Equation (9.7) is given by Equation (9.10).

9.3 Verify that the mean and variance of $k$th sample moment $M_k$ as defined by Equation (9.14) are given by Equations (9.15).

9.4 Let $X_1, X_2, \ldots, X_{10}$ be a sample of size 10 from the standardized normal distribution $N(0, 1)$. Determine probability $P(\overline{X} \le 1)$.

9.5 Let $X_1, X_2, \ldots, X_{10}$ be a sample of size 10 from a uniformly distributed random variable in interval $(0, 1)$.

(a) Determine the pdfs of $X_{(1)}$ and $X_{(10)}$.
(b) Find the probabilities $P[X_{(1)} > 0.5]$ and $P[X_{(10)} \leq 0.5]$.
(c) Determine $E\{X_{(1)}\}$ and $E\{X_{(10)}\}$.

9.6 A sample of size $n$ is taken from a population $X$ with pdf

$$f_X(x) = \begin{cases} e^{-x}, & \text{for } x \geq 0; \\ 0, & \text{elsewhere.} \end{cases}$$

Determine the probability density function of statistic $\overline{X}$. (Hint: use the method of characteristic functions discussed in Chapter 4.)

9.7 Two samples $X_1$ and $X_2$ are taken from an exponential random variable $X$ with unknown parameter $\theta$; that is,

$$f_X(x; \theta) = \frac{1}{\theta} e^{-x/\theta}, \quad x \geq 0.$$

We propose two estimators for $\theta$ in the forms

$$\hat{\Theta}_1 = \overline{X} = \frac{X_1 + X_2}{2},$$

$$\hat{\Theta}_2 = \frac{4}{\pi}(X_1 X_2)^{1/2}.$$

In terms of unbiasedness and minimum variance, which one is the better of the two?

9.8 Let $X_1$ and $X_2$ be a sample of size 2 from a population $X$ with mean $m$ and variance $\sigma^2$.
(a) Two estimators for $m$ are proposed to be

$$\hat{M}_1 = \overline{X} = \frac{X_1 + X_2}{2},$$

$$\hat{M}_2 = \frac{X_1 + 2X_2}{3}.$$

Which is the better estimator?
(b) Consider an estimator for $m$ in the form

$$\hat{M} = aX_1 + (1 - a)X_2, \quad 0 \leq a \leq 1.$$

Determine value $a$ that gives the best estimator in this form.

9.9 It is known that a certain proportion, say $p$, of manufactured parts is defective. From a supply of parts, $n$ are chosen at random and are tested. Define the readings (sample $X_1, X_2, \ldots, X_n$) to be 1 if good and 0 if defective. Then, a good estimator for $p$, $\hat{P}$, is

$$\hat{P} = 1 - \overline{X} = 1 - \frac{1}{n}(X_1 + \cdots + X_n).$$

(a) Is $\hat{P}$ unbiased?
(b) Is $\hat{P}$ consistent?
(c) Show that $\hat{P}$ is an MLE of $p$.

9.10 Let $X$ be a random variable with mean $m$ and variance $\sigma^2$, and let $X_1, X_2, \ldots, X_n$ be independent samples of $X$. Suppose an estimator for $\sigma^2$, $\widehat{\Lambda^2}$ is found from the formula

$$\widehat{\Lambda^2} = \frac{1}{2(n-1)}[(X_2 - X_1)^2 + (X_3 - X_2)^2 + \cdots + (X_n - X_{n-1})^2].$$

Is $\widehat{\Lambda^2}$ an unbiased estimator? Verify your answer.

9.11 The geometrical mean $(X_1 X_2 \cdots X_n)^{1/n}$ is proposed as an estimator for the unknown median of a lognormally distributed random variable $X$. Is it unbiased? Is it unbiased as $n \to \infty$?

9.12 Let $X_1, X_2, X_3$ be a sample of size three from a uniform distribution for which the pdf is

$$f_X(x; \theta) = \begin{cases} \dfrac{1}{\theta}, & \text{for } 0 \le x \le \theta; \\ 0, & \text{elsewhere.} \end{cases}$$

Suppose that $aX_{(1)}$ and $bX_{(3)}$ are proposed as two possible estimators for $\theta$.
(a) Determine $a$ and $b$ such that these estimators are unbiased.
(b) Which one is the better of the two? In the above, $X_{(j)}$ is the $j$th-order statistic.

9.13 Let $X_1, \ldots, X_n$ be a sample from a population whose $k$th moment $\alpha_k = E\{X^k\}$ exists. Show that the $k$th sample moment

$$M_k = \frac{1}{n} \sum_{j=1}^{n} X_j^k$$

is a consistent estimator for $\alpha_k$.

9.14 Let $\theta$ be the parameter to be estimated in each of the distributions given below. For each case, determine the CRLB for the variance of any unbiased estimator for $\theta$.

(a) $f(x; \theta) = \dfrac{1}{\theta} e^{-x/\theta}, x \ge 0$.

(b) $f(x; \theta) = \theta x^{\theta-1}, 0 \le x \le 1, \theta > 0$.

(c) $p(x; \theta) = \theta^x (1 - \theta)^{1-x}, x = 0, 1$.

(d) $p(x; \theta) = \dfrac{\theta^x e^{-\theta}}{x!}, x = 0, 1, 2, \ldots$.

9.15 Determine the CRLB for the variances of $\hat{M}$ and $\widehat{\Sigma^2}$, which are, respectively, unbiased estimators for $m$ and $\sigma^2$ in the normal distribution $N(m, \sigma^2)$.

9.16 The method of moments is based on equating the $k$th sample moment $M_k$ to the $k$th population moment $\alpha_k$; that is

$$M_k = \alpha_k.$$

(a) Verify Equations (9.15).
(b) Show that $M_k$ is a consistent estimator for $\alpha_k$.

9.17 Using the maximum likelihood method and the moment method, determine the respective estimators $\hat{\Theta}$ of $\theta$ and compare their asymptotic variances for the following two cases:

(a) Case 1:

$$f(x;\theta) = \frac{1}{(2\pi)^{1/2}\theta}\exp\left[-\frac{(x-m)^2}{2\theta^2}\right], \text{ where } m \text{ is a known constant.}$$

(b) Case 2:

$$f(x;\theta) = \begin{cases} \dfrac{1}{x(2\pi\theta)^{1/2}}\exp\left[-\dfrac{1}{2\theta}\ln^2 x\right], & \text{for } x > 0; \\ 0, & \text{elsewhere.} \end{cases}$$

9.18 Consider each distribution in Problem 9.14.

(a) Determine an ME for $\theta$ on the basis of a sample of size $n$ by using the first-order moment equation. Determine its asymptotic efficiency (i.e. its efficiency as $n \to \infty$). (Hint: use the second of Equations (9.62) for the asymptotic variance of ME.)

(b) Determine the MLE for $\theta$.

9.19 The number of transistor failures in an electronic computer may be considered as a random variable.

(a) Let $X$ be the number of transistor failures per hour. What is an appropriate distribution for $X$? Explain your answer.

(b) The numbers of transistor failures per hour for 96 hours are recorded in Table 9.3. Estimate the parameter(s) of the distribution for $X$ based on these data by using the method of maximum likelihood.

**Table 9.3**   Data for Problem 9.19

| Hourly failures (No.) | Hours (No.) |
|---|---|
| 0 | 59 |
| 1 | 27 |
| 2 | 9 |
| 3 | 1 |
| >3 | 0 |
| | Total = 96 |

(c) A certain computation requires 20 hours of computing time. Use this model and find the probability that this computation can be completed without a computer breakdown (a breakdown occurs when *two or more* transistors fail).

9.20 Electronic components are tested for reliability. Let $p$ be the probability of an electronic component being successful and $1 - p$ be the probability of component failure. If $X$ is the number of trials at which the first failure occurs, then it has the geometric distribution

$$p_X(k;p) = (1-p)p^{k-1}, \quad k = 1,2,\ldots.$$

Suppose that a sample $X_1, \ldots, X_n$ is taken from population $X$, each $X_j$ consisting of testing $X_j$ components when the first failure occurs.
(a) Determine the MLE of $p$.
(b) Determine the MLE of $P(X > 9)$, the probability that the component will not fail in nine trials. Note:

$$P(X > 9) = \sum_{k=1}^{9} (1 - p)p^{k-1}.$$

9.21 The pdf of a population $X$ is given by

$$f_X(x; \theta) = \begin{cases} \dfrac{2x}{\theta^2}, & \text{for } 0 \leq x \leq \theta; \\ 0, & \text{elsewhere.} \end{cases}$$

Based on a sample of size $n$:
(a) Determine the MLE and ME for $\theta$.
(b) Which one of the two is the better estimator?

9.22 Assume that $X$ has a shifted exponential distribution, with

$$f_X(x; a) = e^{a-x}, \quad x \geq a.$$

On the basis of a sample of size $n$ from $X$, determine the MLE and ME for $a$.

9.23 Let $X_1, X_2, \ldots, X_n$ be a sample of size $n$ from a uniform distribution

$$f(x; \theta) = \begin{cases} 1, & \text{for } \theta - \dfrac{1}{2} \leq x \leq \theta + \dfrac{1}{2}; \\ 0, & \text{elsewhere.} \end{cases}$$

Show that every statistic $h(X_1, \ldots, X_n)$ satisfying

$$X_{(n)} - \frac{1}{2} \leq h(X_1, \ldots, X_n) \leq X_{(1)} + \frac{1}{2}$$

is an MLE for $\theta$, where $X_{(j)}$ is the $j$th-order statistic. Determine an MLE for $\theta$ when the observed sample values are (1.5, 1.4, 2.1, 2.0, 1.9, 2.0, 2.3), with $n = 7$.

9.24 Using the 214 measurements given in Example 9.11 (see Table 9.1), determine the MLE for $\lambda$ in the exponential distribution given by Equation (9.70).

9.25 Let us assume that random variable $X$ in Problem 8.2(j) is Poisson distributed. Using the 58 sample values given (see Figure 8.6), determine the MLE and ME for the mean number of blemishes.

9.26 The time-to-failure $T$ of a certain device has a shifted exponential distribution; that is,

$$f_T(t; t_0, \lambda) = \begin{cases} \lambda e^{-\lambda(t-t_0)}, & \text{for } t \geq t_0; \\ 0, & \text{elsewhere.} \end{cases}$$

Let $T_1, T_2, \ldots, T_n$ be a sample from $T$.
(a) Determine the MLE and ME for $\lambda$ ($\hat{\Lambda}_{ML}$ and $\hat{\Lambda}_{ME}$, respectively) assuming $t_0$ is known.
(b) Determine the MLE and ME for $t_0$ ($\hat{T}_{OML}$ and $\hat{T}_{OME}$, respectively) assuming $\lambda$ is known.
(c) Determine the MLEs and MEs for both $\lambda$ and $t_0$ assuming both are unknown.

9.27 If $X_1, X_2, \ldots, X_n$ is a sample from the gamma distribution; that is,

$$f(x; r, \lambda) = \frac{\lambda^r x^{r-1}}{\Gamma(r)} e^{-\lambda x}, \quad x \geq 0, \ r, \lambda > 0,$$

show that:
(a) If $r$ is known and $\lambda$ is the parameter to be estimated, both the MLE and ME for $\lambda$ are $\hat{\Lambda} = r/\overline{X}$.
(b) If both $r$ and $\lambda$ are to be estimated, then the method of moments and the method of maximum likelihood lead to different estimators for $r$ and $\lambda$. (It is not necessary to determine these estimators.)

9.28 Consider the Buffalo yearly snowfall data, given in Problem 8.2(g) (see Table 8.6) and assume that a normal distribution is appropriate.
(a) Find estimates for the parameters by means of the moment method and the method of maximum likelihood.
(b) Estimate from the model the probability of having another blizzard of 1977 [$P(X > 199.4)$].

9.29 Recorded annual flow $Y$ (in cfs) of a river at a given point are 141, 146, 166, 209, 228, 234, 260, 278, 319, 351, 383, 500, 522, 589, 696, 833, 888, 1173, 1200, 1258, 1340, 1390, 1420, 1423, 1443, 1561, 1650, 1810, 2004, 2013, 2016, 2080, 2090, 2143, 2185, 2316, 2582, 3050, 3186, 3222, 3660, 3799, 3824, 4099, and 6634. Assuming that $Y$ follows a lognormal distribution, determine the MLEs of the distribution parameters.

9.30 Let $X_1$ and $X_2$ be a sample of size 2 from a uniform distribution with pdf

$$f(x; \theta) = \begin{cases} \dfrac{1}{\theta}, & \text{for } 0 \leq x \leq \theta; \\ 0, & \text{elsewhere.} \end{cases}$$

Determine constant $c$ so that the interval

$$0 < \theta < c(X_1 + X_2)$$

is a $[100(1 - \alpha)]\%$ confidence interval for $\theta$.

9.31 The fuel consumption of a certain type of vehicle is approximately normal, with standard deviation 3 miles per gallon. If a sample of 64 vehicles has an average fuel consumption of 16 miles per gallon:
(a) Determine a 95% confidence interval for the mean fuel consumption of all vehicles of this type.
(b) With 95% confidence, what is the possible error if the mean fuel consumption is taken to be 16 miles per gallon?
(c) How large a sample is needed if we wish to be 95% confident that the mean will be within 0.5 miles per gallon of the true mean?

9.32 A total of 93 yearly Buffalo snowfall measurements are given in Problem 8.2(g) (see Table 8.6, page 255). Assume that it is approximately normal with standard deviation $\sigma = 26$ inches. Determine 95% confidence intervals for the mean using measurements of (a) 1909 to 1939, (b) 1909 to 1959, (c) 1909 to 1979, and (d) 1909 to 1999. Display these intervals graphically.

9.33 Let $\overline{X}_1$ and $\overline{X}_2$ be independent sample means from two normal populations $N(m_1, \sigma_1^2)$ and $N(m_2, \sigma_2^2)$, respectively. If $\sigma_1^2$ and $\sigma_2^2$ are known, show that a $[100(1 - \alpha)]\%$ confidence interval for $m_1 - m_2$ is

$$P\left[(\overline{X}_1 - \overline{X}_2) - u_{\alpha/2}\left(\frac{\sigma_1^2}{n_1} + \frac{\sigma_2^2}{n_2}\right)^{1/2} < m_1 - m_2 < (\overline{X}_1 - \overline{X}_2) + u_{\alpha/2}\left(\frac{\sigma_1^2}{n_1} + \frac{\sigma_2^2}{n_2}\right)^{1/2}\right] = 1 - \alpha,$$

where $n_1$ and $n_2$ are, respectively, the sample sizes from $N(m_1, \sigma_1^2)$ and $N(m_2, \sigma_2^2)$, and $u_{\alpha/2}$ is the value of standardized normal random variable $U$ such that $P(U > u_{\alpha/2}) = \alpha/2$.

9.34 Let us assume that random variable $X$ in Problem 8.2(e) has a Poisson distribution with pmf

$$p_X(k; \lambda) = \frac{\lambda^k e^{-\lambda}}{k!}, \quad k = 0, 1, 2, \ldots .$$

Use the sample values of $X$ given in Problem 8.2(e) (see Table 8.5, page 255) and:
(a) Determine MLE $\hat{\lambda}$ for $\lambda$.
(b) Determine a 95% confidence interval for $\lambda$ using asymptotic properties of MLE $\hat{\lambda}$.

9.35 Assume that the lifespan of US males is normally distributed with unknown mean $m$ and unknown variance $\sigma^2$. A sample of 30 mortality histories of US males shows that

$$\overline{x} = \frac{1}{30}\sum_{i=1}^{30} x_i = 71.3 \text{ years},$$

$$s^2 = \frac{1}{29}\sum_{i=1}^{30}(x_i - \overline{x})^2 = 128 \text{ (years)}^2.$$

Determine the observed values of 95% confidence intervals for $m$ and $\sigma^2$.

9.36 The life of light bulbs manufactured in a certain plant can be assumed to be normally distributed. A sample of 15 light bulbs gives the observed sample mean $\overline{x} = 1100$ hours and the observed sample standard deviation $s = 50$ hours.
(a) Determine a 95% confidence interval for the average life.
(b) Determine two-sided and one-sided 95% confidence intervals for its variance.

9.37 A total of 12 of 100 manufactured items examined are found to be defective.
(a) Find a 99% confidence interval for the proportion of defective items in the manufacturing process.

(b) With 99% confidence, what is the possible error if the proportion is estimated to be $12/100 = 0.12$?

9.38 In a public opinion poll such as the one described in Example 9.21, determine the minimum sample size needed for the poll so that with 95% confidence the sample means will be within 0.05 of the true proportion. [Hint: use the fact that $\overline{X}(1 - \overline{X}) \leq 1/4$ in Equation (9.153).]

# 10

# Model Verification

The parameter estimation procedures developed in Chapter 9 presume a distribution for the population. The validity of the model-building process based on this approach thus hinges on the substantiability of the hypothesized distribution. Indeed, if the hypothesized distribution is off the mark, the resulting probabilistic model with parameters estimated by any, however elegant, procedure would, at best, still give a poor representation of the underlying physical or natural phenomenon.

In this chapter, we wish to develop methods of testing or verifying a hypothesized distribution for a population on the basis of a sample taken from the population. Some aspects of this problem were addressed in Chapter 8, in which, by means of histograms and frequency diagrams, a graphical comparison between the hypothesized distribution and observed data was made. In the chemical yield example, for instance, a comparison between the shape of a normal distribution and the frequency diagram constructed from the data, as shown in Figure 8.1, suggested that the normal model is reasonable in that case.

However, the graphical procedure described above is clearly subjective and nonquantitative. On a more objective and quantitative basis, the problem of model verification on the basis of sample information falls within the framework of testing of hypotheses. Some basic concepts in this area of statistical inference are now introduced.

## 10.1 PRELIMINARIES

In our development, statistical hypotheses concern functional forms of the assumed distributions; these distributions may be specified completely with prespecified values for their parameters or they may be specified with parameters yet to be estimated from the sample.

Let $X_1, X_2, \ldots, X_n$ be an independent sample of size $n$ from a population $X$ with a hypothesized probability density function (pdf) $f(x; \boldsymbol{\theta})$ or probability

*Fundamentals of Probability and Statistics for Engineers* T.T. Soong © 2004 John Wiley & Sons, Ltd
ISBNs: 0-470-86813-9 (HB)   0-470-86814-7 (PB)

mass function (pmf) $p(x;\theta)$, where $\theta$ may be specified or unspecified. We denote by *hypothesis H* the hypothesis that the sample represents $n$ values of a random variable with pdf $f(x;\theta)$ or $p(x;\theta)$. This hypothesis is called a *simple hypothesis* when the underlying distribution is completely specified; that is, the parameter values are specified together with the functional form of the pdf or the pmf; otherwise, it is a *composite hypothesis*. To construct a criterion for hypotheses testing, it is necessary that an alternative hypothesis be established against which hypothesis $H$ can be tested. An example of an alternative hypothesis is simply another hypothesized distribution, or, as another example, hypothesis $H$ can be tested against the alternative hypothesis that hypothesis $H$ is not true. In our applications, the latter choice is considered more practical and we shall in general deal with the task of either accepting or rejecting hypothesis $H$ on the basis of a sample from the population.

## 10.1.1   TYPE-I AND TYPE-II ERRORS

As in parameter estimation, errors or risks are inherent in deciding whether a hypothesis $H$ should be accepted or rejected on the basis of sample information. Tests for hypotheses testing are therefore generally compared in terms of the probabilities of errors that might be committed. There are basically two types of errors that are likely to be made – namely, reject $H$ when in fact $H$ is true or, alternatively, accept $H$ when in fact $H$ is false. We formalize the above with Definition 10.1.

**Definition 10.1.** in testing hypothesis $H$, a Type-I error is committed when $H$ is rejected when in fact $H$ is true; a Type-II error is committed when $H$ is accepted when in fact $H$ is false.

In hypotheses testing, an important consideration in constructing statistical tests is thus to control, insofar as possible, the probabilities of making these errors. Let us note that, for a given test, an evaluation of Type-I errors can be made when hypothesis $H$ is given, that is, when a hypothesized distribution is specified. In contrast, the specification of an alternative hypothesis dictates Type-II error probabilities. In our problem, the alternative hypothesis is simply that hypothesis $H$ is not true. The fact that the class of alternatives is so large makes it difficult to use Type-II errors as a criterion. In what follows, methods of hypotheses testing are discussed based on Type-I errors only.

## 10.2   CHI-SQUARED GOODNESS-OF-FIT TEST

As mentioned above, the problem to be addressed is one of testing hypothesis $H$ that specifies the probability distribution for a population $X$ compared with the alternative that the probability distribution of $X$ is not of the stated type on the basis of a sample of size $n$ from population $X$. One of the most popular and most versatile tests devised for this purpose is the chi-squared ($\chi^2$) goodness-of-fit test introduced by Pearson (1900).

## 10.2.1  THE CASE OF KNOWN PARAMETERS

Let us first assume that the hypothesized distribution is completely specified with no unknown parameters. In order to test hypothesis $H$, some statistic $h(X_1, X_2, \ldots, X_n)$ of the sample is required that gives a measure of deviation of the observed distribution as constructed from the sample from the hypothesized distribution.

In the $\chi^2$ test, the statistic used is related to, roughly speaking, the difference between the frequency diagram constructed from the sample and a corresponding diagram constructed from the hypothesized distribution. Let the range space of $X$ be divided into $k$ mutually exclusive intervals $A_1, A_2, \ldots,$ and $A_k$, and let $N_i$ be the number of $X_j$ falling into $A_i, i = 1, 2, \ldots, k$. Then, the *observed* probabilities $P(A_i)$ are given by

$$\text{observed } P(A_i) = \frac{N_i}{n}, \quad i = 1, 2, \ldots, k. \tag{10.1}$$

The *theoretical* probabilities $P(A_i)$ can be obtained from the hypothesized population distribution. Let us denote these by

$$\text{theoretical } P(A_i) = p_i, \quad i = 1, 2, \ldots, k. \tag{10.2}$$

A logical choice of a statistic giving a measure of deviation is

$$\sum_{i=1}^{k} c_i \left( \frac{N_i}{n} - p_i \right)^2, \tag{10.3}$$

which is a natural least-square type deviation measure. Pearson (1900) showed that, if we take coefficient $c_i = n/p_i$, the statistic defined by Expression (10.3) has particularly simple properties. Hence, we choose as our deviation measure

$$D = \sum_{i=1}^{k} \frac{n}{p_i} \left( \frac{N_i}{n} - p_i \right)^2 = \sum_{i=1}^{k} \frac{(N_i - np_i)^2}{np_i}$$

$$= \sum_{i=1}^{k} \frac{N_i^2}{np_i} - n. \tag{10.4}$$

Let us note that $D$ is a statistic since it is a function of $N_i$, which are, in turn, functions of sample $X_1, \ldots, X_n$. The distribution of statistic $D$ is given in Theorem 10.1, attributable to Pearson (1900).

**Theorem 10.1:** assuming that hypothesis $H$ is true, the distribution of $D$ defined by Equation (10.4) approaches a chi-squared distribution with $(k - 1)$ degrees of freedom as $n \to \infty$. Its pdf is given by [see Equation (7.67)]

$$f_D(d) = \begin{cases} \left[ 2^{(k-1)/2} \Gamma\left( \dfrac{k-1}{2} \right) \right]^{-1} d^{(k-3)/2} e^{-d/2}, & \text{for } d \geq 0; \\ 0, & \text{elsewhere.} \end{cases} \tag{10.5}$$

Note that this distribution is independent of the hypothesized distribution.

**Proof of Theorem 10.1:** The complete proof, which can be found in Craméer (1946) and in other advanced texts in statistics, will not be attempted here. To demonstrate its plausibility, we only sketch the proof for the $k = 2$ case.

For $k = 2$, random variable $D$ is

$$D = \frac{(N_1 - np_1)^2}{np_1} + \frac{(N_2 - np_2)^2}{np_2}.$$

Since $N_1 + N_2 = n$, and $p_1 + p_2 = 1$, we can write

$$D = \frac{(N_1 - np_1)^2}{np_1} + \frac{[n - N_1 - n(1 - p_1)]^2}{np_2}$$

$$= (N_1 - np_1)^2 \left( \frac{1}{np_1} + \frac{1}{np_2} \right) = \frac{(N_1 - np_1)^2}{np_1(1 - p_1)}. \tag{10.6}$$

Now, recalling that $N_1$ is the number of, say, successes in $n$ trials, with $p_1$ being the probability of success, it is a binomial random variable with $E\{N_1\} = np_1$ and $\text{var}\{N_1\} = np_1(1 - p_1)$ if hypothesis $H$ is true. As $n$ increases, we have seen in Chapter 7 that $N_1$ approaches a normal distribution by virtue of the central limit theorem (Section 7.2.1). Hence, the distribution of random variable $U$, defined by

$$U = \frac{N_1 - np_1}{[np_1(1 - p_1)]^{1/2}},$$

approaches N(0, 1) as $n \to \infty$. Since

$$D = U^2,$$

following Equation (10.6), random variable $D$ thus approaches a chi-squared distribution with one degree of freedom, and the proof is complete for $k = 2$. The proof for an arbitrary $k$ proceeds in a similar fashion.

By means of Theorem 10.1, a test of hypothesis $H$ considered above can be constructed based on the assignment of a probability of Type-I error. Suppose that we wish to achieve a Type-I error probability of $\alpha$. The $\chi^2$ test suggests that hypothesis $H$ is rejected whenever

$$d = \sum_{i=1}^{k} \frac{n_i^2}{np_i} - n > \chi_{k-1,\alpha}^2, \tag{10.7}$$

and is accepted otherwise, where $d$ is the sample value of $D$ based on sample values $x_i, i = 1, \ldots, n$, and $\chi_{k-1,\alpha}^2$ takes the value such that (see Figure 10.1)

$$P(D > \chi_{k-1,\alpha}^2) = \alpha.$$

Since $D$ has a Chi-squared distribution with $(k - 1)$ degrees of freedom for large $n$, an approximate value for $\chi_{k-1,\alpha}^2$ can be found from Table A.5 in Appendix A for the $\chi^2$ distribution when $\alpha$ is specified.

The probability $\alpha$ of a Type-I error is referred to as the *significance level* in this context. As seen from Figure 10.1, it represents the area under $f_D(d)$ to the right of $\chi_{k-1,\alpha}^2$. Letting $\alpha = 0.05$, for example, the criterion given by Equation (10.7) implies that we reject hypothesis $H$ whenever deviation measure $d$ as calculated from a given set of sample values falls within the 5% region. In other words, we expect to reject $H$ about 5% of the time when in fact $H$ is true. Which significance level should be adopted in a given situation will, of course, depend on the

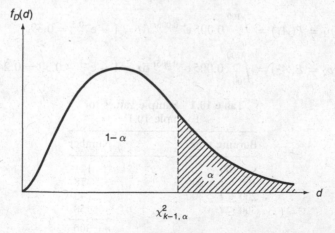

**Figure 10.1** Chi-squared distribution with $(k - 1)$ degrees of freedom

particular case involved. In practice, common values for $\alpha$ are 0.001, 0.01, and 0.05; a value of $\alpha$ between 5% and 1% is regarded as *almost significant*; a value between 1% and 0.1% as *significant*; and a value below 0.1% as *highly significant*.

Let us now give a step-by-step procedure for carrying out the $\chi^2$ test when the distribution of a population $X$ is completely specified.

- Step 1: divide range space $X$ into $k$ mutually exclusive and numerically convenient intervals $A_i, i = 1, 2, \ldots, k$. Let $n_i$ be the number of sample values falling into $A_i$. As a rule, if the number of sample values in any $A_i$ is less than 5, combine interval $A_i$ with either $A_{i-1}$ or $A_{i+1}$.
- Step 2: compute theoretical probabilities $P(A_i) = p_i, i = 1, 2, \ldots, k$, by means of the hypothesized distribution.
- Step 3: construct $d$ as given by Equation (10.7).
- Step 4: choose a value of $\alpha$ and determine from Table A.5 for the $\chi^2$ distribution of $(k - 1)$ degrees of freedom the value of $\chi^2_{k-1, \alpha}$.
- Step 5: reject hypothesis $H$ if $d > \chi^2_{k-1, \alpha}$. Otherwise, accept $H$.

**Example 10.1.** Problem: 300 light bulbs are tested for their burning time $t$ (in hours), and the result is shown in Table 10.1. Suppose that random burning time $T$ is postulated to be exponentially distributed with mean burning time $1/\lambda = 200$ hours; that is, $\lambda = 0.005$, per hour, and

$$f_T(t) = 0.005 \, e^{-0.005t}, \quad t \geq 0. \tag{10.8}$$

Test this hypothesis by using the $\chi^2$ test at the 5% significance level.

Answer: the necessary steps in carrying out the $\chi^2$ test are indicated in Table 10.2. The first column gives intervals $A_i$, which are chosen in this case to be the intervals of $t$ given in Table 10.1. The theoretical probabilities $P(A_i) = p_i$ in the third column are easily calculated by using Equation (10.8). For example,

$$p_1 = P(A_1) = \int_0^{100} 0.005 \, e^{-0.005t} \, dt = 1 - e^{-0.5} = 0.39;$$

$$p_2 = P(A_2) = \int_{100}^{200} 0.005 \, e^{-0.005t} \, dt = 1 - e^{-1} - 0.39 = 0.24.$$

**Table 10.1**   Sample values for
Example 10.1

| Burning time, $t$ | Number |
|---|---|
| $t < 100$ | 121 |
| $100 \leq t < 200$ | 78 |
| $200 \leq t < 300$ | 43 |
| $300 \leq t$ | 58 |
| | $n = 300$ |

**Table 10.2** Table for $\chi^2$ test for Example 10.1

| Interval, $A_i$ | $n_i$ | $p_i$ | $np_i$ | $n_i^2/np_i$ |
|---|---|---|---|---|
| $t < 100$ | 121 | 0.39 | 117 | 125.1 |
| $100 \le t < 200$ | 78 | 0.24 | 72 | 84.5 |
| $200 \le t < 300$ | 43 | 0.15 | 45 | 41.1 |
| $300 \le t$ | 58 | 0.22 | 66 | 51.0 |
| | 300 | 1.00 | 300 | 301.7 |

Note: $n_i$, observed number of occurrences; $p_i$, theoretical $P(A_i)$.

For convenience, the theoretical numbers of occurrences as predicted by the model are given in the fourth column of Table 10.2, which, when compared with the value in the second column, give a measure of goodness of fit of the model to the data. Column 5 ($n_i^2/np_i$) is included in order to facilitate the calculation of $d$. Thus, from Equation (10.7) we have

$$d = \sum_{i=1}^{k} \frac{n_i^2}{np_i} - n = 301.7 - 300 = 1.7.$$

Now, $k = 4$. From Table A.5 for the $\chi^2$ distribution with three degrees of freedom, we find

$$\chi^2_{3,0.05} = 7.815.$$

Since $d < \chi^2_{3,0.05}$, we accept at the 5% significance level the hypothesis that the observed data represent a sample from an exponential distribution with $\lambda = 0.005$.

**Example 10.2.** Problem: a six-year accident record of 7842 California drivers is given in Table 8.2. On the basis of these sample values, test the hypothesis that $X$, the number of accidents in six years per driver, is Poisson-distributed with mean rate $\lambda = 0.08$ per year at the 1% significance level.

Answer: since $X$ is discrete, a natural choice of intervals $A_i$ is those centered around the discrete values, as indicated in the first column of Table 10.3. Note that interval $x > 5$ would be combined with $4 < x \le 5$ if number $n_7$ were less than 5.

The hypothesized distribution for $X$ is

$$p_X(x) = \frac{(\lambda t)^x e^{-\lambda t}}{x!} = \frac{(0.48)^x e^{-0.48}}{x!}, \quad x = 0,1,2,\ldots. \tag{10.9}$$

**Table 10.3**   Table for $\chi^2$ test for Example 10.2

| Interval, $A_i$ | $n_i$ | $p_i$ | $np_i$ | $n_i^2/np_i$ |
|---|---|---|---|---|
| $x \leq 0$ | 5147 | 0.6188 | 4853 | 5459 |
| $0 < x \leq 1$ | 1859 | 0.2970 | 2329 | 1484 |
| $1 < x \leq 2$ | 595 | 0.0713 | 559 | 633 |
| $2 < x \leq 3$ | 167 | 0.0114 | 89 | 313 |
| $3 < x \leq 4$ | 54 | 0.0013 | 10 | 292 |
| $4 < x \leq 5$ | 14 | 0.0001 | 1 | 196 |
| $5 < x$ | 6 | 0.0001 | 1 | 36 |
| | 7842 | 1.0 | 7842 | 8413 |

Note: $n_i$, observed number of occurrences; $p_i$, theoretical $P(A_i)$.

We thus have

$$P(A_i) = p_i = \frac{(0.48)^{i-1} e^{-0.48}}{(i-1)!}, \quad i = 1, 2, \ldots, 6,$$

$$P(A_7) = p_7 = 1 - \sum_{i=1}^{6} p_i,$$

These values are indicated in the third column of Table 10.3.
Column 5 of Table 10.3 gives

$$d = \sum_{i=1}^{k} \frac{n_i^2}{np_i} - n = 8413 - 7842 = 571.$$

With $k = 7$, the value of $\chi_{k-1,\alpha}^2 = \chi_{6,0.01}^2$ is found from Table A.5 to be

$$\chi_{6,0.01}^2 = 16.812.$$

Since $d > \chi_{6,\,0.01}^2$, the hypothesis is rejected at the 1% significance level.

## 10.2.2   THE CASE OF ESTIMATED PARAMETERS

Let us now consider a more common situation in which parameters in the hypothesized distribution also need to be estimated from the data.

A natural procedure for a goodness-of-fit test in this case is first to estimate the parameters by using one of the methods developed in Chapter 9 and then to follow the $\chi^2$ test for known parameters, already discussed in Section 7.2.1. In doing so, however, a complication arises in that theoretical probabilities $p_i$ defined by Equation (10.2) are, being functions of the distribution parameters, functions of the sample. The statistic $D$ now takes the form

$$D = \sum_{i=1}^{k} \frac{n}{\hat{P}_i} \left( \frac{N_i}{n} - \hat{P}_i \right)^2 = \sum_{i=1}^{k} \frac{N_i^2}{n\hat{P}_i} - n, \tag{10.10}$$

where $\hat{P}_i$ is an estimator for $p_i$ and is thus a statistic. We see that $D$ is now a much more complicated function of $X_1, X_2, \ldots, X_n$. The important question to be answered is: what is the new distribution of $D$?

The problem of determining the limiting distribution of $D$ in this situation was first considered by Fisher (1922, 1924), who showed that, as $n \to \infty$, the distribution of $D$ needs to be modified, and the modification obviously depends on the method of parameter estimation used. Fortunately, for a class of important methods of estimation, such as the maximum likelihood method, the modification required is a simple one, namely, statistic $D$ still approaches a chi-squared distribution as $n \to \infty$ but now with $(k - r - 1)$ degrees of freedom, where $r$ is the number of parameters in the hypothesized distribution to be estimated. In other words, it is only necessary to reduce the number of degrees of freedom in the limiting distribution defined by Equation (10.5) by one for *each* parameter estimated from the sample.

We can now state a step-by-step procedure for the case in which $r$ parameters in the distribution are to be estimated from the data.

- Step 1: divide range space $X$ into $k$ mutually exclusive and numerically convenient intervals $A_i, i = 1, \ldots, k$. Let $n_i$ be the number of sample values falling into $A_i$. As a rule, if the number of sample values in any $A_i$ is less than 5, combine interval $A_i$ with either $A_{i-1}$ or $A_{i+1}$.
- Step 2: estimate the $r$ parameters by the method of maximum likelihood from the data.
- Step 3: compute theoretical probabilities $P(A_i) = p_i, i = 1, \ldots, k$, by means of the hypothesized distribution with estimated parameter values.
- Step 4: construct $d$ as given by Equation (10.7).
- Step 5: choose a value of $\alpha$ and determine from Table A.5 for the $\chi^2$ distribution of $(k - r - 1)$ degrees of freedom the value of $\chi^2_{k-r-1,\alpha}$. It is assumed, of course, that $k - r - 1 > 0$.
- Step 6: reject hypothesis $H$ if $d > \chi^2_{k-r-1,\alpha}$. Otherwise, accept $H$.

**Example 10.3.** Problem: vehicle arrivals at a toll gate on the New York State Thruway were recorded. The vehicle counts at one-minute intervals were taken for 106 minutes and are given in Table 10.4. On the basis of these observations, determine whether a Poisson distribution is appropriate for $X$, the number of arrivals per minute, at the 5% significance level.

**Table 10.4**  One-minute arrivals, for Example 10.3

| Vehicles per minute (No.) | Number of occurrences |
|---|---|
| 0 | 0 |
| 1 | 0 |
| 2 | 1 |
| 3 | 3 |
| 4 | 5 |
| 5 | 7 |
| 6 | 13 |
| 7 | 12 |
| 8 | 8 |
| 9 | 9 |
| 10 | 13 |
| 11 | 10 |
| 12 | 5 |
| 13 | 6 |
| 14 | 4 |
| 15 | 5 |
| 16 | 4 |
| 17 | 0 |
| 18 | 1 |
|  | $n = 106$ |

Answer: the hypothesized distribution is

$$p_X(x) = \frac{\lambda^x e^{-\lambda}}{x!}, \quad x = 0, 1, 2, \ldots, \tag{10.11}$$

where parameter $\lambda$ needs to be estimated from the data. Thus, $r = 1$.

To proceed, we first determine appropriate intervals $A_i$ such that $n_i \geq 5$ for all $i$; these are shown in the first column of Table 10.5. Hence, $k = 11$.

The maximum likelihood estimate for $\lambda$ is given by

$$\hat{\lambda} = \overline{x} = \frac{1}{n} \sum_{j=1}^{n} x_j = 9.09.$$

The substitution of this value for parameter $\lambda$ in Equation (10.11) permits us to calculate probabilities $P(A_i) = p_i$. For example,

$$p_1 = \sum_{j=0}^{4} p_X(j) = 0.052,$$

$$p_2 = p_X(5) = 0.058.$$

**Table 10.5**  Table for $\chi^2$ test for Example 10.3

| Interval, $A_i$ | $n_i$ | $p_i$ | $np_i$ | $n_i^2/np_i$ |
|---|---|---|---|---|
| $0 \leq x < 5$ | 9 | 0.052 | 5.51 | 14.70 |
| $5 \leq x < 6$ | 7 | 0.058 | 6.15 | 7.97 |
| $6 \leq x < 7$ | 13 | 0.088 | 9.33 | 18.11 |
| $7 \leq x < 8$ | 12 | 0.115 | 12.19 | 11.81 |
| $8 \leq x < 9$ | 8 | 0.131 | 13.89 | 4.61 |
| $9 \leq x < 10$ | 9 | 0.132 | 13.99 | 5.79 |
| $10 \leq x < 11$ | 13 | 0.120 | 12.72 | 13.29 |
| $11 \leq x < 12$ | 10 | 0.099 | 10.49 | 9.53 |
| $12 \leq x < 13$ | 5 | 0.075 | 7.95 | 3.14 |
| $13 \leq x < 14$ | 6 | 0.054 | 5.72 | 6.29 |
| $14 \leq x$ | 14 | 0.076 | 8.06 | 24.32 |
| | 106 | 1.0 | 106 | 119.56 |

These theoretical probabilities are given in the third column of Table 10.5. From column 5 of Table 10.5, we obtain

$$d = \sum_{i=1}^{k} \frac{n_i^2}{np_i} - n = 119.56 - 106 = 13.56.$$

Table A.5 with $\alpha = 0.05$ and $k - r - 1 = 9$ degrees of freedom gives

$$\chi_{9,0.05}^2 = 16.92.$$

Since $d < \chi_{9,0.05}^2$, the hypothesized distribution with $\lambda = 9.09$ is accepted at the 5% significance level.

**Example 10.4.** Problem: based upon the snowfall data given in Problem 8.2(g) from 1909 to 1979, test the hypothesis that the Buffalo yearly snowfall can be modeled by a normal distribution at 5% significance level.

Answer: for this problem, the assumed distribution for $X$, the Buffalo yearly snowfall, measured in inches, is $N(m, \sigma^2)$ where $m$ and $\sigma^2$ must be estimated from the data. Since the maximum likelihood estimator for $m$ and $\sigma^2$ are $\hat{M} = \overline{X}$, and $\widehat{\Sigma^2} = [(n-1)/n]S^2$, respectively, we have

$$\hat{m} = \overline{x} = \frac{1}{70} \sum_{j=1}^{70} x_j = 83.6,$$

$$\widehat{\sigma^2} = \frac{69}{70} s^2 = \frac{1}{70} \sum_{j=1}^{70} (x_j - 83.6)^2 = 777.4.$$

**Table 10.6**  Table for $\chi^2$ test for Example 10.4

| Interval, $A_i$ | $n_i$ | $p_i$ | $np_i$ | $n_i^2/np_i$ |
|---|---|---|---|---|
| $x \leq 56$ | 13 | 0.161 | 11.27 | 15.00 |
| $56 < x \leq 72$ | 10 | 0.178 | 12.46 | 8.03 |
| $72 < x \leq 88$ | 20 | 0.224 | 15.68 | 25.51 |
| $88 < x \leq 104$ | 13 | 0.205 | 14.35 | 11.78 |
| $104 < x \leq 120$ | 8 | 0.136 | 9.52 | 6.72 |
| $120 < x$ | 6 | 0.096 | 6.72 | 5.36 |
| | 70 | 1.0 | 70 | 72.40 |

With intervals $A_i$ defined as shown in the first column of Table 10.6, theoretical probabilities $P(A_i)$ now can be calculated with the aid of Table A.3. For example, the first two of these probabilities are

$$P(A_1) = P(X \leq 56) = P\left(U \leq \frac{56 - 83.6}{\sqrt{777.4}}\right) = F_U(-0.990)$$

$$= 1 - F_U(0.990) = 1 - 0.8389 = 0.161;$$

$$P(A_2) = P(56 < X \leq 72) = P(-0.990 < U \leq -0.416)$$

$$= [1 - F_U(0.416)] - [1 - F_U(0.990)]$$

$$= 0.339 - 0.161 = 0.178.$$

The information given above allows us to construct Table 10.6. Hence, we have

$$d = \sum_{i=1}^{k} \frac{n_i^2}{np_i} - n = 72.40 - 70 = 2.40.$$

The number of degrees of freedom in this case is $k - r - 1 = 6 - 2 - 1 = 3$. Table A.5 thus gives

$$\chi_{3,0.05}^2 = 7.815.$$

Since $d < \chi_{3,0.05}^2$ normal distribution $N(83.6, 777.4)$ is acceptable at the 5% significance level.

Before leaving this section, let us remark again that statistic $D$ in the $\chi^2$ test is $\chi^2$-distributed only when $n \to \infty$. It is thus a *large sample* test. As a rule, $n > 50$ is considered satisfactory for fulfilling the large-sample requirement.

## 10.3  KOLMOGOROV–SMIRNOV TEST

The so-called *Kolmogorov–Smirnov goodness-of-fit test*, referred to as the K–S test in the rest of this chapter, is based on a statistic that measures the deviation of the observed *cumulative histogram* from the hypothesized cumulative distribution function.

Given a set of sample values $x_1, x_2, \ldots, x_n$ observed from a population $X$, a cumulative histogram can be constructed by (a) arranging the sample values in increasing order of magnitude, denoted here by $x_{(1)}, x_{(2)}, \ldots, x_{(n)}$, (b) determining the *observed* distribution function of $X$ at $x_{(1)}, x_{(2)}, \ldots$, denoted by $F^0[x_{(1)}]$, $F^0[x_{(2)}], \ldots$, from relations $F^0[x_{(i)}] = i/n$, and (c) connecting the values of $F^0[x_{(i)}]$ by straight-line segments.

The test statistic to be used in this case is

$$
\begin{aligned}
D_2 &= \max_{i=1}^{n}\{|F^0[X_{(i)}] - F_X[X_{(i)}]|\} \\
&= \max_{i=1}^{n}\left\{\left|\frac{i}{n} - F_X[X_{(i)}]\right|\right\},
\end{aligned}
\tag{10.12}
$$

where $X_{(i)}$ is the $i$th-order statistic of the sample. Statistic $D_2$ thus measures the maximum of absolute values of the $n$ differences between observed probability distribution function (PDF) and hypothesized PDF evaluated for the observed samples. In the case where parameters in the hypothesized distribution must be estimated, the values for $F_X[X_{(i)}]$ are obtained by using estimated parameter values.

While the distribution of $D_2$ is difficult to obtain analytically, its distribution function at various values can be computed numerically and tabulated. It can be shown that the probability distribution of $D_2$ is independent of the hypothesized distribution and is a function only of $n$, the sample size (e.g. see Massey, 1951).

The execution of the K–S test now follows that of the $\chi^2$ test. At a specified $\alpha$ significance level, the operating rule is to reject hypothesis $H$ if $d_2 > c_{n,\alpha}$; otherwise, accept $H$. Here, $d_2$ is the sample value of $D_2$, and the value of $c_{n,\alpha}$ is defined by

$$
P(D_2 > c_{n,\alpha}) = \alpha.
\tag{10.13}
$$

The values of $c_{n,\alpha}$ for $\alpha = 0.01, 0.05$, and $0.10$ are given in Table A.6 in Appendix A as functions of $n$.

It is instructive to note the important differences between this test and the $\chi^2$ test. Whereas the $\chi^2$ test is a large-sample test, the K–S test is valid for all values of $n$. Furthermore, the K–S test utilizes sample values in their unaltered and unaggregated form, whereas data lumping is necessary in the execution of the $\chi^2$ test. On the negative side, the K–S test is strictly valid only for continuous distributions. We also remark that the values of $c_{n,\alpha}$ given in Table A.6 are based on a completely specified hypothesized distribution. When the parameter values must be estimated, no rigorous method of adjustment is available. In these cases, it can be stated only that the values of $c_{n,\alpha}$ should be somewhat reduced.

The step-by-step procedure for executing the K–S test is now outlined as follows:

- Step 1: rearrange sample values $x_1, x_2, \ldots, x_n$ in increasing order of magnitude and label them $x_{(1)}, x_{(2)}, \ldots, x_{(n)}$.
- Step 2: determine observed distribution function $F^0(x)$ at each $x_{(i)}$ by using $F^0[x_{(i)}] = i/n$.
- Step 3: determine the theoretical distribution function $F_X(x)$ at each $x_{(i)}$ by using the hypothesized distribution. Parameters of the distribution are estimated from the data if necessary.
- Step 4: form the differences $|F^0(x_{(i)}) - F_X(x_{(i)})|$ for $i = 1, 2, \ldots, n$.
- Step 5: calculate

$$d_2 = \max_{i=1}^{n}\{|F^0[x_{(i)}] - F_X[x_{(i)}]|\}.$$

The determination of this maximum value requires enumeration of $n$ quantities. This labor can be somewhat reduced by plotting $F^0(x)$ and $F_X(x)$ as functions of $x$ and noting the location of the maximum by inspection.

- Step 6: choose a value of $\alpha$ and determine from Table A.6 the value of $c_{n,\alpha}$.
- Step 7: reject hypothesis $H$ if $d_2 > c_{n,\alpha}$. Otherwise, accept $H$.

**Example 10.5.** Problem: 10 measurements of the tensile strength of one type of engineering material are made. In dimensionless forms, they are 30.1, 30.5, 28.7, 31.6, 32.5, 29.0, 27.4, 29.1, 33.5, and 31.0. On the basis of this data set, test the hypothesis that the tensile strength follows a normal distribution at the 5% significance level.

Answer: a reordering of the data yields $x_{(1)} = 27.4, x_{(2)} = 28.7, \ldots, x_{(10)} = 33.5$. The determination of $F^0(x_{(i)})$ is straightforward. We have, for example,

$$F^0(27.4) = 0.1, \quad F^0(28.7) = 0.2, \ldots, \quad F^0(33.5) = 1.$$

With regard to the theoretical distribution function, estimates of the mean and variance are first obtained from

$$\hat{m} = \bar{x} = \frac{1}{10}\sum_{j=1}^{10} x_j = 30.3,$$

$$\hat{\sigma}^2 = \left(\frac{n-1}{n}\right)s^2 = \frac{1}{10}\sum_{j=1}^{10}(x_j - 30.3)^2 = 3.14.$$

The values of $F_X[x_{(i)}]$ can now be found based on distribution N(30.3, 3.14) for $X$. For example, with the aid of Table A.3 for standardized normal random variable $U$, we have

$$F_X(27.4) = F_U\left(\frac{27.4 - 30.3}{\sqrt{3.14}}\right) = F_U(-1.64)$$

$$= 1 - F_U(1.64) = 1 - 0.9495 = 0.0505,$$

$$F_X(28.7) = F_U\left(\frac{28.7 - 30.3}{\sqrt{3.14}}\right) = F_U(-0.90)$$

$$= 1 - 0.8159 = 0.1841,$$

and so on.

In order to determine $d_2$, it is constructive to plot $F^0(x)$ and $F_X(x)$ as functions of $x$, as shown in Figure 10.2. It is clearly seen from the figure that the maximum of the differences between $F^0(x)$ and $F_X(x)$ occurs at $x = x_{(4)} = 29.1$. Hence,

$$d_2 = |F^0(29.1) - F_X(29.1)| = 0.4 - 0.2483 = 0.1517.$$

With $\alpha = 0.05$ and $n = 10$, Table A.6 gives

$$c_{10,0.05} = 0.41.$$

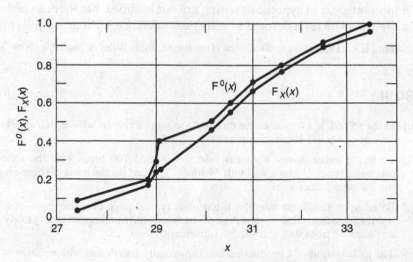

**Figure 10.2** $F^0(x)$ and $F_X(x)$ in Example 10.5.

Since $d_2 < c_{10,0.05}$, we accept normal distribution $N(30.3, 3.14)$ at the 5% significance level.

Let us remark that, since the parameter values were also estimated from the data, it is more appropriate to compare $d_2$ with a value somewhat smaller than 0.41. In view of the fact that the value of $d_2$ is well below 0.41, we are safe in making the conclusion given above.

## REFERENCES

Beard, L.R., 1962, *Statistical Methods in Hydrology*, Army Corps of Engineers, Sacramento, CA.

Cramér, H., 1946, *Mathematical Methods of Statistics*, Princeton University Press, Princeton, NJ.

Fisher, R.A., 1922, "On the Interpretation of $\chi^2$ from Contingency Tables, and Calculations of $P$", *J. Roy. Stat. Soc.* **85** 87–93.

Fisher, R.A., 1924, "The Conditions under which $\chi^2$ Measures the Discrepancy between Observation with Hypothesis", *J. Roy. Stat. Soc.* **87** 442–476.

Massey, F.J., 1951, "The Kolmogorov Test for Goodness of Fit", *J. Am. Stat. Assoc.* **46** 68–78.

Pearson, K., 1900, "On a Criterion that a System of Deviations from the Probable in the Case of a Correlated System of Variables is such that it can be Reasonably Supposed to have Arisen in Random Sampling", *Phil. Mag.* **50** 157–175.

## FURTHER READING AND COMMENTS

We have been rather selective in our choice of topics in this chapter. A number of important areas in hypotheses testing are not included, but they can be found in more complete texts devoted to statistical inference, such as the following:

Lehmann, E.L., 1959, *Testing Statistical Hypotheses*, John Wiley & Sons Inc. New York.

## PROBLEMS

10.1 In the $\chi^2$ test, is a hypothesized distribution more likely to be accepted at $\alpha = 0.05$ than at $\alpha = 0.01$? Explain your answer.

10.2 To test whether or not a coin is fair, it is tossed 100 times with the following outcome: heads 41 times, and tails 59 times. Is it fair on the basis of these tosses at the 5% significance level?

10.3 Based upon telephone numbers listed on a typical page of a telephone directory, test the hypothesis that the last digit of the telephone numbers is equally likely to be any number from 0 to 9 at the 5% significance level.

10.4 The daily output of a production line is normally distributed with mean $m = 8000$ items and standard deviation $\sigma = 1000$ items. A second production line is set up,

**Table 10.7**  Production-line data for Problem 10.4

| Daily output interval | Number of occurrences |
|---|---|
| <4 000 | 3 |
| 4 000–5 000 | 3 |
| 5 000–6 000 | 7 |
| 6 000–7 000 | 16 |
| 7 000–8 000 | 27 |
| 8 000–9 000 | 22 |
| 9 000–10 000 | 11 |
| 10 000–11 000 | 8 |
| 11 000–12 000 | 2 |
| >12 000 | 1 |
| | $n = 100$ |

and 100 daily output readings are taken, as shown in Table 10.7. On the basis of this sample, does the second production line behave in the same statistical manner as the first? Use $\alpha = 0.01$.

10.5 In a given plant, a sample of a given number of production items was taken from each of the five production lines; the number of defective items was recorded, as shown in Table 10.8. Test the hypothesis that the proportion of defects is constant from one production line to another. Use $\alpha = 0.01$.

**Table 10.8**  Production-line data for Problem 10.5

| Production line | Number of defects |
|---|---|
| 1 | 11 |
| 2 | 13 |
| 3 | 9 |
| 4 | 12 |
| 5 | 8 |

10.6 We have rejected in Example 10.2 the Poisson distribution with $\lambda = 0.08$ on the basis of accident data at the 1% significance level. At the same $\alpha$:
(a) Would a Poisson distribution with $\lambda$ estimated from the data be acceptable?
(b) Would a negative binomial distribution be more appropriate?

10.7 The data on the number of arrivals of cars at an intersection in 360 10 s intervals are as shown in Table 10.9.
Three models are proposed:
model 1:

$$p_X(x) = \frac{e^{-1}}{x!}, \quad x = 0, 1, \ldots;$$

**Table 10.9**   Arrival of cars at intersection, for Problem 10.7

| Cars per interval | Number of observations |
|---|---|
| 0 | 139 |
| 1 | 128 |
| 2 | 55 |
| 3 | 25 |
| 4 | 13 |
| | n = 360 |

model 2:

$$p_X(x) = \frac{e^{-\lambda}\lambda^x}{x!}, \quad x = 0, 1, \ldots,$$

where $\lambda$ is estimated from the data;

model 3:

$$p_X(x) = \binom{x+k-1}{k-1}p^k(1-p)^x, \quad x = 0, 1, \ldots,$$

where $k$ and $p$ are estimated from the data.

(a) Use the $\chi^2$ test; are these models acceptable at the 5% significance level?
(b) In you opinion, which is a better model? Explain your answer.
Note: for model 3,

$$m_X = \frac{k(1-p)}{p},$$

$$\sigma_X^2 = \frac{k(1-p)}{p^2}.$$

10.8  Car pooling is encouraged in a city. A survey of 321 passenger vehicles coming into the city gives the car occupancy profile shown in Table 10.10. Suggest a probabilistic model for $X$, the number of passengers per vehicle, and test your hypothesized distribution at $\alpha = 0.05$ on the basis of this survey.

**Table 10.10**   Car occupancy (number of passengers per vehicle, excluding the driver), for Problem 10.8

| Occupancy | Vehicles (No.) |
|---|---|
| 0 | 224 |
| 1 | 47 |
| 2 | 31 |
| 3 | 16 |
| ≥4 | 3 |
| | n = 321 |

10.9   Problem 8.2(c) gives 100 measurements of time gaps (see Table 8.4). On the basis of these data, postulate a likely distribution for $X$ and test your hypothesis at the 5% significance level.

10.10  Consider the data given in Problem 8.2(d) for the sum of two consecutive time gaps. Postulate a likely distribution for $X$ and test your hypothesis at the 5% significance level.

10.11  Problem 8.2(e) gives data for one-minute vehicle arrivals (see Table 8.5). Postulate a likely distribution for $X$ and test your hypothesis at the 5% significance level.

10.12  Problem 8.2(h) gives a histogram for $X$, the peak combustion pressure (see Figure 8.4). On the basis of these data, postulate a likely distribution for $X$ and test your hypothesis at the 5% significance level.

10.13  Suppose that the number of drivers sampled is 200. Based on the histogram given in Problem 8.2(i) (see Figure 8.5), postulate a likely distribution for $X_1$ and test your hypothesis at the 1% significance level.

10.14  Problem 8.2(j) gives a histogram for $X$, the number of blemishes in television tubes (see Figure 8.6). On the basis of this sample, postulate a likely distribution for $X$ and test your hypothesis at the 5% significance level.

10.15  A total of 24 readings of the annual sediment load (in $10^6$ tons) in the Colorado River at the Grand Canyon are (arranged in increasing order of magnitude) 49, 50, 50, 66, 70, 75, 84, 85, 98, 118, 122, 135, 143, 146, 157, 172, 177, 190, 225, 235, 265, 270, 400, 480. Using the Kolmogorov–Smirnov test at the 5% significance level, test the hypothesis that the annual sediment load follows a lognormal distribution (data are taken from Beard, 1962).

10.16  For the snowfall data given in Problem 8.2(g) (see Table 8.6), use the Kolmogorov–Smirnov test and test the normal distribution hypothesis on the basis of snowfall data from 1909–2002.

# 11

# Linear Models and Linear Regression

The tools developed in Chapters 9 and 10 for parameter estimation and model verification are applied in this chapter to a very useful class of models encountered in science and engineering. A commonly occurring situation is one in which a random quantity, $Y$, is a function of one or more independent (and deterministic) variables $x_1, x_2, \ldots$, and $x_m$. For example, wind load ($Y$) acting on a structure is a function of height ($x$); the intensity ($Y$) of strong motion earthquakes is dependent on the distance from the epicenter ($x$); housing price ($Y$) is a function of location ($x_1$) and age ($x_2$); and chemical yield ($Y$) may be related to temperature ($x_1$), pressure ($x_2$), and acid content ($x_3$).

Given a sample of $Y$ values with their associated values of $x_i, i = 1, 2, \ldots, m$, we are interested in estimating on the basis of this sample the relationship between $Y$ and the independent variables $x_1, x_2, \ldots$, and $x_m$. In what follows, we concentrate on some simple cases of the broadly defined problem stated above.

## 11.1 SIMPLE LINEAR REGRESSION

We assume in this section that random variable $Y$ is a function of only one independent variable and that their relationship is linear. By a linear relationship we mean that the mean of $Y$, $E\{Y\}$, is known to be a linear function of $x$, that is,

$$E\{Y\} = \alpha + \beta x. \tag{11.1}$$

The two constants, intercept $\alpha$ and slope $\beta$, are unknown and are to be estimated from a sample of $Y$ values with their associated values of $x$. Note

*Fundamentals of Probability and Statistics for Engineers* T.T. Soong © 2004 John Wiley & Sons, Ltd
ISBNs: 0-470-86813-9 (HB)   0-470-86814-7 (PB)

that $E\{Y\}$ is a function $x$. In any single experiment, $x$ will assume a certain value $x_i$ and the mean of $Y$ will take the value

$$E\{Y_i\} = \alpha + \beta x_i. \tag{11.2}$$

Random variable $Y$ is, of course, itself a function of $x$. If we define a random variable $E$ by

$$E = Y - (\alpha + \beta x), \tag{11.3}$$

we can write

$$\boxed{Y = \alpha + \beta x + E,} \tag{11.4}$$

where $E$ has mean 0 and variance $\sigma^2$, which is identical to the variance of $Y$. The value of $\sigma^2$ is not known in general but it is assumed to be a constant and not a function of $x$.

Equation (11.4) is a standard expression of a *simple linear regression model*. The unknown parameters $\alpha$ and $\beta$ are called *regression coefficients*, and random variable $E$ represents the deviation of $Y$ about its mean. As with simple models discussed in Chapters 9 and 10, simple linear regression analysis is concerned with estimation of the regression parameters, the quality of these estimators, and model verification on the basis of the sample. We note that, instead of a simple sample such as $Y_1, Y_2, \ldots, Y_n$ as in previous cases, our sample in the present context takes the form of pairs $(x_1, Y_1), (x_2, Y_2), \ldots, (x_n, Y_n)$. For each value $x_i$ assigned to $x$, $Y_i$ is an independent observation from population $Y$ defined by Equation (11.4). Hence, $(x_i, Y_i), i = 1, 2, \ldots, n$, may be considered as a sample from random variable $Y$ for given values $x_1, x_2, \ldots,$ and $x_n$ of $x$; these $x$ values need not all be distinct but, in order to estimate both $\alpha$ and $\beta$, we will see that we must have at least two distinct values of $x$ represented in the sample.

## 11.1.1   LEAST SQUARES METHOD OF ESTIMATION

As one approach to point estimation of regression parameters $\alpha$ and $\beta$, the method of least squares suggests that their estimates, $\hat{\alpha}$ and $\hat{\beta}$, be chosen so that the sum of the squared differences between observed sample values $y_i$ and the estimated expected value of $Y, \hat{\alpha} + \hat{\beta}x_i$, is minimized. Let us write

$$e_i = y_i - (\hat{\alpha} + \hat{\beta}x_i). \tag{11.5}$$

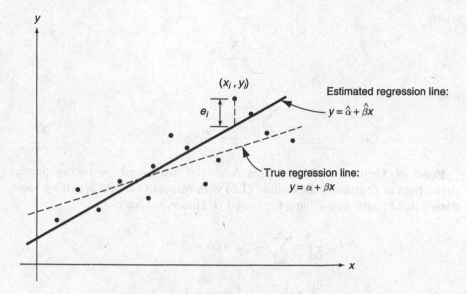

**Figure 11.1**   The least squares method of estimation

The least-square estimates $\hat{\alpha}$ and $\hat{\beta}$, respectively, of $\alpha$ and $\beta$ are found by minimizing

$$Q = \sum_{i=1}^{n} e_i^2 = \sum_{i=1}^{n} [y_i - (\hat{\alpha} + \hat{\beta} x_i)]^2. \tag{11.6}$$

In the above, the sample-value pairs are $(x_1, y_1), (x_2, y_2), \ldots, (x_n, y_n)$, and $e_i, i = 1, 2, \ldots, n$, are called the *residuals*. Figure 11.1 gives a graphical presentation of this procedure. We see that the residuals are the vertical distances between the observed values of $Y, y_i$, and the least-square estimate $\hat{\alpha} + \hat{\beta} x$ of true regression line $\alpha + \beta x$.

The estimates $\hat{\alpha}$ and $\hat{\beta}$ are easily found based on the least-square procedure. The results are stated below as Theorem 11.1.

**Theorem 11.1:** consider the simple linear regression model defined by Equation (11.4). Let $(x_1, y_1), (x_2, y_2), \ldots, (x_n, y_n)$ be observed sample values of $Y$ with associated values of $x$. Then the least-square estimates of $\alpha$ and $\beta$ are

$$\hat{\alpha} = \overline{y} - \hat{\beta}\overline{x}, \tag{11.7}$$

$$\hat{\beta} = \left[ \sum_{i=1}^{n} (x_i - \overline{x})(y_i - \overline{y}) \right] \left[ \sum_{i=1}^{n} (x_i - \overline{x})^2 \right]^{-1}, \tag{11.8}$$

where

$$\bar{x} = \frac{1}{n} \sum_{i=1}^{n} x_i,$$

and

$$\bar{y} = \frac{1}{n} \sum_{i=1}^{n} y_i.$$

**Proof of Theorem 11.1:** estimates $\hat{\alpha}$ and $\hat{\beta}$ are found by taking partial derivatives of $Q$ given by Equation (11.6) with respect to $\hat{\alpha}$ and $\hat{\beta}$, setting these derivatives to zero and solving for $\hat{\alpha}$ and $\hat{\beta}$. Hence, we have

$$\frac{\partial Q}{\partial \hat{\alpha}} = \sum_{i=1}^{n} -2[y_i - (\hat{\alpha} + \hat{\beta} x_i)],$$

$$\frac{\partial Q}{\partial \hat{\beta}} = \sum_{i=1}^{n} -2x_i[y_i - (\hat{\alpha} + \hat{\beta} x_i)].$$

Upon simplifying and setting the above equations to zero, we have the so-called *normal equations*:

$$n\hat{\alpha} + n\bar{x}\hat{\beta} = n\bar{y}, \tag{11.9}$$

$$n\bar{x}\hat{\alpha} + \hat{\beta} \sum_{i=1}^{n} x_i^2 = \sum_{i=1}^{n} x_i y_i. \tag{11.10}$$

Their solutions are easily found to be those given by Equations (11.7) and (11.8).

To ensure that these solutions correspond to the minimum of the sum of squared residuals, we need to verify that

$$\frac{\partial^2 Q}{\partial \hat{\alpha}^2} > 0,$$

and

$$D = \begin{vmatrix} \dfrac{\partial^2 Q}{\partial \hat{\alpha}^2} & \dfrac{\partial^2 Q}{\partial \hat{\alpha}\,\partial \hat{\beta}} \\[2mm] \dfrac{\partial^2 Q}{\partial \hat{\alpha}\,\partial \hat{\beta}} & \dfrac{\partial^2 Q}{\partial \hat{\beta}^2} \end{vmatrix} > 0,$$

at $\hat{\alpha}$ and $\hat{\beta}$. Elementary calculations show that

$$\frac{\partial^2 Q}{\partial \hat{\alpha}^2} = 2n > 0,$$

and

$$D = 4n \sum_{i=1}^{n} (x_i - \bar{x})^2 > 0$$

The proof of this theorem is thus complete. Note that $D$ would be zero if all $x_i$ take the same value. Hence, at least two distinct $x_i$ values are needed for the determination of $\hat{\alpha}$ and $\hat{\beta}$.

It is instructive at this point to restate the foregoing results by using a more compact vector–matrix notation. As we will see, results in vector–matrix form facilitate calculations. Also, they permit easy generalizations when we consider more general regression models.

In terms of observed sample values $(x_1, y_1), (x_2, y_2), \ldots, (x_n, y_n)$, we have a system of observed regression equations

$$y_i = \alpha + \beta x_i + e_i, \quad i = 1, 2, \ldots, n. \tag{11.11}$$

Let

$$C = \begin{bmatrix} 1 & x_1 \\ 1 & x_2 \\ \vdots & \vdots \\ 1 & x_n \end{bmatrix}, \quad y = \begin{bmatrix} y_1 \\ y_2 \\ \vdots \\ y_n \end{bmatrix}, \quad e = \begin{bmatrix} e_1 \\ e_2 \\ \vdots \\ e_n \end{bmatrix},$$

and let

$$\theta = \begin{bmatrix} \alpha \\ \beta \end{bmatrix}.$$

Equations (11.11) can be represented by the vector–matrix equation

$$y = C\theta + e. \tag{11.12}$$

The sum of squared residuals given by Equation (11.6) is now

$$Q = e^T e = (y - C\theta)^T (y - C\theta). \tag{11.13}$$

The least-square estimate of $\boldsymbol{\theta}, \hat{\boldsymbol{\theta}}$, is found by minimizing $Q$. Applying the variational principle discussed in Section 9.3.1.1, we have

$$
\begin{aligned}
\delta Q &= -\delta\boldsymbol{\theta}^T C^T (y - C\boldsymbol{\theta}) - (y - C\boldsymbol{\theta})^T C\delta\boldsymbol{\theta} \\
&= -2\delta\boldsymbol{\theta}^T C^T (y - C\boldsymbol{\theta}).
\end{aligned}
$$

Setting $\delta Q = 0$, the solution for $\hat{\boldsymbol{\theta}}$ is obtained from normal equation

$$C^T (y - C\hat{\boldsymbol{\theta}}) = 0, \tag{11.14}$$

or

$$C^T C\hat{\boldsymbol{\theta}} = C^T y,$$

which gives

$$\boxed{\hat{\boldsymbol{\theta}} = (C^T C)^{-1} C^T y.} \tag{11.15}$$

In the above, the inverse of matrix $C^T C$ exists if there are at least two distinct values of $x_i$ represented in the sample.

We can easily check that Equation (11.15) is identical to Equations (11.7) and (11.8) by noting that

$$
C^T C = \begin{bmatrix} 1 & 1 & \cdots & 1 \\ x_1 & x_2 & \cdots & x_n \end{bmatrix} \begin{bmatrix} 1 & x_1 \\ 1 & x_2 \\ \vdots & \vdots \\ 1 & x_n \end{bmatrix} = \begin{bmatrix} n & n\bar{x} \\ n\bar{x} & \sum\limits_{i=1}^{n} x_i^2 \end{bmatrix},
$$

$$
C^T y = \begin{bmatrix} 1 & 1 & \cdots & 1 \\ x_1 & x_2 & \cdots & x_n \end{bmatrix} \begin{bmatrix} y_1 \\ y_2 \\ \vdots \\ y_n \end{bmatrix} = \begin{bmatrix} n\bar{y} \\ \sum\limits_{i=1}^{n} x_i y_i \end{bmatrix},
$$

and

$$
\begin{aligned}
\hat{\boldsymbol{\theta}} = (C^T C)^{-1} C^T y &= \begin{bmatrix} n & n\bar{x} \\ n\bar{x} & \sum\limits_{i=1}^{n} x_i^2 \end{bmatrix}^{-1} \begin{bmatrix} n\bar{y} \\ \sum\limits_{i=1}^{n} x_i y_i \end{bmatrix} \\
&= \begin{bmatrix} \bar{y} - \left\{ \sum\limits_{i=1}^{n} x_i y_i - n\bar{x}\bar{y} \right\} \left\{ \sum\limits_{i=1}^{n} x_i^2 - n\bar{x}^2 \right\}^{-1} \bar{x} \\ \left\{ \sum\limits_{i=1}^{n} x_i y_i - n\bar{x}\bar{y} \right\} \left\{ \sum\limits_{i=1}^{n} x_i^2 - n\bar{x}^2 \right\}^{-1} \end{bmatrix} \\
&= \begin{bmatrix} \bar{y} - \hat{\beta}\bar{x} \\ \left\{ \sum\limits_{i=1}^{n} (x_i - \bar{x})(y_i - \bar{y}) \right\} \left\{ \sum\limits_{i=1}^{n} (x_i - \bar{x})^2 \right\}^{-1} \end{bmatrix}.
\end{aligned}
$$

**Table 11.1**  Percentage yield, $y_i$, with process temperature, $x_i$, for Example 11.1

|        |    |    |    |    | $i$ |    |    |    |    |    |
|--------|----|----|----|----|----|----|----|----|----|----|
|        | 1  | 2  | 3  | 4  | 5  | 6  | 7  | 8  | 9  | 10 |
| $x$ (°C) | 45 | 50 | 55 | 60 | 65 | 70 | 75 | 80 | 85 | 90 |
| $y$    | 43 | 45 | 48 | 51 | 55 | 57 | 59 | 63 | 66 | 68 |

**Example 11.1.** Problem: it is expected that the average percentage yield, $Y$, from a chemical process is linearly related to the process temperature, $x$, in °C. Determine the least-square regression line for $E\{Y\}$ on the basis of 10 observations given in Table 11.1.

Answer: in view of Equations (11.7) and (11.8), we need the following quantities:

$$\bar{x} = \frac{1}{n}\sum_{i=1}^{n} x_i = \frac{1}{10}(45 + 50 + \cdots + 90) = 67.5,$$

$$\bar{y} = \frac{1}{n}\sum_{i=1}^{n} y_i = \frac{1}{10}(43 + 45 + \cdots + 68) = 55.5,$$

$$\sum_{i=1}^{n}(x_i - \bar{x})^2 = 2062.5,$$

$$\sum_{i=1}^{n}(x_i - \bar{x})(y_i - \bar{y}) = 1182.5.$$

The substitution of these values into Equations (11.7) and (11.8) gives

$$\hat{\beta} = \frac{1182.5}{2062.5} = 0.57,$$
$$\hat{\alpha} = 55.5 - 0.57(67.5) = 17.03.$$

The estimated regression line together with observed sample values is shown in Figure 11.2.

It is noteworthy that regression relationships are valid only for the range of $x$ values represented by the data. Thus, the estimated regression line in Example 11.1 holds only for temperatures between 45 °C and 90 °C. Extrapolation of the result beyond this range can be misleading and is not valid in general.

Another word of caution has to do with the basic linear assumption between $E\{Y\}$ and $x$. Linear regression analysis such as the one performed in Example 11.1 is based on the assumption that the true relationship between $E\{Y\}$ and $x$ is linear. Indeed, if the underlying relationship is nonlinear or nonexistent,

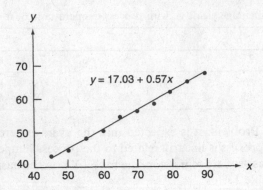

**Figure 11.2**  Estimated regression line and observed data for Example 11.1

linear regression produces meaningless results even if a straight line appears to provide a good fit to the data.

## 11.1.2  PROPERTIES OF LEAST-SQUARE ESTIMATORS

The properties of the estimators for regression coefficients $\alpha$ and $\beta$ can be determined in a straightforward fashion following the vector–matrix expression Equation (11.15). Let $\hat{A}$ and $\hat{B}$ denote, respectively, the estimators for $\alpha$ and $\beta$ following the method of least squares, and let

$$\hat{\Theta} = \begin{bmatrix} \hat{A} \\ \hat{B} \end{bmatrix}. \tag{11.16}$$

We see from Equation (11.15) that

$$\hat{\Theta} = (C^{\mathrm{T}} C)^{-1} C^{\mathrm{T}} Y, \tag{11.17}$$

where

$$Y = \begin{bmatrix} Y_1 \\ \vdots \\ Y_n \end{bmatrix}, \tag{11.18}$$

and $Y_j, j = 1, 2, \ldots, n$, are independent and identically distributed according to Equation (11.4). Thus, if we write

$$Y = C\theta + E, \tag{11.19}$$

then $E$ is a zero-mean random vector with covariance matrix $\Lambda = \sigma^2 I$, $I$ being the $n \times n$ identity matrix.

The mean and variance of estimator $\hat{\Theta}$ are now easily determined. In view of Equations (11.17) and (11.19), we have

$$
\begin{aligned}
E\{\hat{\Theta}\} &= (C^T C)^{-1} C^T E\{Y\} \\
&= (C^T C)^{-1} C^T [C\theta + E\{E\}] \\
&= (C^T C)^{-1} (C^T C)\theta = \theta.
\end{aligned}
\tag{11.20}
$$

Hence, estimators $\hat{A}$ and $\hat{B}$ for $\alpha$ and $\beta$, respectively, are unbiased.

The covariance matrix associated with $\hat{\Theta}$ is given by, as seen from Equation (11.17),

$$
\begin{aligned}
\text{cov}\{\hat{\Theta}\} &= E\{(\hat{\Theta} - \theta)(\hat{\Theta} - \theta)^T\} \\
&= (C^T C)^{-1} C^T \text{cov}\{Y\} C (C^T C)^{-1}.
\end{aligned}
$$

But $\text{cov}\{Y\} = \sigma^2 I$; we thus have

$$
\text{cov}\{\hat{\Theta}\} = \sigma^2 (C^T C)^{-1} C^T C (C^T C)^{-1} = \sigma^2 (C^T C)^{-1}.
\tag{11.21}
$$

The diagonal elements of the matrix in Equation (11.21) give the variances of $\hat{A}$ and $\hat{B}$. In terms of the elements of $C$, we can write

$$
\text{var}\{\hat{A}\} = \left[\sigma^2 \sum_{i=1}^{n} x_i^2\right]\left[n \sum_{i=1}^{n}(x_i - \overline{x})^2\right]^{-1},
\tag{11.22}
$$

$$
\text{var}\{\hat{B}\} = \sigma^2 \left[\sum_{i=1}^{n}(x_i - \overline{x})^2\right]^{-1}.
\tag{11.23}
$$

It is seen that these variances decrease as sample size $n$ increases, according to $1/n$. Thus, it follows from our discussion in Chapter 9 that these estimators are consistent – a desirable property. We further note that, for a fixed $n$, the variance of $\hat{B}$ can be reduced by selecting the $x_i$ in such a way that the denominator of Equation (11.23) is maximized; this can be accomplished by spreading the $x_i$ as far apart as possible. In Example 11.1, for example, assuming that we are free to choose the values of $x_i$, the quality of $\hat{\beta}$ is improved if one-half of the $x$ readings are taken at one extreme of the temperature range and the other half at the other extreme. However, the sampling strategy for minimizing $\text{var}(\hat{A})$ for a fixed $n$ is to make $\overline{x}$ as close to zero as possible.

Are the variances given by Equations (11.22) and (11.23) minimum variances associated with any unbiased estimators for $\alpha$ and $\beta$? An answer to this important question can be found by comparing the results given by Equations (11.22) and (11.23) with the Cramér–Rao lower bounds defined in Section 9.2.2. In order to evaluate these lower bounds, a probability distribution of $Y$ must be made available. Without this knowledge, however, we can still show, in Theorem 11.2, that the least squares technique leads to *linear unbiased minimum-variance estimators* for $\alpha$ and $\beta$; that is, among all unbiased estimators which are *linear* in $Y$, least-square estimators have minimum variance.

**Theorem 11.2:** let random variable $Y$ be defined by Equation (11.4). Given a sample $(x_1, Y_1), (x_2, Y_2), \ldots, (x_n, Y_n)$ of $Y$ with its associated $x$ values, least-square estimators $\hat{A}$ and $\hat{B}$ given by Equation (11.17) are *minimum variance linear unbiased estimators* for $\alpha$ and $\beta$, respectively.

**Proof of Theorem 11.2:** the proof of this important theorem is sketched below with use of vector–matrix notation.

Consider a linear unbiased estimator of the form

$$\Theta^* = [(C^T C)^{-1} C^T + G] Y. \tag{11.24}$$

We thus wish to prove that $G = 0$ if $\Theta^*$ is to be minimum variance.

The unbiasedness requirement leads to, in view of Equation (11.19),

$$GC = 0. \tag{11.25}$$

Consider now the covariance matrix

$$\text{cov}\{\Theta^*\} = E\{(\Theta^* - \theta)(\Theta^* - \theta)^T\}. \tag{11.26}$$

Upon using Equations (11.19), (11.24), and (11.25) and expanding the covariance, we have

$$\text{cov}\{\Theta^*\} = \sigma^2 [(C^T C)^{-1} + G G^T].$$

Now, in order to minimize the variances associated with the components of $\Theta^*$, we must minimize each diagonal element of $G G^T$. Since the $ii$th diagonal element of $G G^T$ is given by

$$(G G^T)_{ii} = \sum_{j=1}^{n} g_{ij}^2,$$

where $g_{ij}$ is the $ij$th element of $G$, we must have

$$g_{ij} = 0, \quad \text{for all } i \text{ and } j.$$

and we obtain

$$G = 0. \tag{11.27}$$

This completes the proof. The theorem stated above is a special case of the *Gauss–Markov theorem*.

Another interesting comparison is that between the least-square estimators for $\alpha$ and $\beta$ and their maximum likelihood estimators with an assigned distribution for random variable $Y$. It is left as an exercise to show that the maximum likelihood estimators for $\alpha$ and $\beta$ are identical to their least-square counterparts under the added assumption that $Y$ is normally distributed.

### 11.1.3 UNBIASED ESTIMATOR FOR $\sigma^2$

As we have shown, the method of least squares does not lead to an estimator for variance $\sigma^2$ of $Y$, which is in general also an unknown quantity in linear regression models. In order to propose an estimator for $\sigma^2$, an intuitive choice is

$$\widehat{\Sigma^2} = k \sum_{i=1}^{n} [Y_i - (\hat{A} + \hat{B}x_i)]^2, \tag{11.28}$$

where coefficient $k$ is to be chosen so that $\widehat{\Sigma^2}$ is unbiased. In order to carry out the expectation of $\widehat{\Sigma^2}$, we note that [see Equation (11.7)]

$$Y_i - \hat{A} - \hat{B}x_i = Y_i - (\overline{Y} - \hat{B}\overline{x}) - \hat{B}x_i$$
$$= (Y_i - \overline{Y}) - \hat{B}(x_i - \overline{x}). \tag{11.29}$$

Hence, it follows that

$$\sum_{i=1}^{n}(Y_i - \hat{A} - \hat{B}x_i)^2 = \sum_{i=1}^{n}(Y_i - \overline{Y})^2 - \hat{B}^2 \sum_{i=1}^{n}(x_i - \overline{x})^2, \tag{11.30}$$

since [see Equation (11.8)]

$$\sum_{i=1}^{n}(x_i - \overline{x})(Y_i - \overline{Y}) = \hat{B} \sum_{i=1}^{n}(x_i - \overline{x})^2. \tag{11.31}$$

Upon taking expectations term by term, we can show that

$$E\{\widehat{\Sigma^2}\} = kE\left\{ \sum_{i=1}^{n}(Y_i - \overline{Y})^2 - \hat{B}^2 \sum_{i=1}^{n}(x_i - \overline{x})^2 \right\}$$
$$= k(n - 2)\sigma^2.$$

Hence, $\widehat{\Sigma^2}$ is unbiased with $k = 1/(n-2)$, giving

$$\widehat{\Sigma^2} = \frac{1}{n-2} \sum_{i=1}^{n} [Y_i - (\hat{A} + \hat{B}x_i)]^2, \qquad (11.32)$$

or, in view of Equation (11.30),

$$\widehat{\Sigma^2} = \frac{1}{n-2} \left[ \sum_{i=1}^{n} (Y_i - \overline{Y})^2 - \hat{B}^2 \sum_{i=1}^{n} (x_i - \overline{x})^2 \right]. \qquad (11.33)$$

**Example 11.2.** Problem: use the results given in Example 11.1 and determine an unbiased estimate for $\sigma^2$.

Answer: we have found in Example 11.1 that

$$\sum_{i=1}^{n} (x_i - \overline{x})^2 = 2062.5,$$

$$\hat{\beta} = 0.57.$$

In addition, we easily obtain

$$\sum_{i=1}^{n} (y_i - \overline{y})^2 = 680.5.$$

Equation (11.33) thus gives

$$\hat{\sigma^2} = \frac{1}{8} [680.5 - (0.57)^2 (2062.5)]$$

$$= 1.30.$$

**Example 11.3.** Problem: an experiment on lung tissue elasticity as a function of lung expansion properties is performed, and the measurements given in Table 11.2 are those of the tissue's Young's modulus ($Y$), in $g\,cm^{-2}$, at varying values of lung expansion in terms of stress ($x$), in $g\,cm^{-2}$. Assuming that $E\{Y\}$ is linearly related to $x$ and that $\sigma_Y^2 = \sigma^2$ (a constant), determine the least-square estimates of the regression coefficients and an unbiased estimate of $\sigma^2$.

**Table 11.2**  Young's modulus, y ($g\,cm^{-2}$), with stress, x ($g\,cm^{-2}$), for Example 11.3

| x | 2 | 2.5 | 3 | 5 | 7 | 9 | 10 | 12 | 15 | 16 | 17 | 18 | 19 | 20 |
|---|---|-----|---|---|---|---|----|----|----|----|----|----|----|----|
| y | 9.1 | 19.2 | 18.0 | 31.3 | 40.9 | 32.0 | 54.3 | 49.1 | 73.0 | 91.0 | 79.0 | 68.0 | 110.5 | 130.8 |

Answer: in this case, we have $n = 14$. The quantities of interest are

$$\bar{x} = \frac{1}{n}\sum_{i=1}^{n} x_i = \frac{1}{14}(2 + 2.5 + \cdots + 20) = 11.11,$$

$$\bar{y} = \frac{1}{n}\sum_{i=1}^{n} y_i = \frac{1}{14}(9.1 + 19.2 + \cdots + 130.8) = 57.59,$$

$$\sum_{i=1}^{n}(x_i - \bar{x})^2 = 546.09,$$

$$\sum_{i=1}^{n}(y_i - \bar{y})^2 = 17,179.54,$$

$$\sum_{i=1}^{n}(x_i - \bar{x})(y_i - \bar{y}) = 2862.12.$$

The substitution of these values into Equations (11.7), (11.8), and (11.33) gives

$$\hat{\beta} = \frac{2862.12}{546.09} = 5.24,$$

$$\hat{\alpha} = 57.59 - 5.24(11.11) = -0.63,$$

$$\widehat{\sigma^2} = \frac{1}{12}[17,179.54 - (5.24)^2(546.09)] = 182.10.$$

The estimated regression line together with the data are shown in Figure 11.3. The estimated standard deviation is $\hat{\sigma} = \sqrt{182.10} = 13.49\,\text{g cm}^{-2}$, and the $1\sigma$-band is also shown in the figure.

## 11.1.4  CONFIDENCE INTERVALS FOR REGRESSION COEFFICIENTS

In addition to point estimators for the slope and intercept in linear regression, it is also easy to construct confidence intervals for them and for $\alpha + \beta x$, the mean of $Y$, under certain distributional assumptions. In what follows, let us assume that $Y$ is normally distributed according to $N(\alpha + \beta x, \sigma^2)$. Since estimators $\hat{A}$, $\hat{B}$, and $\hat{A} + \hat{B}x$ are linear functions of the sample of $Y$, they are also normal random variables. Let us note that, when sample size $n$ is large, $\hat{A}$, $\hat{B}$, and $\hat{A} + \hat{B}x$ are expected to follow normal distributions as a consequence of the central limit theorem (Section 7.2.1), no matter how $Y$ is distributed.

We follow our development in Section 9.3.2 in establishing the desired confidence limits. Based on our experience in Section 9.3.2, the following are not difficult to verify:

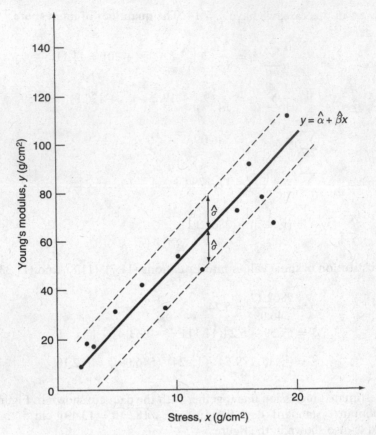

**Figure 11.3**  Estimated regression line and observed data, for Example 11.3

- Result i: let $\widehat{\Sigma^2}$ be the unbiased estimator for $\sigma^2$ as defined by Equation (11.33), and let

$$D = \frac{(n-2)\widehat{\Sigma^2}}{\sigma^2}. \tag{11.34}$$

It follows from the results given in Section 9.3.2.3 that $D$ is a $\chi^2$-distributed random variable with $(n-2)$ degrees of freedom.

- Result ii: consider random variables

$$(\hat{A} - \alpha)\left\{ \left[ \widehat{\Sigma^2} \sum_{i=1}^{n} x_i^2 \right] \left[ n \sum_{i=1}^{n} (x_i - \bar{x})^2 \right]^{-1} \right\}^{-1/2}, \tag{11.35}$$

and

$$(\hat{B} - \beta)\left\{\widehat{\Sigma^2}\left[\sum_{i=1}^{n}(x_i - \overline{x})^2\right]^{-1}\right\}^{-1/2} \tag{11.36}$$

where, as seen from Equations (11.20), (11.22), and (11.23), $\alpha$ and $\beta$ are, respectively, the means of $\hat{A}$ and $\hat{B}$ and the denominators are, respectively, the standard deviations of $\hat{A}$ and $\hat{B}$ with $\sigma^2$ estimated by $\widehat{\Sigma^2}$. The derivation given in Section 9.3.2.2 shows that each of these random variables has a $t$-distribution with $(n - 2)$ degrees of freedom.

- Result iii: estimator $\widehat{E\{Y\}}$ for the mean of $Y$ is normally distributed with mean $\alpha + \beta x$ and variance

$$\begin{aligned}
\text{var}\{\widehat{E\{Y\}}\} &= \text{var}\{\hat{A} + \hat{B}x\} \\
&= \text{var}\{\hat{A}\} + x^2\text{var}\{\hat{B}\} + 2x\text{cov}\{\hat{A}, \hat{B}\} \\
&= \sigma^2\left[\sum_{i=1}^{n}(x_i - \overline{x})^2\right]^{-1}\left(\frac{1}{n}\sum_{i=1}^{n}x_i^2 + x^2 - 2x\overline{x}\right) \\
&= \sigma^2\left\{\frac{1}{n} + (x - \overline{x})^2\left[\sum_{i=1}^{n}(x_i - \overline{x})^2\right]^{-1}\right\}.
\end{aligned} \tag{11.37}$$

Hence, again following the derivation given in Section 9.3.2.2, random variable

$$\left[\widehat{E\{Y\}} - (\alpha + \beta x)\right]\left\{\widehat{\Sigma^2}\left\{\frac{1}{n} + (x_i - \overline{x})^2\left[\sum_{i=1}^{n}(x_i - \overline{x})^2\right]^{-1}\right\}\right\}^{-1/2} \tag{11.38}$$

is also $t$-distributed with $(n - 2)$ degrees of freedom.

Based on the results presented above, we can now easily establish confidence limits for all the parameters of interest. The results given below are a direct consequence of the development in Section 9.3.2.

- Result 1: a $[100(1 - \gamma)]\%$ confidence interval for $\alpha$ is determined by [see Equation (9.141)]

$$L_{1,2} = \hat{A} \mp t_{n-2,\gamma/2}\left\{\left(\widehat{\Sigma^2}\sum_{i=1}^{n}x_i^2\right)\left[n\sum_{i=1}^{n}(x_i - \overline{x})^2\right]^{-1}\right\}^{1/2}. \tag{11.39}$$

- Result 2: a $[100(1 - \gamma)\%$ confidence interval for $\beta$ is determined by [see Equation (9.141)]

$$L_{1,2} = \hat{B} \mp t_{n-2,\gamma/2} \left\{ \widehat{\Sigma^2} \left[ \sum_{i=1}^{n} (x_i - \overline{x})^2 \right]^{-1} \right\}^{1/2} . \qquad (11.40)$$

- Result 3: a $[100(1 - \gamma)]\%$ confidence interval for $E\{Y\} = \alpha + \beta x$ is determined by [see Equation (9.141)]

$$L_{1,2} = E\widehat{\{Y\}} \mp t_{n-2,\gamma/2} \left\{ \widehat{\Sigma^2} \left\{ \frac{1}{n} + (x - \overline{x})^2 \left[ \sum_{i=1}^{n} (x_i - \overline{x})^2 \right]^{-1} \right\} \right\}^{1/2} . \qquad (11.41)$$

- Result 4: a two-sided $[100(1 - \gamma)\%$ confidence interval for $\sigma^2$ is determined by [see Equation (9.144)]

$$L_1 = \frac{(n - 2)\widehat{\Sigma^2}}{\chi^2_{n-2,\gamma/2}},$$
$$L_2 = \frac{(n - 2)\widehat{\Sigma^2}}{\chi^2_{n-2,1-\gamma/2}}. \qquad (11.42)$$

If a one-sided confidence interval for $\sigma^2$ is desired, it is given by [see Equation (9.145)]

$$L_1 = \frac{(n - 2)\widehat{\Sigma^2}}{\chi^2_{n-1,\gamma}}. \qquad (11.43)$$

A number of observations can be made regarding these confidence intervals. In each case, both the position and the width of the interval will vary from sample to sample. In addition, the confidence interval for $\alpha + \beta x$ is shown to be a function of $x$. If one plots the observed values of $L_1$ and $L_2$ they form a *confidence band* about the estimated regression line, as shown in Figure 11.4. Equation (11.41) clearly shows that the narrowest point of the band occurs at $x = \overline{x}$; it becomes broader as $x$ moves away from $\overline{x}$ in either direction.

**Example 11.4.** Problem: in Example 11.3, assuming that $Y$ is normally distributed, determine a 95% confidence band for $\alpha + \beta x$.

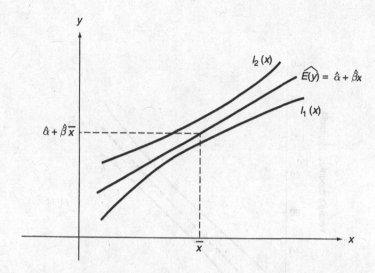

**Figure 11.4**   Confidence band for $E\{Y\} = \alpha + \beta x$

Answer: equation (11.41) gives the desired confidence limits, with $n = 14$, $\gamma = 0.05$, and

$$\widehat{E\{y\}} = \hat{\alpha} + \hat{\beta}x = -0.63 + 5.24x,$$

$$t_{n-2,\gamma/2} = t_{12,0.025} = 2.179, \quad \text{from Table A.4,}$$

$$\bar{x} = 11.11,$$

$$\sum_{i=1}^{n}(x_i - \bar{x})^2 = 546.09,$$

$$\widehat{\sigma^2} = 182.10.$$

The observed confidence limits are thus given by

$$l_{1,2} = (-0.63 + 5.24x) \mp 2.179\left\{182.10\left[\frac{1}{14} + \frac{(x - 11.11)^2}{546.09}\right]\right\}^{1/2}.$$

This result is shown graphically in Figure 11.5.

## 11.1.5   SIGNIFICANCE TESTS

Following the results given above, tests of hypotheses about the values of $\alpha$ and $\beta$ can be carried out based upon the approach discussed in Chapter 10. Let us demonstrate the underlying ideas by testing hypothesis $H_0: \beta = \beta_0$ against hypothesis $H_1: \beta \neq \beta_0$, where $\beta_0$ is some specified value.

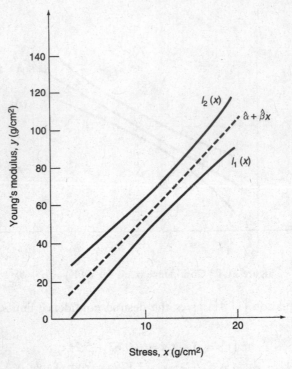

**Figure 11.5**  The 95% confidence band for $E\{Y\}$, for Example 11.4

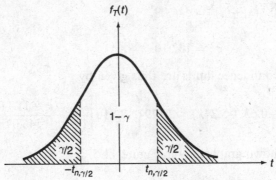

**Figure 11.6**  Probability density function of $T = (\hat{B} - \beta_0)\left\{\widehat{\Sigma^2}[\sum_{i=1}^{n}(x_i - \overline{x})^2]^{-1}\right\}^{-1/2}$

Using $\hat{B}$ as the test statistic, we have shown in Section 11.1.4 that the random variable defined by Equation (11.36) has a $t$-distribution with $n - 2$ degrees of freedom. Suppose we wish to achieve a Type-I error probability of $\gamma$. We would reject $H_0$ if $|\hat{\beta} - \beta_0|$ exceeds (see Figure 11.6)

$$t_{n-2,\gamma/2}\left\{\widehat{\sigma^2}\left[\sum_{i=1}^n (x_i - \overline{x})^2\right]^{-1}\right\}^{1/2}. \tag{11.44}$$

Similarly, significance tests about the value of $\alpha$ can be easily carried out with use of $\hat{A}$ as the test statistic.

An important special case of the above is the test of $H_0: \beta = 0$ against $H_1: \beta \neq 0$. This particular situation corresponds essentially to the significance test of linear regression. Accepting $H_0$ is equivalent to concluding that there is no reason to accept a linear relationship between $E\{Y\}$ and $x$ at a specified significance level $\gamma$. In many cases, this may indicate the lack of a causal relationship between $E\{Y\}$ and independent variable $x$.

**Example 11.5.** Problem: it is speculated that the starting salary of a clerk is a function of the clerk's height. Assume that salary ($Y$) is normally distributed and its mean is linearly related to height ($x$); use the data given in Table 11.3 to test the assumption that $E\{Y\}$ and $x$ are linearly related at the 5% significance level.

**Table 11.3**  Salary, $y$ (in $10\,000$), with height, $x$ (in feet), for Example 11.5

| $x$ | 5.7 | 5.7 | 5.7 | 5.7 | 6.1 | 6.1 | 6.1 | 6.1 |
|---|---|---|---|---|---|---|---|---|
| $y$ | 2.25 | 2.10 | 1.90 | 1.95 | 2.40 | 1.95 | 2.10 | 2.25 |

Answer: in this case, we wish to test $H_0: \beta = 0$ against $H_1: \beta \neq 0$, with $\gamma = 0.05$. From the data in Table 11.3, we have

$$\hat{\beta} = \left[\sum_{i=1}^n (x_i - \overline{x})(y_i - \overline{y})\right]\left[\sum_{i=1}^n (x_i - \overline{x})^2\right]^{-1}$$

$$= 0.31,$$

$$t_{n-2,\gamma/2} = t_{6,0.025} = 2.447, \quad \text{from Table A.4,}$$

$$\widehat{\sigma^2} = \frac{1}{n-2}\left[\sum_{i=1}^n (y_i - \overline{y})^2 - \hat{\beta}^2 \sum_{i=1}^n (x_i - \overline{x})^2\right] = 0.02,$$

$$\sum_{i=1}^n (x_i - \overline{x})^2 = 0.32.$$

According to Equation (11.44), we have

$$t_{6,0.025}\left\{\widehat{\sigma^2}\left[\sum_{i=1}^n (x_i - \overline{x})^2\right]^{-1}\right\}^{1/2} = 0.61.$$

Since $\hat{\beta} = 0.31 < 0.61$, we accept $H_0$. That is, we conclude that the data do not indicate a linear relationship between $E\{Y\}$ and $x$; the probability that we are wrong in accepting $H_0$ is 0.05.

In closing, let us remark that we are often called on to perform tests of *simultaneous hypotheses*. For example, one may wish to test $H_0: \alpha = 0$ and $\beta = 1$ against $H_1: \alpha \neq 0$ or $\beta \neq 1$ or both. Such tests involve both estimators $\hat{A}$ and $\hat{B}$ and hence require their joint distribution. This is also often the case in multiple linear regression, to be discussed in the next section. Such tests customarily involve $F$-distributed test statistics, and we will not pursue them here. A general treatment of simultaneous hypotheses testing can be found in Rao (1965), for example.

## 11.2 MULTIPLE LINEAR REGRESSION

The vector–matrix approach proposed in the preceding section provides a smooth transition from simple linear regression to linear regression involving more than one independent variable. In multiple linear regression, the model takes the form

$$E\{Y\} = \beta_0 + \beta_1 x_1 + \beta_2 x_2 + \cdots + \beta_m x_m. \tag{11.45}$$

Again, we assume that the variance of $Y$ is $\sigma^2$ and is independent of $x_1, x_2, \ldots$, and $x_m$. As in simple linear regression, we are interested in estimating $(m + 1)$ regression coefficients $\beta_0, \beta_1, \ldots$, and $\beta_m$, obtaining certain interval estimates, and testing hypotheses about these parameters on the basis of a sample of $Y$ values with their associated values of $(x_1, x_2, \ldots, x_m)$. Let us note that our sample of size $n$ in this case takes the form of arrays $(x_{11}, x_{21}, \ldots, x_{m1}, Y_1)$, $(x_{12}, x_{22}, \ldots, x_{m2}, Y_2)$, $\ldots$, $(x_{1n}, x_{2n}, \ldots, x_{mn}, Y_n)$. For each set of values $x_{ki}, k = 1, 2, \ldots, m$, of $x_i$, $Y_i$ is an independent observation from population $Y$ defined by

$$Y = \beta_0 + \beta_1 x_1 + \cdots + \beta_m x_m + E. \tag{11.46}$$

As before, $E$ is the random error, with mean 0 and variance $\sigma^2$.

### 11.2.1 LEAST SQUARES METHOD OF ESTIMATION

To estimate the regression coefficients, the method of least squares will again be employed. Given observed sample-value sets $(x_{1i}, x_{2i}, \ldots, x_{mi}, y_i), i = 1, 2, \ldots, n$, the system of observed regression equations in this case takes the form

$$y_i = \beta_0 + \beta_1 x_{1i} + \cdots + \beta_m x_{mi} + e_i, \quad i = 1, 2, \ldots, n. \tag{11.47}$$

If we let

$$
C = \begin{bmatrix} 1 & x_{11} & x_{21} & \cdots & x_{m1} \\ 1 & x_{12} & x_{22} & \cdots & x_{m2} \\ \vdots & \vdots & \vdots & \vdots & \vdots \\ 1 & x_{1n} & x_{2n} & \cdots & x_{mn} \end{bmatrix}, \quad y = \begin{bmatrix} y_1 \\ y_2 \\ \vdots \\ y_n \end{bmatrix}, \quad e = \begin{bmatrix} e_1 \\ e_2 \\ \vdots \\ e_n \end{bmatrix},
$$

and

$$
\theta = \begin{bmatrix} \beta_0 \\ \beta_1 \\ \vdots \\ \beta_m \end{bmatrix},
$$

Equation (11.47) can be represented by vector–matrix equation:

$$
y = C\theta + e. \tag{11.48}
$$

Comparing Equation (11.48) with Equation (11.12) in simple linear regression, we see that the observed regression equations in both cases are identical except that the $C$ matrix is now an $n \times (m+1)$ matrix and $\theta$ is an $(m+1)$-dimensional vector. Keeping this dimension difference in mind, the results obtained in the case of simple linear regression based on Equation (11.12) again hold in the multiple linear regression case. Thus, without further derivation, we have for the solution of least-square estimates $\hat{\theta}$ of $\theta$ [see Equation (11.15)]

$$
\boxed{\hat{\theta} = (C^{\mathrm{T}}C)^{-1}C^{\mathrm{T}}y.} \tag{11.49}
$$

The existence of matrix inverse $(C^{\mathrm{T}}C)^{-1}$ requires that there are at least $(m+1)$ distinct sets of values of $(x_{1i}, x_{2i}, \ldots, x_{mi})$ represented in the sample. It is noted that $C^{\mathrm{T}}C$ is a $(m+1) \times (m+1)$ symmetric matrix.

**Example 11.6.** Problem: the average monthly electric power consumption ($Y$) at a certain manufacturing plant is considered to be linearly dependent on the average ambient temperature ($x_1$) and the number of working days in a month ($x_2$). Consider the one-year monthly data given in Table 11.4. Determine the least-square estimates of the associated linear regression coefficients.

**Table 11.4** Average monthly power consumption $y$ (in thousands of kwh), with number of working days in the month, $x_2$, and average ambient temperature, $x_1$, (in °F) for Example 11.6

| $x_1$ | 20 | 26 | 41 | 55 | 60 | 67 | 75 | 79 | 70 | 55 | 45 | 33 |
|-------|-----|-----|-----|-----|-----|-----|-----|-----|-----|-----|-----|-----|
| $x_2$ | 23 | 21 | 24 | 25 | 24 | 26 | 25 | 25 | 24 | 25 | 25 | 23 |
| $y$ | 210 | 206 | 260 | 244 | 271 | 285 | 270 | 265 | 234 | 241 | 258 | 230 |

Answer: in this case, $C$ *is a* $12 \times 3$ matrix, and

$$C^T C = \begin{bmatrix} 12 & 626 & 290 \\ 626 & 36{,}776 & 15{,}336 \\ 290 & 15{,}336 & 7{,}028 \end{bmatrix},$$

$$C^T y = \begin{bmatrix} 2{,}974 \\ 159{,}011 \\ 72{,}166 \end{bmatrix}.$$

We thus have, upon finding the inverse of $C^T C$ by using either matrix inversion formulae or readily available matrix inversion computer programs,

$$\hat{\theta} = (C^T C)^{-1} C^T y = \begin{bmatrix} -33.84 \\ 0.39 \\ 10.80 \end{bmatrix},$$

or

$$\hat{\beta}_0 = -33.84, \quad \hat{\beta}_1 = 0.39, \quad \hat{\beta}_3 = 10.80.$$

The estimated regression equation based on the data is thus

$$\widehat{E\{y\}} = \hat{\beta}_0 + \hat{\beta}_1 x_1 + \hat{\beta}_2 x_2$$
$$= -33.84 + 0.39 x_1 + 10.80 x_2.$$

Since Equation (11.48) is identical to its counterpart in the case of simple linear regression, much of the results obtained therein concerning properties of least-square estimators, confidence intervals, and hypotheses testing can be duplicated here with, of course, due regard to the new definitions for matrix $C$ and vector $\theta$.

Let us write estimator $\hat{\Theta}$ for $\theta$ in the form

$$\hat{\Theta} = (C^T C)^{-1} C^T Y. \tag{11.50}$$

We see immediately that

$$E\{\hat{\Theta}\} = (C^T C)^{-1} C^T E\{Y\} = \theta. \tag{11.51}$$

Hence, least-square estimator $\hat{\Theta}$ is again unbiased. It also follows from Equation (11.21) that the covariance matrix for $\hat{\Theta}$ is given by

$$\text{cov}\{\hat{\Theta}\} = \sigma^2 (C^T C)^{-1}. \tag{11.52}$$

Confidence intervals for the regression parameters in this case can also be established following similar procedures employed in the case of simple linear regression. Concerning hypotheses testing, it was mentioned in Section 11.1.5 that testing of simultaneous hypotheses is more appropriate in multiple linear regression, and that we will not pursue it here.

## 11.3  OTHER REGRESSION MODELS

In science and engineering, one often finds it necessary to consider regression models that are nonlinear in the independent variables. Common examples of this class of models include

$$Y = \beta_0 + \beta_1 x + \beta_2 x^2 + E, \tag{11.53}$$

$$Y = \beta_0 \exp(\beta_1 x + E), \tag{11.54}$$

$$Y = \beta_0 + \beta_1 x_1 + \beta_2 x_2 + \beta_{11} x_1^2 + \beta_{22} x_2^2 + \beta_{12} x_1 x_2 + E, \tag{11.55}$$

$$Y = \beta_0 \beta_1^x + E. \tag{11.56}$$

Polynomial models such as Equation (11.53) or Equation (11.55) are still linear regression models in that they are linear in the unknown parameters $\beta_0, \beta_1, \beta_2, \ldots$, [etc. Hence, they can be estimated by using multiple linear regression techniques. Indeed, let $x_1 = x$, and $x_2 = x^2$ in Equation (11.53), it takes the form of a multiple linear regression model with two independent variables and can thus be analyzed as such. Similar equivalence can be established between Equation (11.55) and a multiple linear regression model with five independent variables.

Consider the exponential model given by Equation (11.54). Taking logarithms of both sides, we have

$$\ln Y = \ln \beta_0 + \beta_1 x + E. \tag{11.57}$$

In terms of random variable $\ln Y$, Equation (11.57) represents a linear regression equation with regression coefficients $\ln \beta_0$ and $\beta_1$. Linear regression techniques again apply in this case. Equation (11.56), however, cannot be conveniently put into a linear regression form.

**Example 11.7.** Problem: on average, the rate of population increase ($Y$) associated with a given city varies with $x$, the number of years after 1970. Assuming that

$$E\{Y\} = \beta_0 + \beta_1 x + \beta_2 x^2,$$

compute the least-square estimates for $\beta_0$, $\beta_1$, and $\beta_2$ based on the data presented in Table 11.5.

**Table 11.5**  Population increase, $y$, with number of years after 1970, $x$, for Example 11.7

| $x$ | 0 | 1 | 2 | 3 | 4 | 5 |
|------|------|------|------|------|------|------|
| $y(\%)$ | 1.03 | 1.32 | 1.57 | 1.75 | 1.83 | 2.33 |

Answer: let $x_1 = x$, $x_2 = x^2$, and let

$$\theta = \begin{bmatrix} \beta_0 \\ \beta_1 \\ \beta_2 \end{bmatrix}.$$

The least-square estimate for $\theta, \hat{\theta}$, is given by Equation (11.49), with

$$C = \begin{bmatrix} 1 & 0 & 0 \\ 1 & 1 & 1 \\ 1 & 2 & 4 \\ 1 & 3 & 9 \\ 1 & 4 & 16 \\ 1 & 5 & 25 \end{bmatrix}$$

and

$$y = \begin{bmatrix} 1.03 \\ 1.32 \\ 1.57 \\ 1.75 \\ 1.83 \\ 2.33 \end{bmatrix}$$

Thus

$$\hat{\theta} = (C^T C)^{-1} C^T y = \begin{bmatrix} 6 & 15 & 55 \\ 15 & 55 & 225 \\ 55 & 225 & 979 \end{bmatrix}^{-1} \begin{bmatrix} 9.83 \\ 28.68 \\ 110.88 \end{bmatrix}$$

$$= \begin{bmatrix} 1.07 \\ 0.20 \\ 0.01 \end{bmatrix},$$

or

$$\hat{\beta}_0 = 1.07, \quad \hat{\beta}_1 = 0.20, \quad \text{and} \quad \hat{\beta}_2 = 0.01.$$

Let us note in this example that, since $x_2 = x_1^2$, matrix $C$ is constrained in that its elements in the third column are the squared values of their corresponding elements in the second column. It needs to be cautioned that, for high-order polynomial regression models, constraints of this type may render matrix $C^T C$ ill-conditioned and lead to matrix-inversion difficulties.

## REFERENCE

Rao, C.R., 1965, *Linear Statistical Inference and Its Applications*, John Wiley & Sons Inc., New York.

## FURTHER READING

Some additional useful references on regression analysis are given below.

Anderson, R.L., and Bancroft, T.A., 1952, *Statistical Theory in Research*, McGraw-Hill, New York.
Bendat, J.S., and Piersol, A.G., 1966, *Measurement and Analysis of Random Data*, John Wiley & Sons Inc., New York.
Draper, N., and Smith, H., 1966, *Applied Regression Analysis*, John Wiley & Sons Inc., New York.
Graybill, F.A., 1961, *An Introduction to Linear Statistical Models, Volume 1*. McGraw-Hill, New York.

## PROBLEMS

11.1 A special case of simple linear regression is given by

$$Y = \beta x + E.$$

Determine:
(a) The least-square estimator $\hat{B}$ for $\beta$;
(b) The mean and variance of $\hat{B}$;
(c) An unbiased estimator for $\sigma^2$, the variance of $Y$.

11.2 In simple linear regression, show that the maximum likelihood estimators for $\alpha$ and $\beta$ are identical to their least-square estimators when $Y$ is normally distributed.

11.3 Determine the maximum likelihood estimator for variance $\sigma^2$ of $Y$ in simple linear regression assuming that $Y$ is normally distributed. Is it a biased estimator?

11.4 Since data quality is generally not uniform among data points, it is sometimes desirable to estimate the regression coefficients by minimizing the sum of *weighted* squared residuals; that is, $\hat{\alpha}$ and $\hat{\beta}$ in simple linear regression are found by minimizing

$$\sum_{i=1}^{n} w_i e_i^2,$$

where $w_i$ are assigned weights. In vector–matrix notation, show that estimates $\hat{\alpha}$ and $\hat{\beta}$ now take the form

$$\hat{\theta} = \begin{bmatrix} \hat{\alpha} \\ \hat{\beta} \end{bmatrix} = (C^{\mathrm{T}} W C)^{-1} C^{\mathrm{T}} W y,$$

where

$$W = \begin{bmatrix} w_1 & & & 0 \\ & w_2 & & \\ & & \ddots & \\ 0 & & & w_n \end{bmatrix}.$$

11.5 (a) In simple linear regression [Equation (11.4)]. use vector–matrix notation and show that the unbiased estimator for $\sigma^2$ given by Equation (11.33) can be written in the form

$$\widehat{\Sigma^2} = \frac{1}{n-2} [(\boldsymbol{Y} - C\hat{\boldsymbol{\Theta}})^{\mathrm{T}} (\boldsymbol{Y} - C\hat{\boldsymbol{\Theta}})].$$

(b) In multiple linear regression [Equation (11.46)], show that an unbiased estimator for $\sigma^2$ is given by

$$\widehat{\Sigma^2} = \frac{1}{n-m-1} [(\boldsymbol{Y} - C\hat{\boldsymbol{\Theta}})^{\mathrm{T}} (\boldsymbol{Y} - C\hat{\boldsymbol{\Theta}})].$$

11.6 Given the data in Table 11.6:

**Table 11.6**  Data for Problem 11.6

| $x$ | 0 | 1 | 2 | 3 | 4 | 5 | 6 | 7 | 8 | 9 |
|-----|-----|-----|-----|-----|-----|-----|-----|-----|-----|-----|
| $y$ | 3.2 | 3.1 | 3.9 | 4.7 | 4.3 | 4.4 | 4.8 | 5.3 | 5.9 | 6.0 |

(a) Determine the least-square estimates of $\alpha$ and $\beta$ in the linear regression equation

$$Y = \alpha + \beta x + E.$$

(b) Determine an unbiased estimate of $\sigma^2$, the variance of $Y$.
(c) Estimate $E\{Y\}$ at $x = 5$.
(d) Determine a 95% confidence interval for $\beta$.
(e) Determine a 95% confidence band for $\alpha + \beta x$.

11.7 In transportation studies, it is assumed that, on average, peak vehicle noise level ($Y$) is linearly related to the logarithm of vehicle speed ($v$). Some measurements taken for a class of light vehicles are given in Table 11.7. Assuming that

$$Y = \alpha + \beta \log_{10} v + E,$$

**Table 11.7**  Noise level, $y$ (in dB) with vehicle
speed, $v$ (in $km\,h^{-1}$), for Problem 11.7

| $v$ | 20 | 30 | 40 | 50 | 60 | 70 | 80 | 90 | 100 |
|---|---|---|---|---|---|---|---|---|---|
| $y$ | 55 | 63 | 68 | 70 | 72 | 78 | 74 | 76 | 79 |

determine the estimated regression line for $Y$ as a function of $\log_{10} v$.

11.8  An experimental study of nasal deposition of particles was carried out and
showed a linear relationship between $E\{Y\}$ and $\ln d^2 f$, where $Y$ is the fraction
of particles of aerodynamic diameter, $d$ (in mm), that is deposited in the nose
during an inhalation of $f(l\,min^{-1})$. Consider the data given in Table 11.8 (four
readings are taken at each value of $\ln d^2 f$). Estimate the regression parameters in
the linear regression equation

$$E\{Y\} = \alpha + \beta \ln d^2 f,$$

and estimate $\sigma^2$, the variance of $Y$.

**Table 11.8**  Fraction of particles inhaled of diameter $d$ (in mm), with $\ln d^2 f$
($f$ is inhalation, in $l\,min^{-1}$), for Problem 11.8

| $\ln d^2 f$ | 1.6 | 1.7 | 2.0 | 2.8 | 3.0 | 3.0 | 3.6 |
|---|---|---|---|---|---|---|---|
| $y$ | 0.39 | 0.41 | 0.42 | 0.61 | 0.83 | 0.79 | 0.98 |
|  | 0.30 | 0.28 | 0.34 | 0.51 | 0.79 | .0.69 | 0.88 |
|  | 0.21 | 0.20 | 0.22 | 0.47 | 0.70 | 0.63 | 0.87 |
|  | 0.12 | 0.10 | 0.18 | 0.39 | 0.61 | 0.59 | 0.83 |

11.9  For a study of the stress–strain history of soft biological tissues, experimental
results relating dynamic moduli of aorta ($D$) to stress frequency ($\omega$) are given in
Table 11.9.
  (a) Assuming that $E\{D\} = \alpha + \beta\omega$, and $\sigma_D^2 = \sigma^2$, estimate regression coefficients
      $\alpha$ and $\beta$.
  (b) Determine a one-sided 95% confidence interval for the variance of $D$.
  (c) Test if the slope estimate is significantly different from zero at the 5%
      significance level.

**Table 11.9**  The dynamic modulus of aorta, d (normalized) with frequency,
$\omega$ (in Hz), for Problem 11.9

| $\omega$ | 1 | 2 | 3 | 4 | 5 | 6 | 7 | 8 | 9 | 10 |
|---|---|---|---|---|---|---|---|---|---|---|
| $d$ | 1.60 | 1.51 | 1.40 | 1.57 | 1.60 | 1.59 | 1.80 | 1.59 | 1.82 | 1.59 |

11.10  Given the data in Table 11.10
  (a) Determine the least-square estimates of $\beta_0, \beta_1$, and $\beta_2$ assuming that

$$E\{Y\} = \beta_0 + \beta_1 x_1 + \beta_2 x_2.$$

**Table 11.10**   Data for Problem 11.10

| $x_1$ | $-1$ | $-1$ | 1 | 1 | 2 | 2 | 3 | 3 |
|-------|------|------|-----|-----|-----|-----|-----|-----|
| $x_2$ | 1 | 2 | 3 | 4 | 5 | 6 | 7 | 8 |
| $y$ | 2.0 | 3.1 | 4.8 | 4.9 | 5.4 | 6.8 | 6.9 | 7.5 |

(b)  Estimate $E\{Y\}$ at $x_1 = x_2 = 2$.

11.11  In Problem 11.7, when vehicle weight is taken into account, we have the multiple linear regression equation

$$Y = \beta_0 + \beta_1 \log_{10} v + \beta_2 \log_{10} w + E,$$

where $w$ is vehicle unladen weight in Mg. Use the data given in Table 11.11 and estimate the regression parameters in this case.

**Table 11.11**   Noise level, y (in dB), with vehicle weight (unladen, in Mg) and vehicle speed (in $km\,h^{-1}$), for Problem 11.11

| $v$ | 20 | 40 | 60 | 80 | 100 | 120 |
|-----|------|------|------|------|------|------|
| $w$ | 1.0 | 1.0 | 1.7 | 3.0 | 1.0 | 0.7 |
| $y$ | 54 | 59 | 78 | 91 | 78 | 67 |

11.12  Given the data in Table 11.12:

**Table 11.12**   Data for Problem 11.12

| $x$ | 0 | 1 | 2 | 3 | 4 | 5 | 6 | 7 |
|-----|-----|-----|-----|-----|-----|-----|-----|-----|
| $y$ | 3.2 | 2.8 | 5.1 | 7.3 | 7.6 | 5.9 | 4.1 | 1.8 |

(a)  Determine the least-square estimates of $\beta_0$, $\beta_1$, and $\beta_2$ assuming that

$$E\{Y\} = \beta_0 + \beta_1 x + \beta_2 x^2.$$

(b)  Estimate $E\{Y\}$ at $x = 3$.

11.13  A large number of socioeconomic variables are important to account for mortality rate. Assuming a multiple linear regression model, one version of the model for mortality rate ($Y$) is expressed by

$$Y = \beta_0 + \beta_1 x_1 + \beta_2 x_2 + \beta_3 x_3 + \beta_4 x_4 + E,$$

where

$x_1$ = mean annual precipitation in inches,
$x_2$ = education in terms of median school years completed for those over 25 years old
$x_3$ = percentage of area population that is nonwhite,
$x_4$ = relative pollution potential of $SO_2$ (sulfur dioxide).

**Table 11.13** Data for Problem 11.13

| $x_1$ | 13 | 11 | 21 | 30 | 35 | 27 | 27 | 40 |
|---|---|---|---|---|---|---|---|---|
| $x_2$ | 9 | 10.5 | 11 | 10 | 9 | 12.3 | 9 | 9 |
| $x_3$ | 1.5 | 7 | 21 | 27 | 30 | 6 | 27 | 33 |
| $x_4$ | 4 | 21 | 64 | 67 | 17 | 28 | 82 | 101 |
| $y$ | 795 | 841 | 820 | 1050 | 1010 | 970 | 980 | 1090 |

Some available data are presented in Table 11.13. Determine the least-square estimate of the regression parameters.

# Appendix A: Tables

## A.1 BINOMIAL MASS FUNCTION

**Table A.1** Binomial mass function: a table of

$$p_X(k) = \binom{n}{k} p^k (1-p)^{n-k},$$

for $n = 2$ to $10$, $p = 0.01$ to $0.50$

| n | k | p | | | | | | | | | | | | |
|---|---|---|---|---|---|---|---|---|---|---|---|---|---|---|
| | | 0.01 | 0.05 | 0.10 | 0.15 | 0.20 | 0.25 | 0.30 | $\frac{1}{3}$ | 0.35 | 0.40 | 0.45 | 0.49 | 0.50 |
| 2 | 0 | 0.9801 | 0.9025 | 0.8100 | 0.7225 | 0.6400 | 0.5625 | 0.4900 | 0.4444 | 0.4225 | 0.3600 | 0.3025 | 0.2601 | 0.2500 |
| | 1 | 0.0198 | 0.0950 | 0.1800 | 0.2550 | 0.3200 | 0.3750 | 0.4200 | 0.4444 | 0.4550 | 0.4800 | 0.4950 | 0.4998 | 0.5000 |
| | 2 | 0.0001 | 0.0025 | 0.0100 | 0.0225 | 0.0400 | 0.0625 | 0.0900 | 0.1111 | 0.1225 | 0.1600 | 0.2025 | 0.2401 | 0.2500 |
| 3 | 0 | 0.9703 | 0.8574 | 0.7290 | 0.6141 | 0.5120 | 0.4219 | 0.3430 | 0.2963 | 0.2746 | 0.2160 | 0.1664 | 0.1327 | 0.1250 |
| | 1 | 0.0294 | 0.1354 | 0.2430 | 0.3251 | 0.3840 | 0.4219 | 0.4410 | 0.4444 | 0.4436 | 0.4320 | 0.4084 | 0.3823 | 0.3750 |
| | 2 | 0.0003 | 0.0071 | 0.0270 | 0.0574 | 0.0960 | 0.1406 | 0.1890 | 0.2222 | 0.2389 | 0.2880 | 0.3341 | 0.3674 | 0.3750 |
| | 3 | 0.0000 | 0.0001 | 0.0010 | 0.0034 | 0.0080 | 0.0156 | 0.0270 | 0.0370 | 0.0429 | 0.0640 | 0.0911 | 0.1176 | 0.1250 |
| 4 | 0 | 0.9606 | 0.8145 | 0.6561 | 0.5220 | 0.4096 | 0.3164 | 0.2401 | 0.1975 | 0.1785 | 0.1296 | 0.0915 | 0.0677 | 0.0625 |
| | 1 | 0.0388 | 0.1715 | 0.2916 | 0.3685 | 0.4096 | 0.4219 | 0.4116 | 0.3951 | 0.3845 | 0.3456 | 0.2995 | 0.2600 | 0.2500 |
| | 2 | 0.0006 | 0.0135 | 0.0486 | 0.0975 | 0.1536 | 0.2109 | 0.2646 | 0.2963 | 0.3105 | 0.3456 | 0.3675 | 0.3747 | 0.3750 |
| | 3 | 0.0000 | 0.0005 | 0.0036 | 0.0115 | 0.0256 | 0.0469 | 0.0756 | 0.0988 | 0.1115 | 0.1536 | 0.2005 | 0.2400 | 0.2500 |
| | 4 | 0.0000 | 0.0000 | 0.0001 | 0.0005 | 0.0016 | 0.0039 | 0.0081 | 0.0123 | 0.0150 | 0.0256 | 0.0410 | 0.0576 | 0.0625 |
| 5 | 0 | 0.9510 | 0.7738 | 0.5905 | 0.4437 | 0.3277 | 0.2373 | 0.1681 | 0.1317 | 0.1160 | 0.0778 | 0.0503 | 0.0345 | 0.0312 |
| | 1 | 0.0480 | 0.2036 | 0.3280 | 0.3915 | 0.4096 | 0.3955 | 0.3602 | 0.3292 | 0.3124 | 0.2592 | 0.2059 | 0.1657 | 0.1562 |
| | 2 | 0.0010 | 0.0214 | 0.0729 | 0.1382 | 0.2048 | 0.2637 | 0.3087 | 0.3292 | 0.3364 | 0.3456 | 0.3369 | 0.3185 | 0.3125 |
| | 3 | 0.0000 | 0.0011 | 0.0081 | 0.0244 | 0.0512 | 0.0879 | 0.1323 | 0.1646 | 0.1811 | 0.2304 | 0.2757 | 0.3060 | 0.3125 |
| | 4 | 0.0000 | 0.0000 | 0.0004 | 0.0022 | 0.0064 | 0.0146 | 0.0284 | 0.0412 | 0.0488 | 0.0768 | 0.1128 | 0.1470 | 0.1562 |
| | 5 | 0.0000 | 0.0000 | 0.0000 | 0.0001 | 0.0003 | 0.0010 | 0.0024 | 0.0041 | 0.0053 | 0.0102 | 0.0185 | 0.0283 | 0.0312 |
| 6 | 0 | 0.9415 | 0.7351 | 0.5314 | 0.3771 | 0.2621 | 0.1780 | 0.1176 | 0.0878 | 0.0754 | 0.0467 | 0.0277 | 0.0176 | 0.0156 |
| | 1 | 0.0571 | 0.2321 | 0.3543 | 0.3993 | 0.3932 | 0.3560 | 0.3025 | 0.2634 | 0.2437 | 0.1866 | 0.1359 | 0.1014 | 0.0938 |
| | 2 | 0.0014 | 0.0305 | 0.0984 | 0.1762 | 0.2458 | 0.2966 | 0.3241 | 0.3292 | 0.3280 | 0.3110 | 0.2780 | 0.2437 | 0.2344 |
| | 3 | 0.0000 | 0.0021 | 0.0146 | 0.0415 | 0.0819 | 0.1318 | 0.1852 | 0.2195 | 0.2355 | 0.2765 | 0.3032 | 0.3121 | 0.3125 |

*Fundamentals of Probability and Statistics for Engineers* T.T. Soong © 2004 John Wiley & Sons, Ltd
ISBNs: 0-470-86813-9 (HB)   0-470-86814-7 (PB)

**Table A.1**    Continued

| n | k | 0.01 | 0.05 | 0.10 | 0.15 | 0.20 | 0.25 | 0.30 | $\frac{1}{3}$ | 0.35 | 0.40 | 0.45 | 0.49 | 0.50 |
|---|---|------|------|------|------|------|------|------|------|------|------|------|------|------|
|   | 4 | 0.0000 | 0.0001 | 0.0012 | 0.0055 | 0.0154 | 0.0330 | 0.0595 | 0.0823 | 0.0951 | 0.1382 | 0.1861 | 0.2249 | 0.2344 |
|   | 5 | 0.0000 | 0.0000 | 0.0001 | 0.0004 | 0.0015 | 0.0044 | 0.0102 | 0.0165 | 0.0205 | 0.0369 | 0.0609 | 0.0864 | 0.0938 |
|   | 6 | 0.0000 | 0.0000 | 0.0000 | 0.0000 | 0.0001 | 0.0002 | 0.0007 | 0.0014 | 0.0018 | 0.0041 | 0.0083 | 0.0139 | 0.0156 |
| 7 | 0 | 0.9321 | 0.6983 | 0.4783 | 0.3206 | 0.2097 | 0.1335 | 0.0824 | 0.0585 | 0.0490 | 0.0280 | 0.0152 | 0.0090 | 0.0078 |
|   | 1 | 0.0659 | 0.2573 | 0.3720 | 0.3960 | 0.3670 | 0.3115 | 0.2471 | 0.2048 | 0.1848 | 0.1306 | 0.0872 | 0.0603 | 0.0547 |
|   | 2 | 0.0020 | 0.0406 | 0.1240 | 0.2097 | 0.2753 | 0.3115 | 0.3177 | 0.3073 | 0.2985 | 0.2613 | 0.2140 | 0.1740 | 0.1641 |
|   | 3 | 0.0000 | 0.0036 | 0.0230 | 0.0617 | 0.1147 | 0.1730 | 0.2269 | 0.2561 | 0.2679 | 0.2903 | 0.2918 | 0.2786 | 0.2734 |
|   | 4 | 0.0000 | 0.0002 | 0.0026 | 0.0109 | 0.0287 | 0.0577 | 0.0972 | 0.1280 | 0.1442 | 0.1935 | 0.2388 | 0.2676 | 0.2734 |
|   | 5 | 0.0000 | 0.0000 | 0.0002 | 0.0012 | 0.0043 | 0.0115 | 0.0250 | 0.0384 | 0.0466 | 0.0774 | 0.1172 | 0.1543 | 0.1641 |
|   | 6 | 0.0000 | 0.0000 | 0.0000 | 0.0001 | 0.0004 | 0.0013 | 0.0036 | 0.0064 | 0.0084 | 0.0172 | 0.0320 | 0.0494 | 0.0547 |
|   | 7 | 0.0000 | 0.0000 | 0.0000 | 0.0000 | 0.0000 | 0.0001 | 0.0002 | 0.0005 | 0.0006 | 0.0016 | 0.0037 | 0.0068 | 0.0078 |
| 8 | 0 | 0.9227 | 0.6634 | 0.4305 | 0.2725 | 0.1678 | 0.1001 | 0.0576 | 0.0390 | 0.0319 | 0.0168 | 0.0084 | 0.0046 | 0.0039 |
|   | 1 | 0.0746 | 0.2793 | 0.3826 | 0.3847 | 0.3355 | 0.2670 | 0.1977 | 0.1561 | 0.1373 | 0.0896 | 0.0548 | 0.0352 | 0.0312 |
|   | 2 | 0.0026 | 0.0515 | 0.1488 | 0.2376 | 0.2936 | 0.3115 | 0.2965 | 0.2731 | 0.2587 | 0.2090 | 0.1569 | 0.1183 | 0.1094 |
|   | 3 | 0.0001 | 0.0054 | 0.0331 | 0.0839 | 0.1468 | 0.2076 | 0.2541 | 0.2731 | 0.2786 | 0.2787 | 0.2568 | 0.2273 | 0.2188 |
|   | 4 | 0.0000 | 0.0004 | 0.0046 | 0.0185 | 0.0459 | 0.0865 | 0.1361 | 0.1707 | 0.1875 | 0.2322 | 0.2627 | 0.2730 | 0.2734 |
|   | 5 | 0.0000 | 0.0000 | 0.0004 | 0.0026 | 0.0092 | 0.0231 | 0.0467 | 0.0683 | 0.0808 | 0.1239 | 0.1719 | 0.2098 | 0.2188 |
|   | 6 | 0.0000 | 0.0000 | 0.0000 | 0.0002 | 0.0011 | 0.0038 | 0.0100 | 0.0171 | 0.0217 | 0.0413 | 0.0703 | 0.1008 | 0.1094 |
|   | 7 | 0.0000 | 0.0000 | 0.0000 | 0.0000 | 0.0001 | 0.0004 | 0.0012 | 0.0024 | 0.0033 | 0.0079 | 0.0164 | 0.0277 | 0.0312 |
|   | 8 | 0.0000 | 0.0000 | 0.0000 | 0.0000 | 0.0000 | 0.0000 | 0.0001 | 0.0002 | 0.0002 | 0.0007 | 0.0017 | 0.0033 | 0.0039 |
| 9 | 0 | 0.9135 | 0.6302 | 0.3874 | 0.2316 | 0.1342 | 0.0751 | 0.0404 | 0.0260 | 0.0207 | 0.0101 | 0.0046 | 0.0023 | 0.0020 |
|   | 1 | 0.0830 | 0.2985 | 0.3874 | 0.3679 | 0.3020 | 0.2253 | 0.1556 | 0.1171 | 0.1004 | 0.0605 | 0.0339 | 0.0202 | 0.0176 |
|   | 2 | 0.0034 | 0.0629 | 0.1722 | 0.2597 | 0.3020 | 0.3003 | 0.2668 | 0.2341 | 0.2162 | 0.1612 | 0.1110 | 0.0776 | 0.0703 |
|   | 3 | 0.0001 | 0.0077 | 0.0446 | 0.1069 | 0.1762 | 0.2336 | 0.2668 | 0.2731 | 0.2716 | 0.2508 | 0.2119 | 0.1739 | 0.1641 |
|   | 4 | 0.0000 | 0.0006 | 0.0074 | 0.0283 | 0.0661 | 0.1168 | 0.1715 | 0.2048 | 0.2194 | 0.2508 | 0.2600 | 0.2506 | 0.2461 |
|   | 5 | 0.0000 | 0.0000 | 0.0008 | 0.0050 | 0.0165 | 0.0389 | 0.0735 | 0.1024 | 0.1181 | 0.1672 | 0.2128 | 0.2408 | 0.2461 |
|   | 6 | 0.0000 | 0.0000 | 0.0001 | 0.0006 | 0.0028 | 0.0087 | 0.0210 | 0.0341 | 0.0424 | 0.0743 | 0.1160 | 0.1542 | 0.1641 |
|   | 7 | 0.0000 | 0.0000 | 0.0000 | 0.0000 | 0.0003 | 0.0012 | 0.0039 | 0.0073 | 0.0098 | 0.0212 | 0.0407 | 0.0635 | 0.0703 |
|   | 8 | 0.0000 | 0.0000 | 0.0000 | 0.0000 | 0.0000 | 0.0001 | 0.0004 | 0.0009 | 0.0013 | 0.0035 | 0.0083 | 0.0153 | 0.0176 |
|   | 9 | 0.0000 | 0.0000 | 0.0000 | 0.0000 | 0.0000 | 0.0000 | 0.0000 | 0.0001 | 0.0001 | 0.0003 | 0.0008 | 0.0016 | 0.0020 |
| 10 | 0 | 0.9044 | 0.5987 | 0.3487 | 0.1969 | 0.1074 | 0.0563 | 0.0282 | 0.0173 | 0.0135 | 0.0060 | 0.0025 | 0.0012 | 0.0010 |
|   | 1 | 0.0914 | 0.3151 | 0.3874 | 0.3474 | 0.2684 | 0.1877 | 0.1211 | 0.0867 | 0.0725 | 0.0403 | 0.0207 | 0.0114 | 0.0098 |
|   | 2 | 0.0042 | 0.0746 | 0.1937 | 0.2759 | 0.3020 | 0.2816 | 0.2335 | 0.1951 | 0.1757 | 0.1209 | 0.0736 | 0.0495 | 0.0439 |
|   | 3 | 0.0001 | 0.0105 | 0.0574 | 0.1298 | 0.2013 | 0.2503 | 0.2668 | 0.2601 | 0.2522 | 0.2150 | 0.1665 | 0.1267 | 0.1172 |
|   | 4 | 0.0000 | 0.0010 | 0.0112 | 0.0401 | 0.0881 | 0.1460 | 0.2001 | 0.2276 | 0.2377 | 0.2508 | 0.2384 | 0.2130 | 0.2051 |
|   | 5 | 0.0000 | 0.0001 | 0.0015 | 0.0085 | 0.0264 | 0.0584 | 0.1029 | 0.1366 | 0.1536 | 0.2007 | 0.2340 | 0.2456 | 0.2461 |
|   | 6 | 0.0000 | 0.0000 | 0.0001 | 0.0012 | 0.0055 | 0.0162 | 0.0368 | 0.0569 | 0.0689 | 0.1115 | 0.1596 | 0.1966 | 0.2051 |
|   | 7 | 0.0000 | 0.0000 | 0.0000 | 0.0001 | 0.0008 | 0.0031 | 0.0090 | 0.0163 | 0.0212 | 0.0425 | 0.0746 | 0.1080 | 0.1172 |
|   | 8 | 0.0000 | 0.0000 | 0.0000 | 0.0000 | 0.0001 | 0.0004 | 0.0014 | 0.0030 | 0.0043 | 0.0106 | 0.0229 | 0.0389 | 0.0439 |
|   | 9 | 0.0000 | 0.0000 | 0.0000 | 0.0000 | 0.0000 | 0.0000 | 0.0001 | 0.0003 | 0.0005 | 0.0016 | 0.0042 | 0.0083 | 0.0098 |
|   | 10 | 0.0000 | 0.0000 | 0.0000 | 0.0000 | 0.0000 | 0.0000 | 0.0000 | 0.0000 | 0.0000 | 0.0001 | 0.0003 | 0.0008 | 0.0010 |

## A.2  POISSON MASS FUNCTION

**Table A.2**  Poisson mass function: a table of

$$p_k(0, t) = \frac{(\lambda t)^k e^{-\lambda t}}{k!},$$

for $k = 0$ to $24$, $\lambda t = 0.1$ to $10$

| $\lambda t$ | \multicolumn{13}{c}{$k$} |
|---|---|
| | 0 | 1 | 2 | 3 | 4 | 5 | 6 | 7 | 8 | 9 | 10 | 11 | 12 |
| 0.1 | 0.9048 | 0.0905 | 0.0045 | 0.0002 | 0.0000 | | | | | | | | |
| 0.2 | 0.8187 | 0.1637 | 0.0164 | 0.0011 | 0.0001 | 0.0000 | | | | | | | |
| 0.3 | 0.7408 | 0.2222 | 0.0333 | 0.0033 | 0.0002 | 0.0000 | | | | | | | |
| 0.4 | 0.6703 | 0.2681 | 0.0536 | 0.0072 | 0.0007 | 0.0001 | 0.0000 | | | | | | |
| 0.5 | 0.6065 | 0.3033 | 0.0758 | 0.0126 | 0.0016 | 0.0002 | 0.0000 | | | | | | |
| 0.6 | 0.5488 | 0.3293 | 0.0988 | 0.0198 | 0.0030 | 0.0004 | 0.0000 | | | | | | |
| 0.7 | 0.4966 | 0.3476 | 0.1217 | 0.0284 | 0.0050 | 0.0007 | 0.0001 | 0.0000 | | | | | |
| 0.8 | 0.4493 | 0.3595 | 0.1438 | 0.0383 | 0.0077 | 0.0012 | 0.0002 | 0.0000 | | | | | |
| 0.9 | 0.4066 | 0.3659 | 0.1647 | 0.0494 | 0.0111 | 0.0020 | 0.0003 | 0.0000 | | | | | |
| 1.0 | 0.3679 | 0.3679 | 0.1839 | 0.0613 | 0.0153 | 0.0031 | 0.0005 | 0.0001 | 0.0000 | | | | |
| 1.1 | 0.3329 | 0.3662 | 0.2014 | 0.0738 | 0.0203 | 0.0045 | 0.0008 | 0.0001 | 0.0000 | | | | |
| 1.2 | 0.3012 | 0.3614 | 0.2169 | 0.0867 | 0.0260 | 0.0062 | 0.0012 | 0.0002 | 0.0000 | | | | |
| 1.3 | 0.2725 | 0.3543 | 0.2303 | 0.0998 | 0.0324 | 0.0084 | 0.0018 | 0.0003 | 0.0001 | 0.0000 | | | |
| 1.4 | 0.2466 | 0.3452 | 0.2417 | 0.1128 | 0.0395 | 0.0111 | 0.0026 | 0.0005 | 0.0001 | 0.0000 | | | |
| 1.5 | 0.2231 | 0.3347 | 0.2510 | 0.1255 | 0.0471 | 0.0141 | 0.0035 | 0.0008 | 0.0001 | 0.0000 | | | |
| 1.6 | 0.2019 | 0.3230 | 0.2584 | 0.1378 | 0.0551 | 0.0176 | 0.0047 | 0.0011 | 0.0002 | 0.0000 | | | |
| 1.7 | 0.1827 | 0.3106 | 0.2640 | 0.1496 | 0.0636 | 0.0216 | 0.0061 | 0.0015 | 0.0003 | 0.0001 | 0.0000 | | |
| 1.8 | 0.1653 | 0.2975 | 0.2678 | 0.1607 | 0.0723 | 0.0260 | 0.0078 | 0.0020 | 0.0005 | 0.0001 | 0.0000 | | |
| 1.9 | 0.1496 | 0.2842 | 0.2700 | 0.1710 | 0.0812 | 0.0309 | 0.0098 | 0.0027 | 0.0006 | 0.0001 | 0.0000 | | |
| 2.0 | 0.1353 | 0.2707 | 0.2707 | 0.1804 | 0.0902 | 0.0361 | 0.0120 | 0.0034 | 0.0009 | 0.0002 | 0.0000 | | |
| 2.2 | 0.1108 | 0.2438 | 0.2681 | 0.1966 | 0.1082 | 0.0476 | 0.0174 | 0.0055 | 0.0015 | 0.0004 | 0.0001 | 0.0000 | |
| 2.4 | 0.0907 | 0.2177 | 0.2613 | 0.2090 | 0.1254 | 0.0602 | 0.0241 | 0.0083 | 0.0025 | 0.0007 | 0.0002 | 0.0000 | |
| 2.6 | 0.0743 | 0.1931 | 0.2510 | 0.2176 | 0.1414 | 0.0735 | 0.0319 | 0.0118 | 0.0038 | 0.0011 | 0.0003 | 0.0001 | 0.0000 |
| 2.8 | 0.0608 | 0.1703 | 0.2384 | 0.2225 | 0.1557 | 0.0872 | 0.0407 | 0.0163 | 0.0057 | 0.0018 | 0.0005 | 0.0001 | 0.0000 |
| 3.0 | 0.0498 | 0.1494 | 0.2240 | 0.2240 | 0.1680 | 0.1008 | 0.0504 | 0.0216 | 0.0081 | 0.0027 | 0.0008 | 0.0002 | 0.0001 |
| 3.2 | 0.0408 | 0.1304 | 0.2087 | 0.2226 | 0.1781 | 0.1140 | 0.0608 | 0.0278 | 0.0111 | 0.0040 | 0.0013 | 0.0004 | 0.0001 |
| 3.4 | 0.0334 | 0.1135 | 0.1929 | 0.2186 | 0.1858 | 0.1264 | 0.0716 | 0.0348 | 0.0148 | 0.0056 | 0.0019 | 0.0006 | 0.0002 |
| 3.6 | 0.0273 | 0.0984 | 0.1771 | 0.2125 | 0.1912 | 0.1377 | 0.0826 | 0.0425 | 0.0191 | 0.0076 | 0.0028 | 0.0009 | 0.0003 |
| 3.8 | 0.0224 | 0.0850 | 0.1615 | 0.2046 | 0.1944 | 0.1477 | 0.0936 | 0.0508 | 0.0241 | 0.0102 | 0.0039 | 0.0013 | 0.0004 |
| 4.0 | 0.0183 | 0.0733 | 0.1465 | 0.1954 | 0.1954 | 0.1563 | 0.1042 | 0.0595 | 0.0298 | 0.0132 | 0.0053 | 0.0019 | 0.0006 |
| 5.0 | 0.0067 | 0.0337 | 0.0842 | 0.1404 | 0.1755 | 0.1755 | 0.1462 | 0.1044 | 0.0653 | 0.0363 | 0.0181 | 0.0082 | 0.0034 |
| 6.0 | 0.0025 | 0.0149 | 0.0446 | 0.0892 | 0.1339 | 0.1606 | 0.1606 | 0.1377 | 0.1033 | 0.0688 | 0.0413 | 0.0225 | 0.0113 |
| 7.0 | 0.0009 | 0.0064 | 0.0223 | 0.0521 | 0.0912 | 0.1277 | 0.1490 | 0.1490 | 0.1304 | 0.1014 | 0.0710 | 0.0452 | 0.0264 |
| 8.0 | 0.0003 | 0.0027 | 0.0107 | 0.0286 | 0.0573 | 0.0916 | 0.1221 | 0.1396 | 0.1396 | 0.1241 | 0.0993 | 0.0722 | 0.0481 |
| 9.0 | 0.0001 | 0.0011 | 0.0050 | 0.0150 | 0.0337 | 0.0607 | 0.0911 | 0.1171 | 0.1318 | 0.1318 | 0.1186 | 0.0970 | 0.0728 |
| 10.0 | 0.0000 | 0.0005 | 0.0023 | 0.0076 | 0.0189 | 0.0378 | 0.0631 | 0.0901 | 0.1126 | 0.1251 | 0.1251 | 0.1137 | 0.0948 |

**Table A.2**   Continued

| $\lambda t$ | k | | | | | | | | | | | |
|------|--------|--------|--------|--------|--------|--------|--------|--------|--------|--------|--------|--------|
|      | 13     | 14     | 15     | 16     | 17     | 18     | 19     | 20     | 21     | 22     | 23     | 24     |
| 5.0  | 0.0013 | 0.0005 | 0.0002 |        |        |        |        |        |        |        |        |        |
| 6.0  | 0.0052 | 0.0022 | 0.0009 | 0.0003 | 0.0001 |        |        |        |        |        |        |        |
| 7.0  | 0.0142 | 0.0071 | 0.0033 | 0.0014 | 0.0006 | 0.0002 | 0.0001 |        |        |        |        |        |
| 8.0  | 0.0296 | 0.0169 | 0.0090 | 0.0045 | 0.0021 | 0.0009 | 0.0004 | 0.0002 | 0.0001 |        |        |        |
| 9.0  | 0.0504 | 0.0324 | 0.0194 | 0.0109 | 0.0058 | 0.0029 | 0.0014 | 0.0006 | 0.0003 | 0.0001 |        |        |
| 10.0 | 0.0729 | 0.0521 | 0.0347 | 0.0217 | 0.0128 | 0.0071 | 0.0037 | 0.0019 | 0.0009 | 0.0004 | 0.0002 | 0.0001 |

From Parzen, E., 1960, *Modern Probability and Its Applications*, John Wiley & Sons, with permission.

## A.3  STANDARDIZED NORMAL DISTRIBUTION FUNCTION

**Table A.3**  Standardized normal distribution function: a table of

$$F_U(u) = \frac{1}{(2\pi)^{1/2}} \int_{-\infty}^{u} e^{-x^2/2}\, dx,$$

for $u = 0.0$ to $3.69$

| u | 0.00 | 0.01 | 0.02 | 0.03 | 0.04 | 0.05 | 0.06 | 0.07 | 0.08 | 0.09 |
|---|------|------|------|------|------|------|------|------|------|------|
| 0.0 | 0.5000 | 0.5040 | 0.5080 | 0.5120 | 0.5160 | 0.5199 | 0.5239 | 0.5279 | 0.5319 | 0.5359 |
| 0.1 | 0.5398 | 0.5438 | 0.5478 | 0.5517 | 0.5557 | 0.5596 | 0.5636 | 0.5675 | 0.5714 | 0.5733 |
| 0.2 | 0.5793 | 0.5832 | 0.5871 | 0.5910 | 0.5948 | 0.5987 | 0.6026 | 0.6064 | 0.6103 | 0.6141 |
| 0.3 | 0.6179 | 0.6217 | 0.6255 | 0.6293 | 0.6331 | 0.6368 | 0.6406 | 0.6443 | 0.6480 | 0.6517 |
| 0.4 | 0.6554 | 0.6591 | 0.6628 | 0.6664 | 0.6700 | 0.6736 | 0.6772 | 0.6808 | 0.6844 | 0.6879 |
| 0.5 | 0.6915 | 0.6950 | 0.6985 | 0.7019 | 0.7054 | 0.7088 | 0.7123 | 0.7157 | 0.7190 | 0.7224 |
| 0.6 | 0.7257 | 0.7291 | 0.7324 | 0.7357 | 0.7389 | 0.7422 | 0.7454 | 0.7486 | 0.7517 | 0.7549 |
| 0.7 | 0.7580 | 0.7611 | 0.7642 | 0.7673 | 0.7704 | 0.7734 | 0.7764 | 0.7794 | 0.7823 | 0.7852 |
| 0.8 | 0.7881 | 0.7910 | 0.7939 | 0.7967 | 0.7995 | 0.8023 | 0.8051 | 0.8078 | 0.8106 | 0.8133 |
| 0.9 | 0.8159 | 0.8186 | 0.8212 | 0.8238 | 0.8264 | 0.8289 | 0.8315 | 0.8340 | 0.8365 | 0.8389 |
| 1.0 | 0.8413 | 0.8438 | 0.8461 | 0.8485 | 0.8508 | 0.8531 | 0.8554 | 0.8577 | 0.8599 | 0.8621 |
| 1.1 | 0.8643 | 0.8665 | 0.8686 | 0.8708 | 0.8729 | 0.8749 | 0.8770 | 0.8790 | 0.8810 | 0.8830 |
| 1.2 | 0.8849 | 0.8869 | 0.8888 | 0.8907 | 0.8925 | 0.8944 | 0.8962 | 0.8980 | 0.8997 | 0.9015 |
| 1.3 | 0.9032 | 0.9049 | 0.9066 | 0.9082 | 0.9099 | 0.9115 | 0.9131 | 0.9147 | 0.9162 | 0.9177 |
| 1.4 | 0.9192 | 0.9207 | 0.9222 | 0.9236 | 0.9251 | 0.9265 | 0.9279 | 0.9292 | 0.9306 | 0.9319 |
| 1.5 | 0.9332 | 0.9345 | 0.9357 | 0.9370 | 0.9382 | 0.9394 | 0.9406 | 0.9418 | 0.9429 | 0.9441 |
| 1.6 | 0.9452 | 0.9463 | 0.9474 | 0.9484 | 0.9495 | 0.9505 | 0.9515 | 0.9525 | 0.9535 | 0.9545 |
| 1.7 | 0.9554 | 0.9564 | 0.9573 | 0.9482 | 0.9591 | 0.9599 | 0.9608 | 0.9616 | 0.9625 | 0.9633 |
| 1.8 | 0.9641 | 0.9649 | 0.9656 | 0.9664 | 0.9671 | 0.9678 | 0.9686 | 0.9693 | 0.9699 | 0.9706 |
| 1.9 | 0.9713 | 0.9719 | 0.9726 | 0.9732 | 0.9738 | 0.9744 | 0.9750 | 0.9756 | 0.9761 | 0.9767 |
| 2.0 | 0.9772 | 0.9778 | 0.9783 | 0.9788 | 0.9793 | 0.9798 | 0.9803 | 0.9808 | 0.9812 | 0.9817 |
| 2.1 | 0.9821 | 0.9826 | 0.9830 | 0.9834 | 0.9838 | 0.9842 | 0.9846 | 0.9850 | 0.9854 | 0.9857 |
| 2.2 | 0.9861 | 0.9864 | 0.9868 | 0.9871 | 0.9875 | 0.9878 | 0.9881 | 0.9884 | 0.9887 | 0.9890 |
| 2.3 | 0.9893 | 0.9896 | 0.9898 | 0.9901 | 0.9904 | 0.9906 | 0.9909 | 0.9911 | 0.9913 | 0.9916 |
| 2.4 | 0.9918 | 0.9920 | 0.9922 | 0.9925 | 0.9927 | 0.9929 | 0.9931 | 0.9932 | 0.9934 | 0.9936 |
| 2.5 | 0.9938 | 0.9940 | 0.9941 | 0.9943 | 0.9945 | 0.9946 | 0.9948 | 0.9949 | 0.9951 | 0.9952 |
| 2.6 | 0.9953 | 0.9955 | 0.9956 | 0.9957 | 0.9959 | 0.9960 | 0.9961 | 0.9962 | 0.9963 | 0.9964 |
| 2.7 | 0.9965 | 0.9966 | 0.9967 | 0.9968 | 0.9969 | 0.9970 | 0.9971 | 0.9972 | 0.9973 | 0.9974 |
| 2.8 | 0.8874 | 0.9975 | 0.9976 | 0.9977 | 0.9977 | 0.9978 | 0.9979 | 0.9979 | 0.9980 | 0.9981 |
| 2.9 | 0.9981 | 0.9982 | 0.9982 | 0.9983 | 0.9984 | 0.9984 | 0.9985 | 0.9985 | 0.9986 | 0.9986 |
| 3.0 | 0.9987 | 0.9987 | 0.9987 | 0.9988 | 0.9988 | 0.9989 | 0.9989 | 0.9989 | 0.9990 | 0.9990 |
| 3.1 | 0.9990 | 0.9991 | 0.9991 | 0.9991 | 0.9992 | 0.9992 | 0.9992 | 0.9992 | 0.9993 | 0.9993 |
| 3.2 | 0.9993 | 0.9993 | 0.9994 | 0.9994 | 0.9994 | 0.9994 | 0.9994 | 0.9995 | 0.9995 | 0.9995 |
| 3.3 | 0.9995 | 0.9995 | 0.9995 | 0.9996 | 0.9996 | 0.9996 | 0.9996 | 0.9996 | 0.9996 | 0.9997 |
| 3.4 | 0.9997 | 0.9997 | 0.9997 | 0.9997 | 0.9997 | 0.9997 | 0.9997 | 0.9997 | 0.9997 | 0.9998 |
| 3.6 | 0.9998 | 0.9998 | 0.9999 | 0.9999 | 0.9999 | 0.9999 | 0.9999 | 0.9999 | 0.9999 | 0.9999 |

From Parzen, E., 1960, *Modern Probability and Its Applications*, John Wiley & Sons, with permission.

## A.4   STUDENT'S $t$ DISTRIBUTION WITH $n$ DEGREES OF FREEDOM

**Table A.4**  Student's distribution with $n$ degrees of
Freedom: a table of $t_{n,\alpha}$ in $P(T > t_{n,\alpha}) = \alpha$, for $\alpha = 0.005$
to 0.10, $n = 1, 2, \ldots$

| $n$ | $\alpha$ | | | | |
|---|---|---|---|---|---|
| | 0.10 | 0.05 | 0.025 | 0.01 | 0.005 |
| 1 | 3.078 | 6.314 | 12.706 | 31.821 | 63.657 |
| 2 | 1.886 | 2.920 | 4.303 | 6.965 | 9.925 |
| 3 | 1.638 | 2.353 | 3.182 | 4.541 | 5.841 |
| 4 | 1.533 | 2.132 | 2.776 | 3.747 | 4.604 |
| 5 | 1.476 | 2.015 | 2.571 | 3.365 | 4.032 |
| 6 | 1.440 | 1.943 | 2.447 | 3.143 | 3.707 |
| 7 | 1.415 | 1.895 | 2.365 | 2.998 | 3.499 |
| 8 | 1.397 | 1.860 | 2.306 | 2.896 | 3.355 |
| 9 | 1.383 | 1.833 | 2.262 | 2.821 | 3.250 |
| 10 | 1.372 | 1.812 | 2.228 | 2.764 | 3.169 |
| 11 | 1.363 | 1.796 | 2.201 | 2.718 | 3.106 |
| 12 | 1.356 | 1.782 | 2.179 | 2.681 | 3.055 |
| 13 | 1.350 | 1.771 | 2.160 | 2.650 | 3.012 |
| 14 | 1.345 | 1.761 | 2.145 | 2.624 | 2.977 |
| 15 | 1.341 | 1.753 | 2.131 | 2.602 | 2.947 |
| 16 | 1.337 | 1.746 | 2.120 | 2.583 | 2.921 |
| 17 | 1.333 | 1.740 | 2.110 | 2.567 | 2.898 |
| 18 | 1.330 | 1.734 | 2.101 | 2.552 | 2.878 |
| 19 | 1.328 | 1.729 | 2.093 | 2.539 | 2.861 |
| 20 | 1.325 | 1.725 | 2.086 | 2.528 | 2.845 |
| 21 | 1.323 | 1.721 | 2.080 | 2.518 | 2.831 |
| 22 | 1.321 | 1.717 | 2.074 | 2.508 | 2.819 |
| 23 | 1.319 | 1.714 | 2.069 | 2.500 | 2.807 |
| 24 | 1.318 | 1.711 | 2.064 | 2.492 | 2.979 |
| 25 | 1.316 | 1.708 | 2.060 | 2.485 | 2.787 |
| 26 | 1.315 | 1.706 | 2.056 | 2.479 | 2.779 |
| 27 | 1.314 | 1.703 | 2.052 | 2.473 | 2.771 |
| 28 | 1.313 | 1.701 | 2.048 | 2.467 | 2.763 |
| 29 | 1.311 | 1.699 | 2.045 | 2.462 | 2.756 |
| $\infty$ | 1.282 | 1.645 | 1.960 | 2.326 | 2.576 |

From Fisher, R.A., 1925, *Statistical Methods for Research Workers*,
14th edn, Hafner Press. Reproduced by permission of The University of
Adelaide, Australia.

## A.5 CHI-SQUARED DISTRIBUTION WITH $n$ DEGREES OF FREEDOM

**Table A.5** Chi-squared distribution with $n$ degrees of freedom: a table of $\chi^2_{n,\alpha}$ in $P(D > \chi^2_{n,\alpha}) = \alpha$, for $\alpha = 0.005$ to $0.995$, $n = 1$ to $30$

| $n$ | $\alpha$ | | | | | | | |
|---|---|---|---|---|---|---|---|---|
| | 0.995 | 0.99 | 0.975 | 0.95 | 0.05 | 0.025 | 0.01 | 0.005 |
| 1 | $0.0^4393$ | $0.0^3157$ | $0.0^3982$ | $0.0^2393$ | 3.841 | 5.024 | 6.635 | 7.879 |
| 2 | 0.0100 | 0.0201 | 0.0506 | 0.103 | 5.991 | 7.378 | 9.210 | 10.597 |
| 3 | 0.717 | 0.115 | 0.216 | 0.352 | 7.815 | 9.348 | 11.346 | 12.838 |
| 4 | 0.207 | 0.297 | 0.484 | 0.711 | 9.488 | 11.143 | 13.277 | 14.860 |
| 5 | 0.412 | 0.554 | 0.831 | 1.145 | 11.070 | 12.832 | 15.086 | 16.750 |
| 6 | 0.676 | 0.872 | 1.237 | 1.635 | 12.592 | 14.449 | 16.812 | 18.548 |
| 7 | 0.989 | 1.239 | 1.690 | 2.167 | 14.067 | 16.013 | 18.475 | 20.278 |
| 8 | 1.344 | 1.646 | 2.180 | 2.733 | 15.507 | 17.535 | 20.090 | 21.955 |
| 9 | 1.735 | 2.088 | 2.700 | 3.325 | 16.919 | 19.023 | 21.666 | 23.589 |
| 10 | 2.156 | 2.558 | 3.247 | 3.940 | 18.307 | 20.483 | 23.209 | 25.188 |
| 11 | 2.603 | 3.053 | 3.816 | 4.575 | 19.675 | 21.920 | 24.725 | 26.757 |
| 12 | 3.074 | 3.571 | 4.404 | 5.226 | 21.026 | 23.337 | 26.217 | 28.300 |
| 13 | 3.565 | 4.107 | 5.009 | 5.892 | 22.362 | 24.736 | 27.688 | 29.819 |
| 14 | 4.075 | 4.660 | 5.628 | 6.571 | 23.685 | 26.119 | 29.141 | 31.319 |
| 15 | 4.601 | 5.229 | 6.262 | 7.261 | 24.996 | 27.488 | 30.578 | 32.801 |
| 16 | 5.142 | 5.812 | 6.908 | 7.962 | 26.296 | 28.845 | 32.000 | 34.267 |
| 17 | 5.697 | 6.408 | 7.564 | 8.672 | 27.587 | 30.191 | 33.409 | 35.718 |
| 18 | 6.265 | 7.015 | 8.231 | 9.390 | 28.869 | 31.526 | 34.805 | 37.156 |
| 19 | 6.844 | 7.633 | 8.907 | 10.117 | 30.144 | 32.852 | 36.191 | 38.582 |
| 20 | 7.434 | 8.260 | 9.591 | 10.851 | 31.410 | 34.170 | 37.566 | 39.997 |
| 21 | 8.034 | 8.897 | 10.283 | 11.591 | 32.671 | 35.479 | 38.932 | 41.401 |
| 22 | 8.643 | 9.542 | 10.982 | 12.338 | 33.924 | 36.781 | 40.289 | 42.796 |
| 23 | 9.260 | 10.196 | 11.689 | 13.091 | 35.172 | 38.076 | 41.638 | 44.181 |
| 24 | 9.886 | 10.856 | 12.401 | 13.848 | 36.415 | 39.364 | 42.980 | 45.558 |
| 25 | 10.520 | 11.524 | 13.120 | 14.611 | 37.652 | 40.646 | 44.314 | 46.928 |
| 26 | 11.160 | 12.198 | 13.844 | 15.379 | 38.885 | 41.923 | 45.642 | 48.290 |
| 27 | 11.808 | 12.879 | 14.573 | 16.151 | 40.113 | 43.194 | 46.963 | 49.645 |
| 28 | 12.461 | 13.565 | 15.308 | 16.928 | 41.337 | 44.461 | 48.278 | 50.993 |
| 29 | 13.121 | 14.256 | 16.047 | 17.708 | 42.557 | 45.722 | 49.588 | 52.336 |
| 30 | 13.787 | 14.953 | 16.791 | 18.493 | 43.773 | 46.979 | 50.892 | 53.672 |

From Pearson, E.S. and Hartley, H.O., 1954, *Biometrika Tables for Statisticians, Volume 1*, Cambridge University Press, with permission.

## A.6  $D_2$ DISTRIBUTION WITH SAMPLE SIZE $n$

**Table A.6**  $D_2$ distribution with sample size $n$: a table of $c_{n,\alpha}$ in $P(D_2 > c_{n,\alpha}) = \alpha$, for $\alpha = 0.01$ to $0.10$, $n = 5, 10, \ldots$

| $n$ | $\alpha$ | | |
|---|---|---|---|
|  | 0.10 | 0.05 | 0.01 |
| 5 | 0.51 | 0.56 | 0.67 |
| 10 | 0.37 | 0.41 | 0.49 |
| 15 | 0.30 | 0.34 | 0.40 |
| 20 | 0.26 | 0.29 | 0.35 |
| 25 | 0.24 | 0.26 | 0.32 |
| 30 | 0.22 | 0.24 | 0.29 |
| 40 | 0.19 | 0.21 | 0.25 |
| Large $n$ | $\dfrac{1.22}{\sqrt{n}}$ | $\dfrac{1.36}{\sqrt{n}}$ | $\dfrac{1.63}{\sqrt{n}}$ |

From Lindgren, B.W., 1962, *Statistical Theory*, Macmillan, with permission.

## REFERENCES

Parzen, E., 1960, *Modern Probability Theory and Its Applications*, John Wiley & sons Inc., New York.

Fisher, R.A., 1970, *Statistical Methods for Research Workers*, 14th edn, Hafner Press, New York (1st edn, 1925).

Pearson, E.S., and Hartley, H.O., 1954, *Biometrika Tables for Statisticians, Volume 1.* Cambridge University Press, Cambridge.

Lindgren, B.W., 1962, *Statistical Theory*, Macmillan, New York.

# Appendix B: Computer Software

A large number of computer software packages and spreadsheets are now available that can be used to generate probabilities such as those provided in Tables A.1–A.6 as well as to perform other statistical calculations. For example, some statistical functions available in Microsoft® Excel™ 2000 are listed below, which can be used to carry out many probability calculations and to do many exercises in the text.

AVEDEV: gives the average of the absolute deviations of data points from their mean

AVERAGE: gives the average of its arguments

AVERAGEA: gives the average of its arguments, including numbers, text, and logical values

BETADIST: gives the beta probability distribution function

BETAINV: gives the inverse of the beta probability distribution function

BINOMDIST: gives the individual term binomial probability

CHIDIST: gives the one-tailed probability of the Chi-squared distribution

CHIINV: gives the inverse of the one-tailed probability of the Chi-squared distribution

CHITEST: gives the test for independence

CONFIDENCE: gives the confidence interval for a population mean

CORREL: gives the correlation coefficient between two data sets

COUNT: counts how many numbers are in the list of arguments

COUNTA: counts how many values are in the list of arguments

COVAR: gives covariance, the average of the products of paired deviations

CRITBINOM: gives the smallest value for which the binomial distribution function is less than or equal to the criterion value

DEVSQ: gives the sum of squares of deviations

EXPONDIST: gives the exponential distribution

FORECAST: gives a value along a linear trend

FREQUENCY: gives a frequency distribution as a vertical array

Fundamentals of Probability and Statistics for Engineers T.T. Soong © 2004 John Wiley & Sons, Ltd
ISBNs: 0-470-86813-9 (HB)   0-470-86814-7 (PB)

GAMMADIST: gives the gamma distribution

GAMMAINV: gives the inverse of the gamma distribution function

GAMMALN: gives the natural logarithm of the gamma function

GEOMEAN: gives the geometric mean

GROWTH: gives values along an exponential trend

HYPGEOMDIST: gives the hypergeometric distribution

INTERCEPT: gives the intercept of the linear regression line

KURT: gives the kurtosis of a data set

LARGE: gives the $k$th largest value in a data set

LINEST: gives the parameters of a linear trend

LOGEST: gives the parameters of an exponential trend

LOGINV: gives the inverse of the lognormal distribution

LOGNORMDIST: gives the lognormal distribution function

MAX: gives the maximum value in a list of arguments

MAXA: gives the maximum value in a list of arguments, including numbers, text, and logical values

MEDIAN: gives the median of the given numbers

MIN: gives the minimum value in a list of arguments

MINA: gives the smallest value in a list of arguments, including numbers, text, and logical values

MODE: gives the most common value in a data set

NEGBINOMDIST: gives the negative binomial distribution

NORMDIST: gives the normal distribution function

NORMINV: gives the inverse of the normal distribution function

NORMSDIST: gives the standardized normal distribution function

NORMSINV: gives the inverse of the standardized normal distribution function

PERCENTILE: gives the $k$th percentile of values in a range

PERCENTRANK: gives the percentage rank of a value in a data set

PERMUT: gives the number of permutations for a given number of objects

POISSON: gives the Poisson distribution

PROB: gives the probability that values in a range are between two limits

QUARTILE: gives the quartile of a data set

RANK: gives the rank of a number in a list of numbers

SKEW: gives the skewness of a distribution

SLOPE: gives the slope of the linear regression line

SMALL: gives the $k$th smallest value in a data set

STANDARDIZE: gives a normalized value

STDEV: estimates standard deviation based on a sample

STDEVA: estimates standard deviation based on a sample, including numbers, text, and logical values

STDEVP: calculates standard deviation based on the entire population

STDEVPA: calculates standard deviation based on the entire population, including numbers, text, and logical values

STEYX: gives the standard error of the predicted $y$ value for each $x$ in the regression

TDIST: gives the Student's $t$-distribution

TINV: gives the inverse of the Student's $t$-distribution

TREND: gives values along a linear trend

TRIMMEAN: gives the mean of the interior of a data set

TTEST: gives the probability associated with a Student's $t$-test

VAR: estimates variance based on a sample

VARA: estimates variance based on a sample, including numbers, text, and logical values

VARP: calculates variance based on the entire population

VARPA: calculates variance based on the entire population, including numbers, text, and logical values

WEIBULL: gives the Weibull distribution

# Appendix C: Answers to Selected Problems

## CHAPTER 2

2.1 (a) Incorrect, (b) Correct, (c) Correct, (d) Correct, (e) Correct, (f) Correct

2.4 (a) $\{1, 2, \ldots, 10\}$, (b) $\{1, 3, 4, 5, 6\}$, (c) $\{2, 7\}$, (d) $\{2, 4, 6, 7, 8, 9, 10\}$, (e) $\{1, 2, \ldots, 10\}$, (f) $\{1, 3, 4, 5, 6\}$, (g) $\{1, 5\}$

2.7 (a) $\overline{A}\,\overline{B}\,\overline{C}$, (b) $A\overline{B}\,\overline{C}$, (c) $(A\overline{B}\,\overline{C}) \cup (\overline{A}\,\overline{B}\,C) \cup (\overline{A}\,B\overline{C})$, (d) $A \cup B \cup C$, (e) $(AB\overline{C}) \cup (A\overline{B}\,C)$, (f) $\overline{A}BC$, (g) $(AB) \cup (BC) \cup (CA)$, (h) $\overline{ABC}$, (i) $ABC$

2.9 (a) $A \cup B$, (b) $A\overline{B} \cup \overline{A}B$

2.11 (a) 0.00829, (b) 0.00784, (c) 0.00829

2.14 (a) 0.553, (b) 0.053, (c) 0.395

2.16 0.9999

2.18 (a) 0.8865, (b) $[1 - (1 - p_A)(1 - p_C)][1 - (1 - p_B)(1 - p_D)]$

2.20 No

2.22 No, (a) $P(A) = P(B) = 0.5$, (b) Impossible

2.23 Under condition of mutual exclusiveness: (a) false, (b) true, (c) false, (d) true, (e) false
Under condition of independence: (a) true, (b) false, (c) false, (d) false, (e) true

2.24 (a) Approximately $10^{-5}$, (b) Yes, (c) 0.00499

2.26 (a) $\dfrac{t_1 - t_0}{t}$, (b) $\dfrac{t_1 - t_0}{t - t_0}$

2.28 (a) 0.35, (b) 0.1225, (c) 0.65

2.30 (a) 0.08, (b) 0.375

2.32 (a) 0.351, (b) 0.917, (c) 0.25

2.34 (a) 0.002, (b) 0.086, (c) 0.4904

*Fundamentals of Probability and Statistics for Engineers* T.T. Soong © 2004 John Wiley & Sons, Ltd
ISBNs: 0-470-86813-9 (HB)   0-470-86814-7 (PB)

## CHAPTER 3

3.1   (a) $a = 1$,

$$p(x) = \begin{cases} 1, & \text{for } x = 5 \\ 0, & \text{elsewhere} \end{cases}$$

(c) $a = 2$, $p(x)\dfrac{1}{2^x}$, for $x = 1, 2, \ldots$

(e) $a > 0$,

$$f(x) = \begin{cases} ax^{a-1}, & \text{for } 0 \le x \le 1 \\ 0, & \text{elsewhere} \end{cases}$$

or $a = 0$,

$$p(x) = \begin{cases} 1, & \text{for } x = 0 \\ 0, & \text{elsewhere} \end{cases}$$

(g) $a = 1/2$, neither pdf nor pmf exists

3.2   (a) $1, 1/3, 63/64, 1 - e^{-6a}, 1, 1, (2 - e^{-1/3})/2$

      (b) $1, 1, 127/128, e^{-a/2} - e^{-7a}, 1 - (1/2)^a, 1/2, (e^{-1/4} - e^{-7/2})/2$

3.4   (a)

$$F_X(x) = \begin{cases} 0, & \text{for } x < 90 \\ 0.1x - 9, & \text{for } 90 \le x < 100 \\ 1, & \text{for } x \ge 100 \end{cases}$$

(b)

$$F_X(x) = \begin{cases} 0, & \text{for } x < 0 \\ 2x - x^2, & \text{for } 0 \le x \le 1 \\ 1, & \text{for } x > 1 \end{cases}$$

(c) $F_X(x) = \dfrac{1}{\pi}\tan^{-1} x + \dfrac{1}{2}$, for $-\infty < x < \infty$

3.6   $2/3$

3.9

$$F_X(x) = \begin{cases} 0, & \text{for } x < 0 \\ \dfrac{x}{b}, & \text{for } 0 \le x \le b \\ 1, & \text{for } x > b \end{cases}$$

$$f_X(x) = \begin{cases} \dfrac{1}{b}, & \text{for } 0 \le x \le b \\ 0, & \text{elsewhere} \end{cases}$$

3.11   $3/a$

3.12   (b) $1/6$

3.13 (a): (i) $p_X(x) = \begin{cases} 0.6, & \text{for } x = 1 \\ 0.4, & \text{for } x = 2 \end{cases}$ $p_Y(y) = \begin{cases} 0.6, & \text{for } y = 1 \\ 0.4, & \text{for } y = 2 \end{cases}$

(iii) $f_X(x) = \begin{cases} e^{-x}, & \text{for } x \geq 0 \\ 0, & \text{elsewhere} \end{cases}$ $f_Y(y) = \begin{cases} e^{-y}, & \text{for } y \geq 0 \\ 0, & \text{elsewhere} \end{cases}$

(b): (i) No, (iii) Yes

3.17 $[F_X(x) - F_X(100)]/[1 - F_X(100)],\ x \geq 100$

3.19 (a) 0.087, (b) 0.3174, (c) 0.274

3.22 0.0039

3.26 (a)

$$p_{X_3}(x) = \begin{cases} 0.016, & \text{for } x = 1 \\ 0.035, & \text{for } x = 2 \\ 0.080, & \text{for } x = 3 \\ 0.125, & \text{for } x = 4 \\ 0.415, & \text{for } x = 5 \\ 0.192, & \text{for } x = 6 \\ 0.137, & \text{for } x = 7 \end{cases}$$

(b) Table of $p_{X_4 X_3}(i, j)$

| $i$ | $j$ | | | | | | |
|---|---|---|---|---|---|---|---|
| | 1 | 2 | 3 | 4 | 5 | 6 | 7 |
| 1 | 0.006 | 0.004 | 0.003 | 0.003 | 0.004 | 0.000 | 0.000 |
| 2 | 0.002 | 0.009 | 0.008 | 0.005 | 0.010 | 0.002 | 0.001 |
| 3 | 0.003 | 0.008 | 0.015 | 0.014 | 0.031 | 0.008 | 0.005 |
| 4 | 0.001 | 0.004 | 0.015 | 0.027 | 0.051 | 0.017 | 0.011 |
| 5 | 0.002 | 0.007 | 0.029 | 0.054 | 0.196 | 0.075 | 0.050 |
| 6 | 0.001 | 0.002 | 0.005 | 0.015 | 0.071 | 0.060 | 0.032 |
| 7 | 0.000 | 0.001 | 0.005 | 0.008 | 0.052 | 0.030 | 0.038 |

## CHAPTER 4

4.1 (a) 5, 0; (c) 2, 2; (e) $a/(a+1)$, $a/[(a+1)^2(a+2)]$; (g) 1, 3

4.3 2.44 min

4.6 (a) 1/2, (b) 2, 4; (c) 0, 1

4.12 (a) $(1-p)/\lambda$, (b) $1/\lambda$

4.14 24 min

4.16 $P(|X-1| \leq 0.75) \geq 0.41$ by the Chebyshev inequality, $P(|X-1| \leq 0.75)$ $= 0.75$

4.19 (a) $P(55 \leq X \leq 85) \geq 0$

(b) $P(55 \leq X \leq 85) \geq 5/9$, much more improved bound

4.20 3/4

4.23 1.53

4.25 $\sigma^2_{X_2}[(\sigma^2_{X_1} + \sigma^2_{X_2})^{1/2}(\sigma^2_{X_2} + \sigma^2_{X_3})^{1/2}]^{-1}$

4.27 (a) $m_{X_1} + m_{X_2}, \sigma^2_{X_1} + \sigma^2_{X_2}$

  (b) $\sigma_{X_2}/(\sigma^2_{X_1} + \sigma^2_{X_2})^{1/2}$; it approaches one if $\sigma^2_{X_2} \gg \sigma^2_{X_1}$

4.28 $n, 2n$

4.30 (a) $\phi_X(t) = e^{5jt}, 5, 0$

  (c) $\phi_X(t) = \sum_{k=1}^{\infty} \frac{1}{2^k} e^{jtk}, 2, 2$

  (e) $\phi_X(t) = \frac{2e^{jt}}{jt}\left(1 - \frac{1}{jt}\right) + \frac{2}{(jt)^2}, \frac{2}{3}, \frac{1}{18}$

## CHAPTER 5

5.1 (a)
$$F_Y(y) = \begin{cases} 0, & \text{for } y < 8 \\ \dfrac{1}{3}, & \text{for } 8 \le y \le 17 \\ 1, & \text{for } y > 17 \end{cases}$$

  (b)
$$F_Y(y) = \begin{cases} 0, & \text{for } y < 8 \\ \dfrac{y-8}{9}, & \text{for } 8 \le y \le 17 \\ 1, & \text{for } y > 17 \end{cases}$$

5.3
$$f_Y(y) = \begin{cases} 0, & \text{for } y < -1 \\ \dfrac{y+1}{9}, & \text{for} -1 \le y < 2 \\ \dfrac{5-y}{9}, & \text{for } 2 \le y < 5 \\ 0, & \text{for } y \ge 5 \end{cases}$$

5.5
$$f_Y(y) = \begin{cases} \dfrac{1}{y(2\pi)^{1/2}} e^{-\ln^2 y/2}, & \text{for } y \ge 0 \\ 0, & \text{elsewhere} \end{cases}$$

5.9
$$f_X(x) = \begin{cases} \dfrac{2}{\pi(a^2 - x^2)^{1/2}}, & \text{for } 0 \le x \le a \\ 0, & \text{elsewhere} \end{cases}$$

5.10 (a)

$$f_W(w) = \begin{cases} \dfrac{0.19}{2a(w/a)^{1/2}}\left(\dfrac{(w/a)^{1/2}}{36.6}\right)^{-7.96} \exp\left[-\left(\dfrac{(w/a)^{1/2}}{36.6}\right)^{-6.96}\right], & \text{for } w > 0 \\ 0, & \text{elsewhere} \end{cases}$$

$m_W = 1.71a \times 10^3$, $\sigma_W^2 = 8.05a^2 \times 10^5$

(b) Same as (a)

5.12 $Y$ is discrete and

$$p_Y(y) = \begin{cases} \displaystyle\int_0^\infty f_X(x)\,\mathrm{d}x, & \text{for } y = 1 \\ \displaystyle\int_{-\infty}^0 f_X(x)\,\mathrm{d}x, & \text{for } y = 0 \end{cases}$$

5.14 (a)

$$f_A(a) = \begin{cases} \dfrac{1}{0.08 r_0 (\pi a)^{1/2}}, & \text{for } 4\pi(0.99r_0)^2 \le a \le 4\pi(1.01r_0)^2 \\ 0, & \text{elsewhere} \end{cases}$$

(b)

$$f_V(v) = \begin{cases} \dfrac{1}{0.08\pi r_0}\left(\dfrac{3v}{4\pi}\right)^{-2/3}, & \text{for } \dfrac{4}{3}\pi(0.99r_0)^3 \le v \le \dfrac{4}{3}\pi(1.01r_0)^3 \\ 0, & \text{elsewhere} \end{cases}$$

5.16 (a)

$$f_Y(y) = \begin{cases} \dfrac{2+y}{4}, & \text{for } -2 < y \le 0 \\ \dfrac{2-y}{4}, & \text{for } 0 < y \le 2 \\ 0, & \text{elsewhere} \end{cases}$$

(b) Same as (a)

5.21

$$f_T(t) = \begin{cases} (a_1 + a_2 + \cdots + a_n)e^{-(a_1+a_2+\cdots+a_n)t}, & \text{for } t > 0 \\ 0, & \text{elsewhere} \end{cases}$$

5.23 $f_Y(y) = \displaystyle\int_{-\infty}^\infty f_{X_2}(x_2)[f_{X_1}(x_2 + y) + f_{X_1}(x_2 - y)]\mathrm{d}x_2, \quad -\infty < y < \infty$

5.25

$$f_Y(y) = \begin{cases} y e^{-y^2/2}, & \text{for } y \ge 0 \\ 0, & \text{elsewhere} \end{cases}$$

5.27

$$f_Y(y) = \begin{cases} \dfrac{3y^2}{(1+y)^4}, & \text{for } y \geq 0 \\ 0, & \text{elsewhere} \end{cases}$$

5.29

$$f_{R\Phi}(r,\phi) = \begin{cases} \dfrac{r}{2\pi\sigma^2}\,e^{-r^2/2\sigma^2}, & \text{for } r \geq 0, \text{and} -\pi \leq \phi \leq \pi \\ 0, & \text{elsewhere} \end{cases}$$

$$f_R(r) = \begin{cases} \dfrac{r}{\sigma^2}e^{-r^2/2\sigma^2}, & \text{for } r \geq 0 \\ 0, & \text{elsewhere} \end{cases}$$

$$f_\Phi(\phi) = \begin{cases} \dfrac{1}{2\pi}, & \text{for } -\pi \leq \phi \leq \pi \\ 0, & \text{elsewhere} \end{cases}$$

$R$ and $\Phi$ are independent

## CHAPTER 6

6.3  (a) 0.237,   (b) 3.75

6.5  0.611, 4.2

6.8  (a) $1 - \sum\limits_{k=0}^{m} \binom{n}{k} p^k (1-p)^{n-k}$

   (b) $\sum\limits_{k=m+1}^{n} (k-m)\binom{n}{k} p^k (1-p)^{n-k}$

6.10  0.584

6.12  $\sum\limits_{k=r}^{s+r-1} \binom{k-1}{r-1} p^r (1-p)^{k-r}$

6.14  0.096

6.17  (a) 0.349,   (b) 5

6.26  0.93

6.28  $1.4 \times 10^{-24}$

6.30  $p_X(k) = 22.5^k\, e^{-22.5}/k!, \;\; k = 0, 1, 2, \ldots$

6.32  $p_k(0, t) = \left(\dfrac{t^v}{w}\right)^k \dfrac{\exp(-t^v/w)}{k!}, \;\; k = 0, 1, 2, \ldots$

## CHAPTER 7

7.1  0.847

7.3  (a) 0.9,  (b) 0.775

7.6  (a) $4.566 \times 10^{-3}$,  (b) 0.8944,  (c) 0.383,  (d) 0.385

7.9  $X_2$ is preferred in both cases

7.14  0.0062

7.20  (a) 0.221

(b)

$$f_Y(y) = \begin{cases} \dfrac{1}{(0.294)(2\pi)^{1/2}(y-a)} \exp\left[-\dfrac{1}{0.172}\ln^2\left(\dfrac{y-a}{0.958b}\right)\right], & \text{for } y \geq a \\ 0, & \text{elsewhere} \end{cases}$$

$$m_Y = a + b$$

7.22  0.153

7.30  (a) 0.056,  (b) 0.989

7.34  0.125, 0, 0, 0.875. No partial failure is possible

7.36

$$f_S(s) = \begin{cases} n(n-1)\displaystyle\int_{-\infty}^{\infty}[F_X(y) - F_X(y-s)]^{n-2}f_X(y-s)f_X(y)dy, & \text{for } s \geq 0 \\ 0, & \text{elsewhere} \end{cases}$$

## CHAPTER 8

8.2  (a)  (i)  Type-1 asymptotic maximum-value distribution is suggested

(ii)  $\alpha \cong 0.025$,  $u \cong 46.92$

(d)  (i)  Gamma is suggested,  (ii)  $\lambda \cong 0.317$

(f)  (i)  Poisson is suggested,  (ii)  $v \cong 45.81$

(h)  (i)  Normal is suggested,  (ii)  $m \cong 2860$, $\sigma \cong 202.9$

(j)  (i)  Poisson is suggested,  (ii)  $v \cong 7.0$

(i)  (i)  Lognormal is suggested,  (ii)  $\theta_X \cong 76.2$, $\sigma_{\ln X}^2 \cong 0.203$

## CHAPTER 9

9.1  1.75, 27.96

9.5  (a)

$$f_Y(y) = \begin{cases} 10y^9, & \text{for } 0 \leq y \leq 1 \\ 0, & \text{elsewhere} \end{cases}$$

$$f_Z(z) = \begin{cases} 10(1-z)^9, & \text{for } 0 \leq z \leq 1 \\ 0, & \text{elsewhere} \end{cases}$$

(c) 0.091, 0.91

9.7   $\hat{\Theta}_1$ is better

9.11  It is biased, but unbiased as $n \to \infty$

9.14  (a) $\theta^2/n$,   (b) $\theta^2/n$,   (c) $\theta(1-\theta)/n$,   (d) $\theta/n$

9.15  $\sigma^2/n, 2\sigma^4/n$

9.20  (a) $1 - \dfrac{1}{\overline{X}}$,

  (b) $1 - \displaystyle\sum_{k=1}^{9} \frac{1}{\overline{X}} \left(1 - \frac{1}{\overline{X}}\right)^{k-1}$

9.22  $\hat{A}_{ML} = X_{(1)}$,  $\hat{A}_{ME} = \overline{X} - 1$

9.24  $\hat{\lambda} = 0.13 \text{ sec}^{-1}$

9.26  (a) $\hat{\Lambda}_{ML} = \hat{\Lambda}_{ME} = (\overline{T} - t_0)^{-1}$
  (b) $\hat{T}_{0ML} = T_{(1)}, \hat{T}_{0ME} = \overline{T} - (1/\lambda)$
  (c)

$$\hat{\Lambda}_{ML} = [\overline{T} - T_{(1)}]^{-1}, \hat{T}_{0ML} = T_{(1)}$$
$$\hat{\Lambda}_{ME} = (M_2 - \overline{T}^2)^{-1/2}, \hat{T}_{0ME} = \overline{T} - (M_2 - \overline{T}^2)^{1/2}$$

9.30  $(2\alpha)^{-1/2}$

9.32  (a) $l_{1,2} = 63.65, 81.55$,   (b) $l_{1,2} = 70.57, 84.43$,   (c) $l_{1,2} = 77.74, 89.46$

9.34  (a) 9.16,   (b) $l_{1,2} = 8.46, 9.86$

9.36  (a) $l_{1,2} = 1072, 1128$,   (b) $l_{1,2} = 1340, 6218$ and $l_1 = 1478$

9.38  384


## CHAPTER 10

10.1   More likely to be accepted at $\alpha = 0.01$

10.3   Hypothesis is accepted

10.5   Hypothesis is accepted

10.8   Poisson hypothesis is rejected

10.10  Gamma hypothesis is accepted

10.12  Normal hypothesis is accepted

10.14  Poisson hypothesis is accepted

10.16  Hypothesis is accepted


## CHAPTER 11

11.1  (a)

$$\hat{B} = \left(\sum_{i=1}^{n} x_i Y_i\right)\left(\sum_{i=1}^{n} x_i^2\right)^{-1}$$

(b)

$$E\{\hat{B}\} = \beta, \quad \text{var}\{\hat{B}\} = \sigma^2 \left( \sum_{i=1}^{n} x_i^2 \right)^{-1},$$

(c)

$$\widehat{\Sigma^2} = \frac{1}{n-1} \sum_{i=1}^{n} (Y_i - \hat{B}x_i)^2$$

11.3 $\widehat{\Sigma^2} = \frac{1}{n} \sum_{i=1}^{n} (Y_i - \hat{A} - \hat{B}x_i)^2$, $E\{\widehat{\Sigma^2}\} = \frac{n-2}{n} \sigma^2$, hence biased.

11.7 $14.98 + 32.14 \log_{10} v$

11.9 (a) $\hat{\alpha} = 1.486$ and $\hat{\beta} = 0.022$, (b) $l_1 = 0.006$, (c) Not significantly different from zero

11.11 $\hat{\beta}_0 = 66.18$, $\hat{\beta}_1 = 0.42$, $\hat{\beta}_2 = 46.10$

11.13 $\hat{\beta}_0 = 717.18$, $\hat{\beta}_1 = 10.84$, $\hat{\beta}_2 = -3.78$, $\hat{\beta}_3 = -1.57$, $\hat{\beta}_4 = 0.38$

# Subject Index